Mathematics of Social Choice and Finance

Vladimir A. Dobrushkin
Department of Mathematics—University of Rhode Island

Kendall Hunt
publishing company

www.kendallhunt.com
Send all inquiries to:
4050 Westmark Drive
Dubuque, IA 52004-1840

ISBN 978-1-4652-5099-5

Contents

Preface

This book started its life as a collection of lecture notes for the undergraduate course "Mathematics of Social Choice and Finance" taught at the mathematical department of the University of Rhode Island. This course is intended for students majoring in the liberal arts or other fields that have only a general mathematical requirement. The book is not appropriate for readers who want a broad and sophisticated introduction to large fields of mathematical applications. It is intended for students without any college-level mathematical background who are willing to work through mathematical arguments to understand more deeply the important topics presented in the book.

As a rule, people do not have an accurate picture of mathematics and its role in human life. We do not expect that the reader is intrigued by the romance of the subject or has had a pleasant experience working on mathematical problems. As you read this book, we hope you discover the importance of mathematics in real life and admire its beauty, elegance, and strength. Because we live in an increasingly complex world where information is growing rapidly, it is more important than ever that people have appropriate skills to be successful in their lives. It is now difficult to find a professional human activity that does not require utilizing information technologies. All students know how to use computers during their education: submitting applications, homework, and projects via the Internet becomes a custom at any university, institute, or college. Moreover, students communicate with each other, instructors, parents, and their friends via email and social media, search the web for new information, and do a lot of other activities on computers. Due to the digital world we are all part of finite mathematics, from a theoretical point of view, because this generation should possess confidence and comprehension in utilizing the information technologies.

This book is intended to develop some skills needed in a real digital world—reading complex texts; using quantitative data; and utilizing information technology with calculators; entering, collecting, recording, tabulating, manipulating, and retrieving of data; and searching the Internet. Since mathematics uses penetrating techniques of thought that sharpen problem solving skills, we have designed the course to help students learn to think clearly, logically, and analytically, and to understand the importance and practical applications of math in everyday life, science, and technology.

There are many ways to teach mathematics and they all depend on the objectives and background of students. Much time in school is spent learning basic math and its important tools. This book will show you how to develop valuable skills by working on particular problems from real life. This book has no modest goals—it contains three chapters that could be read independently. While the text is written for students majoring in liberal arts, students from other fields, including mathematics, computer science, engineering, and others will also benefit from the book. They will find this book to be a valuable resource that can help them fill in the gaps or tie together loose ends that might have accumulated during their years of study or field experience.

Given this target audience, our goal is to impart the highly technical information and issues involving mathematics in a manner that is accessible to nearly all levels of readers. The topics discussed in the book are explained fully and are of sufficient rigor for the target audience. Given its format and level of coverage, the text can be equated to a snorkeling adventure: we primarily stay at the surface to examine the features and concepts of mathematics. Occasionally, we hold our joint breath and dive under the surface to explore a particular concept more fully. We do not, however, analyze in detail the underlying mathematical foundations of any particular topic. Instead, we offer a glimpse into this world through various examples.

To be successful in the course, you should have some mathematical background. The *prerequisite* is a basic high school algebra background. Since the course requires numerical calculations, we recommend the use of a calculator (with memory). All students, including those with only an algebra background, will benefit from our presentation.

Organizational Structure

We have decided to explore three topic that are important for liberal arts students—voting methods, apportionment problems, and everyday finance.

- **Social Choice**

  The design of an electoral system is fundamental to any democracy. It is important to know how a group of people makes a choice when individual preferences may differ. The book presents different approaches from voting theory accompanied with many examples and problems. An abundance of exercises is one of the strengths of the book. Going through them, the reader will develop essential skills: how to collect data, analyze data, retrieve data, and how to use the information to find a consensus of different opinions. It will also help the reader to develop a critical point of view on many "obvious" thoughts and sharpen the way one looks at our world.

- **Apportionment Problems**

  A fair division is one of the oldest problems in the world: how something that must be shared by a set of competing parties can be divided among them in a way that parties accept as a fair division. The founding fathers were concerned that every citizen in the country should be fairly represented in government; therefore, they placed the requirements for state representatives at the very beginning of the U.S. Constitution. Apportionment problems are part of our individual lives and professional activities, so the book provides several methods for solving allocation problems.

- **Everyday Finance**

  By understanding and applying the financial principles that you will learn in this chapter, you can avoid financial pitfalls and use your knowledge of the mathematics of how money works to your advantage. Our society depends on a vast network of financial obligations between individuals, families, companies, banks, government's, and even countries. You will learn how to borrow, invest, and save money, how to secure your retirement, and many other issues. The chapter presents a thorough treatment of simple and compound interest, present and future values of ordinary annuities, mortgages, and amortized loans.

Each chapter concludes with a review that contains all formulas and concepts covered. As a result, students are able to grasp the material and definitions covered in the chapter.

The text uses only standard notations and abbreviations (*et al.* (et alii from Latin) means "and others," or "and co-workers;" *i.e.* (from Latin "id est") meaning that is, that is to say, or in other words; *e.g.* stands for the Latin phrase "exempli gratia," which means for example; and *etc.* (et cetera) means "and the others," "and other things," "and the rest"). However, we find it convenient to type □ at the end of exercises and examples (unless a new one serves as a delimiter); the symbol ◁ indicates the end of definitions. Since we will deal with numbers throughout the course, sometimes rational numbers are written not as fractions, but in scientific notation, where we put a line above a repeated string of numbers. For instance, $1.209090909\ldots$ is shorten to $1.2\overline{09}$.

To the Instructor

This textbook is designed for use in a one-semester course. Ample material is included to provide you with a great amount of flexibility for its use. Since all three chapters in the book are independent of each other, it makes instructor's task easier to teach these topics.

The text has many exercises from different areas, so an instructor may want to use the book as a supplementary resource for teaching. A solution manual that is available to instructors contains answers to all the problems.

This book is very flexible and it can be used in lectures or in class working through the problems that accompany the material being covered. Since examples and exercises are integrated into the text, we suggest taking advantage of their presence and embracing students with interactive discussion of the material.

To the Student

Mathematics of Social Choice and Finance serves as an excellent foundation on which you can build a solid understanding of mathematics and its application in the real world. It will help you to utilize mathematical terminology and concepts. Such knowledge will give you control over the part of your life that includes numbers and logic.

Acknowledgments

I am thankful to Professors Ray Beauregard, Nick Kuchura, and Alexander Rozenblyum, who made a lot of suggestions to the final version of the text and the exposition of the material, and to Yelena Kashina for editorial assistance. Many thanks to all of them. Additional impetus and help has been provided by the professional staff of our publisher, Kendall Hunt, particularly Robert Largent, Sarah Flynn, and Amy Wagner.

Finally, I thank my family for putting up with me while I was engaged in the writing of this book.

Vladimir Dobrushkin
University of Rhode Island, Kingston, RI
June, 2014

Chapter 1

Social Choice

One of the fundamental tenets of a democracy is the right to vote. The **Voting Rights Act**, adopted initially in 1965 and extended in 1970, 1975, and 1982, is generally considered the most successful piece of civil rights legislation ever adopted by the U. S. Congress. This act provides specific protections for voting and grants equal voting rights to every citizen of the United States. As citizens of a democratic country, we exercise our voting rights by participating in presidential elections, state elections, local elections, and many others. In addition to such political elections, we are constantly involved with making group decisions that affect our professional, financial, environmental, and everyday aspects of our life such as where to hold the company holiday party, electing the board of trustees, voting on a location of a casino, and many others.

Since we all have different opinions, a social choice of how groups can best arrive at decisions determined by voting. A winner in every election represents the best social choice of a particular population, or an option that people are willing to accept because it is at least "close" to their ideal preference. We consider the problem about utilization of individual choices in such a way as to make a choice for society as a whole. Hence voting is a method to resolve conflicting tendencies. Individual choices should be expressed as general as possible, and the social choice should be a composite of different choices each applying to a specific constituency or it may apply to society as a whole.

Although modern literature on social choice began with Duncan Black in the 1940–50's, major discoveries were originally made—and lost—much earlier, for example by Charles Dodgson (Lewis Carroll) and, most significantly for present purposes, by Borda and Condorcet in the late 18th century (1770 – 80's).

However, voting is only the first part of determining the best social choice. The second part is the *counting*, which is the heart of the democratic process. This is a subject of a voting system that contains rules for valid voting, and how votes are aggregated to yield a final result. The study of formally defined voting systems is called voting theory, a subfield of political science, economics, or mathematics. A central political and decision science issue

1

is to understand how election outcomes can change with a voting method or the slate of candidates.

This chapter examines the underlying mathematical structures and symmetries of elections to explain why different voting procedures can give dramatically different outcomes even if no one changes her or his vote. Mathematical theory of voting can create alternative voting methods that may then be applied to distinct elections as well as to the everyday functioning of the legislative branch.

We present four of the most famous winning selection methods and some of their variations—plurality methods, point distribution systems, head-to-head comparisons, and approval voting. Most of the methods are extended for ranking candidates. All these methods may fail to determine a single winner, or finish in a tie between two or more alternatives. We do not discuss tie-breaking methods because they can lead us to rather complicated issues.

1.1 Preference Ranking

It is customary to reserve the word "candidates" or "competitors" to humans, the word "platforms" for political considerations, and the word "alternatives" or "options" for other choices. For example, if a department wants to determine a monthly meeting day, then there are five alternatives (Monday, Tuesday, Wednesday, Thursday, or Friday), but if the chair is to be elected, then we say that there are a certain number of candidates.

We could consider only a single winning method that determines only one option out of the number of alternatives running in an election for the same position. Choosing a governor or mayor would be an example of this. A multiple winner method will select more than one candidate; for example, an election that fills five seats in a legislature, with one electorate choosing from a field of, say, forty candidates. Such methods are out of our consideration as well as methods using the nonequal weight in voting as, for instance, a shareholder power, which depends on the number of shares.

When we all talk about voting, there are some "naive" properties of fairness that most of the people assume to hold:

1. All voters are treated equally, namely, every voter has the same power as anybody else.

2. All candidates or alternatives are treated equally, that is, an election procedure should not depend on the name of the candidates, just how the ballots are marked.

Later in §1.6, we consider other criteria of voting systems to be fair.

Different voting systems have different forms for allowing the individual to express his or her choice. People usually express their opinion through ballots—which can come in many forms. Some voting systems include additional options on the ballot, such as write-in candidates, a none of the above option, or a no confidence in that candidate option. Single-winner systems can be classified based on their ballot type. One vote systems are where a voter picks one choice at a time.

It was noted many years ago that a one-choice ballot does not express the will of a voter explicitly. In many cases, a voter would like to say that one candidate is most wanted, another is decent, but the third one is disliked. Such a possibility to express a position of a voter is given in a preference ballot. However, it may happen that some alternatives mean the same value to a voter, so indifference among choices is resolved in some way by each voter (may be by randomly ordering them).

In ranked voting systems, each voter ranks the candidates in order of preference. There are a number of different but equivalent ways to design a ballot that allows a voter to specify a set of rank preferences:

- Written numbers: The voter writes a "1" beside their first choice, a "2" beside their second choice, and so on. This is the most common ballot design. Hand-written numeric rankings are compact and easy to hand count.

- Column marks: The voter places marks in columns. These ballots can be easily counted by optical scanners. However considerations of space may limit the number of preferences a voter can express.

- Written names: The numbers are written on the ballot paper and the voter must write the names of candidates beside them.

- Touch screen: When voting is done with a computer aid, a touch screen can be used. In this case, a computer can help the voter to arrange her or his choices in any way.

Below are examples of ballots where a voter identifies a single alternative or ranks some of them:

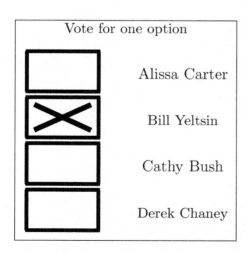

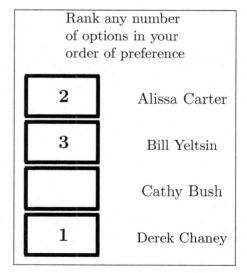

The above right ballot gives an example of a **preference ballot** because it expresses preference ranking of alternatives. A ballot in which ties are not allowed is called a **linear ballot**. In what follows, we consider only linear preference ballots. It does not mean that ties cannot happen in an election; we only insist that a voter avoids ties in their ranking.

Example 1.1.1. Four candidates (A, B, C, and D) are running for the president position of a corporation, and nine board members voted for these candidates. They were asked to rank their choices that resulted in the following ballots:

Ballot			Ballot			Ballot			Ballot			Ballot			Ballot			Ballot			Ballot			Ballot
A	1		A	2		A	1		A	4		A	1		A	4		A	2		A	1		A	3
B	2		B	3		B	4		B	2		B	2		B	2		B	4		B	4		B	1
C	3		C	1		C	3		C	3		C	3		C	1		C	1		C	2		C	4
D	4		D	4		D	2		D	1		D	4		D	3		D	3		D	3		D	2

The committee summarized the results in Table 4.

Number of votes:	2	1	1	1	1	1	1	1
A	1	1	1	2	2	3	4	4
B	2	4	4	3	4	1	3	2
C	3	3	2	1	1	4	2	1
D	4	2	3	4	3	2	1	3

Table 4: Ranking summary of preference voting.

Preferential voting (or preference voting) is a type of ballot structure used in several electoral systems in which voters rank a list or group of candidates in order of preference. Ballots with options allow voters to express their preferences in a more precise manner than ballots with a single choice. Voters' preferences usually depend on available information.

Example 1.1.2. Our next example shows that linearity in ballots is sometimes hard to achieve. Suppose that a club consisting of 300 members is going to celebrate its anniversary. Its members are asked how much they are willing to budget for the upcoming club party. Various club members suggest different amounts. A few suggest $0 because they do not want the club party. Others suggest extravagant amounts. Most people are somewhere in between.

In view of disagreements, the club members nominate a committee of three activists to come up with a decision according to a voting procedure. After recording all possible choices, there are about 150 of them, the committee realizes that they are put into a corner.

The first step to resolve this issue is to express members' opinions in a preference ballot with a reasonable number of options—four or five of them. However, how to arrange options on a line, with each person's favorite choice being a point on the line, is not clear. Moreover, the committee does not know how to maintain what mathematicians call "linearity." This means that you can arrange the options on a line, with each person's favorite choice being a point on the line. Then, if it is linear, an alternative is always preferred if it is closer to the voter's optimal cost.

This is easier to understand by considering an example. If someone thinks that $8.45 should be spent, we might expect that they would prefer spending $10.99 over $25.00 and $5.55 over $0, because these options are closer; this is what linear preferences mean. It is less clear that the person would prefer $5 over $9, just because 5 is closer to 9 than 5 is to 0—linear preferences does not assume this.

In fact, people's preferences might not be linear. Someone might say, if we are going to do it, let us do it right. Either $100 or nothing. Others may claim that it is not the right time for any party because of economic troubles. For others the day of party may affect the amount of contribution. So we see that non-linear situations may easily occur in everyday life. □

Some elections permit **write-in** candidates where a voter writes the names of candidates against the corresponding places. However, this type of ballot may be difficult to implement, especially if the number of options is extensive. It is a custom to represent a ranking of alternatives as a vertical list with the most preferable competitor on top and the least wanted alternative on the bottom. Instead of column representation, the preference ranking might be expressed as a row, with the most wanted candidate at the left and with least wanted alternative at the right. To exercise a maximum amount of freedom, some elections permit ties. For instance, if there are three options, A, B, and C, they may be ranked so that B is first, A is second, and C is third; or A and C may tie for the second/third place. This is expressed in the following self-explanatory notation:

$$\begin{bmatrix} B \\ A \\ C \end{bmatrix} \text{ equivalent to } \langle B, A, C \rangle \qquad \text{or} \qquad \begin{bmatrix} B \\ A \sim C \end{bmatrix} \text{ equivalent to } \langle B, A \sim C \rangle .$$

The left ballot shows that B is the most wanted candidate, but C is the least wanted. Every ballot can be written not only in a vector form (as a vector row or vector column), but also as a list $B \succ A \succ C$ or $B \succ A \sim C$, where $B \succ A$ means that alternative A is less preferable than B, and notation $A \sim C$ shows a tie between alternatives A and C.

As an example, consider the determination of the most valuable (to his team) player in the National Hockey League (NHL). Such a player is awarded annually with the most prestigious prize—the Hart Memorial Trophy. The winner is selected in a poll of the Professional Hockey Writers' Association in all NHL cities at the end of the regular season. The original Hart Trophy was donated to the NHL in 1923 by Dr. David A. Hart, father of Cecil Hart, former manager-coach of the Montreal Canadiens. Selection of the winner is made in two steps. First, three most valuable players are selected as nominees. After that, the winner is determined from these nominees using write-in ballots. At the end of the 2008/9 season, a typical ballot might look like the following table:

First choice:	Alexander Ovechkin, Washington Capitals
Second choice:	Evgeni Malkin, Pittsburgh Penguins
Third choice:	Pavel Datsyuk, Detroit Red Wings

Example 1.1.3.   We reconsider Example 1.1.1 assuming that the nine members wrote their preferences on the ballots. Then the preference schedule takes on the form of Table 6.

□

Number of votes	2	1	1	1	1	1	1	1
First choice:	A	A	A	C	C	B	D	C
Second choice:	B	D	C	A	A	D	C	B
Third choice:	C	C	D	B	D	A	B	D
Fourth choice:	D	B	B	D	B	C	A	A

Table 6: Summary of write-in voting.

It is very important to keep in mind that if even one of the candidates drops out of (or is eliminated from) the election, the voter's preference remains. This means that if a voter prefers alternative A over alternative B and prefers alternative B over alternative C, then the voter prefers A to C. Therefore, when you compare two candidates, say A and C, you have to look at where A and C are placed and which one is above the other. It is the same as if you eliminate all candidates, but A and C. For instance, let us consider a ballot in either preference ranking form or in equivalent write-in form:

Ballot	
A	1
B	2
C	3
D	4

or

Ballot	
1st	A
2nd	B
3d	C
4th	D

after elimination ⟹

Ballot	
A	1
C	3

Ballot	
1st	A
3d	C

So we see that the result of comparison between two candidates, A and C, depends only on places where they are written in the ballot. Since the first choice is better than the third one, we know that candidate A is preferable to candidate C.

When there are only two candidates, it is reasonable to apply the **majority rule:**
The person (or alternative) who attracts more than half of the votes (at least 50% plus one vote) is the winner.

Obviously, dictatorship has no role in our study. Of course, we cannot ignore the possibility of a tie. In the real world, the number of voters is often very large so ties seldom occur. However, generally speaking, ties may happen in any election, under some conditions (for instance, an election with an even number of voters in a two-candidate run). A result known as the May theorem states: among all two-candidate voting systems that never result in a tie, the only fair voting method is the majority rule.

At first glance, there should always be a majority in any group of people. However, this is not always the case, there could be several conflicting majorities. One might try to explain this phenomenon by saying that the voters have behaved irrationally, or that their opinions are nonsensical. Most likely, different choices were picked according to different voter's objectives. Actually, the unique majority may not exist or its determination may depend on the formulation of the options on the ballots. It is clear that sometimes voters' choices can not be arranged in a left to right, high to low, line. But when we can order them, there are many different ways to arrange candidates according to different issues.

It is common to think of "the majority" as the largest group of people. In reality, there may be several groups of people within one population that share similar opinions, with some overlap between different majorities. A concentrated majority may be able to dominate a more dispersed majority that has failed to coordinate and unite.

Often people assume that the purpose of an election is to discover, and act on, the majority will. Since the majority might be in conflict, this is not always possible. It so happens that the election goal is to find the best choice that is as close to a majority rule as possible. Another example of elections that lead to outcomes far away from an average choice is given by the actual distribution of ideological attitudes of Democratic and Republican members of the U.S. House of Representatives in recent decades. It was noticed that each party leader tended to be more extreme than the median members of the party, in the direction of the party mode, who in turn are more extreme than the median voter in the population as a whole. As a result, the median voter may face a choice between two polarized options, neither of which reflects the voter's position.

Problems.

1. The ballots of preferential voting are given below. Make a ranking summary of the voting and convert the table into a write-in schedule.

Ballot			Ballot			Ballot			Ballot			Ballot			Ballot			Ballot			Ballot	
A	1		A	3		A	3		A	4		A	1		A	4		A	2		A	3
B	2		B	4		B	4		B	2		B	2		B	2		B	4		B	4
C	3		C	1		C	1		C	3		C	3		C	1		C	1		C	1
D	4		D	2		D	2		D	1		D	4		D	3		D	3		D	2

2. Thirteen ballots of write-in voting are given below. Make a summary of the voting and convert the table into a ranking schedule.

Ballot			Ballot			Ballot			Ballot			Ballot			Ballot			Ballot	
1st	A		1st	B		1st	C		1st	D		1st	A		1st	B		1st	C
2nd	B		2nd	C		2nd	D		2nd	A		2nd	B		2nd	C		4th	D
3d	C		3d	D		3d	A		3d	B		3d	C		3d	D		3d	A
4th	D		4th	A		4th	B		4th	C		4th	D		4th	A		4th	B

Ballot		Ballot		Ballot		Ballot		Ballot		Ballot	
1st	D	1st	A	1st	B	1st	C	1st	D	1st	A
2nd	A	2nd	B	2nd	C	2nd	D	2nd	A	2nd	B
3d	B	3d	C	3d	D	3d	A	3d	B	3d	C
4th	C	4th	D	4th	A	4th	B	4th	C	4th	D

3. An election is held between five candidates: A, B, C, D, and E. The preference ranking for the election is as follows:

Number of votes:	2	1	2	3	1	4	2	1
A	1	1	5	2	5	3	4	4
B	2	4	4	3	4	1	3	5
C	3	5	2	1	1	4	2	1
D	4	2	3	5	3	2	5	3
E	5	3	1	4	2	5	1	2

  (a) Convert the table into a write-in schedule.

  (b) How many people voted in this election?

  (c) How many first-place votes are needed for a majority?

  (d) Which candidate had the most first-place votes?

  (e) Which candidate had the most last-place votes?

4. An election is held between five candidates: A, B, C, D, and E. The summary of write-in voting is as follows:

Number of votes:	2	3	1	3	1	2	1	5
First choice:	A	E	A	C	C	B	E	C
Second choice:	B	D	E	A	A	D	C	B
Third choice:	C	C	D	E	E	A	B	D
Fourth choice:	D	B	B	D	B	C	A	E
Fifth choice:	E	A	C	B	D	E	D	A

  (a) Convert the table into a ranking schedule.

  (b) How many people voted in this election?

  (c) How many first-place votes are needed for a majority?

  (d) Which candidate had the most first-place votes?

  (e) Which candidate had the most last-place votes?

5. An election is held to choose the Chair of the Computer Science department at a university. The candidates are professors Agu, Beck, Ciaraldi, Fisler, and Gennert (A, B, C, F, and G, for short). The preference schedule is presented here.

Number of votes:	1	3	6	3	4	2	1	5
First choice:	A	B	F	C	G	B	A	G
Second choice:	B	A	G	A	A	F	C	B
Third choice:	C	C	A	F	E	A	B	D
Fourth choice:	F	G	B	G	B	C	G	E
Fifth choice:	G	F	C	B	G	G	F	A

    (a) How many people voted in this election?

    (b) How many first-place votes are needed for a majority?

    (c) Which candidate had the most first-place votes?

    (d) Which candidate had the most last-place votes?

6. A local Thai restaurant conducted a survey among students about their favorite appetizer in the restaurant. The most favorable alternatives were Chicken Fingers, Edamame, French Fries, Nime Chow, and Spring Rolls (C, E, F, N, and S, for short). The preference schedule is presented here.

Number of votes:	4	3	2	5	7	2	3	4
Chicken Fingers	1	2	3	4	5	4	3	1
Edamame	2	1	5	3	4	2	1	5
French Fries	3	4	1	5	2	1	2	3
Nime Chow	4	5	2	1	3	3	5	4
Spring Rolls	5	3	4	2	1	5	4	2

    (a) How many people voted in this election?

    (b) How many first-place votes are needed for a majority?

    (c) Which appetizer had the most first-place votes?

    (d) Which appetizer had the most last-place votes?

7. An election is held between five candidates: A, B, C, D, and E. The preference ranking for the election is as follows:

Number of votes:	2	3	2	2	1	4	2	3
A	1	3	5	2	4	3	5	4
B	2	2	4	3	1	5	4	3
C	3	4	3	1	5	2	1	5
D	4	5	2	4	3	1	3	1
E	5	1	1	5	2	4	2	2

Suppose that the election rules are that when there is a candidate with a majority of the first-place votes, she or he is the winner. Otherwise, all candidates with 25% or less of the first-place votes are eliminated and the ballots are recounted.

    (a) Which candidates are eliminated in this election?

    (b) Find the ranking schedule for the recount.

    (c) Which candidate is the majority winner after the recount?

8. An election is held between five candidates: A, B, C, D, and E. The summary of write-in voting is as follows:

Number of votes:	1	2	3	4	1	2	1	5
First choice:	A	E	A	C	C	B	E	D
Second choice:	B	A	D	A	D	E	C	E
Third choice:	C	D	B	B	B	C	D	A
Fourth choice:	D	C	E	E	A	D	B	B
Fifth choice:	E	B	C	D	E	A	A	C

Suppose that the election rules are that when there is a candidate with a majority of the first-place votes, she or he is the winner. Otherwise, all candidates with 25% or less of the first-place votes are eliminated and the ballots are recounted.

    (a) Which candidates are eliminated in this election?

    (b) Find the summary of write-in voting for the recount.

    (c) Which candidate is the majority winner after the recount?

9. A party is holding its annual convention. The 500 voting delegates are choosing among three possible party platforms: F (a free democratic platform), G (a green platform), and SD (a social democratic platform). Twenty-three percent of the delegates prefer SD to F and F to G. Thirty-one percent of the delegates like G the most and SD the least. The rest of delegates like F the most and G the least. Write out the preference schedule for this election.

10. A condominium association is holding its annual election for president. The three candidates are A, B, and C. 25% of the voters like B the most and C the least; 35% of the voters like A the most and B the least. Of the remaining voters there are three voters that prefer C to A and A to B, and five voters that prefer A to B and B to C. Write out the preference schedule for this election.

11. In a competition between three candidates—A, B, and C—a certain number of voters cast their ballots. Of them, five voters prefer A to B, five voters prefer A to C, seven voters prefer B to A, and four voters prefer B to C. How many voters participated in this competition?

1.2   The Plurality Methods

The most prevalent single-winner voting method, by far, is **plurality** (also called "relative majority," or "winner-takes-all") in which each voter picks up only one choice, and the choice that receives the most votes wins, even if it receives less than the majority of the votes.

Therefore, in an election, where the winner is determined by a plurality method, voters do not need to put the candidates in order—the only information that is taken into account is the voter's only choice. It is believed that on average there will only be two viable candidates for any given election under the plurality system. The vast majority of elections for political office in Europe and the United States is based on a plurality system.

Example 1.2.1. Plurality Method
There are 12 soccer teams in a local competition. The teams must vote for the game day out of four available choices: Monday (M for short), Tuesday (T), Thursday (R), and Sunday (S). The captains of the teams decide to use the plurality method, and their ballots read as follows: T, M, S, T, R, M, M, T, R, S, M, R. The tally of the original ballots shows that Monday gets four votes, whereas all others get less: Tuesday and Thursday have three votes each, but Sunday has only two votes. Therefore, Monday is chosen as the group's selection.

                                                                           □

A plurality method attracts its attention because of its simplicity in applications, and it does not require any explanation—it is clear to every voter. The ballots are also very easy to count. The counting process can even be done without any equipment or a voting machine. All elections conducted under the plurality methods satisfy the fairness conditions (on page 2). Moreover, they support another criterion known as the majority rule.

Majority Criterion:
If an alternative cast a majority of the first-place votes, then this alternative should be the winner of the election. ∎

Terrible tragedies in democracy's history were often sparked by the plurality rule. Since the plurality method in a multi-competitor election usually determines a winner that has failed to attract the majority of voters, it causes the election to be vulnerable to attacks from rival factions. Any plurality elected president that is not a clear majority winner has a weak mandate to set and enforce policies. Let us recall two examples, one is known from U.S. history. Abraham Lincoln became U.S. president in 1860 with less than 40% of the vote. This gave secessionists a powerful argument to deny his authority—hastening the Civil War. Another well-known example is that of Salvador Allende who became president of Chile in 1970 with 36% of the vote. That was followed in 1973 by a right-wing coup and 17 years of military dictatorship.

Other than satisfying the basic fairness criteria, the plurality method is generally considered a very poor method of an election if the number of alternatives exceeds two. The following flaws support this observation: the plurality method is very easy to manipulate and is susceptible to insincere voting, as the following two examples show.

Example 1.2.2. Manipulation
Suppose that in an election, two candidates, A and B, compete for an office. We also assume that they are representatives of two different parties, so their supporters will never switch from one candidate to another. Since there are many companies on the market (for instance, Gallup company) that can predict and estimate with very good accuracy the outcome in any election, these two candidates know that one of them, say A, would attract 60% of voters, but the other one would have about 40%.

In efforts to help his cause, candidate B hires another candidate—usually called a "clone" candidate—denoted by C, that is near-identical to candidate A. Then supporters of candidate A will split approximately in half between candidate A and candidate C. This will result in both of them getting approximately 30% of the votes, and candidate B will get 40% of the votes. So candidate B will become the winner in the election according to the plurality system. This example shows that the plurality system makes it fairly likely that an extreme candidate with a small but intense base of support can defeat a slate of candidates with broad support and similar views (see also Example 1.2.7, page 16). □

Another problem with the practical application of the plurality method is caused by **insincere voting**. With more than two alternatives contest in the election, the voter is faced with the dilemma: either support his or her attractive but "weak" candidate or vote for another more popular candidate. Since plurality does not allow a voter to express his or her support for more than one candidate, it forces the voter to make a single choice. Therefore, a voter may get a feeling of "wasting" his or her vote by supporting the attractive candidate, which may lead to switching to another serious competitor. In closely contested elections, a few insincere ballots may completely change the outcome of an election.

Example 1.2.3. Insincere Voting

Consider an election between four competitors, say A, B, C, and D. The results of preference rankings are summarized in the following table:

Competitors	Number of Votes				
	1	3	4	5	6
A	1	1	4	2	1
B	3	2	2	3	4
C	2	4	1	1	3
D	4	3	3	4	2

or

Ranking	Number of Votes				
	1	3	4	5	6
First choice	A	A	C	C	A
Second choice	C	B	B	A	D
Third choice	B	D	D	B	C
Fourth choice	D	C	A	D	B

Since candidate A has 10 first-place votes, but candidate C has only 9 first-place votes, the former one is the winner. Now suppose that the voter (his or her preference ranking is given in the first column) switches the first two places. This results in the following outcome:

Competitors	Number of Votes				
	1	3	4	5	6
A	2	1	4	2	1
B	3	2	2	3	4
C	1	4	1	1	3
D	4	3	3	4	2

or

Ranking	Number of Votes				
	1	3	4	5	6
First choice	C	A	C	C	A
Second choice	A	B	B	A	D
Third choice	B	D	D	B	C
Fourth choice	D	C	A	D	B

This will give candidate C 10 first-place votes, while A will have only 9 supporters. The result of such an election will be different—candidate C wins the election.

Example 1.2.4. Assurance to Win

Suppose that 100 votes are cast in an election among four candidates: Daly, Howarth, Russo, and Stewart. The election is to be decided by plurality. After the first 75 votes are counted, the tallies are as follows:

Daly   17   Howarth   12   Russo   21   Stewart   25

What is the minimal number of the remaining votes every candidate can receive to be assured of a win?

Solution To start, we identify the leader—Stewart—and the next to the leader—Russo. There are four votes difference between these two most wanted candidates so far. Since there are twenty-five votes left uncounted, we can give four of them to Russo, which makes a tie between these two competitors. Then twenty-one votes will be left at our disposal. In the worst-case scenario for the other 2 candidates (Daly and Howarth), assume that all other votes are distributed between Russo and Stewart. Then to win the election, Stewart needs more than half of the remaining votes $(21/2 = 10.5)$, which leads to eleven votes. Alternatively, the second place behind Stewart needs eleven votes plus four to catch up, so Russo needs fifteen votes.

We can also solve this problem algebraically. Let x be the number of votes Stewart needs in order not to lose (to ensure at least a tie) the election. If Russo gets all the votes that do not belong to Stewart and the race ends with a tie between these two most wanted candidates, we should have

$$25 + x = 21 + (25 - x).$$

Solving this equation for x, we obtain

$$2x = 21 \quad \implies \quad x = \frac{21}{2} = 10.5.$$

Therefore, if Stewart gets at least 11 votes out of remaining 25 votes, it would ensure his win in the election.

Now we consider two other candidates. Starting with Daly, we see that he needs 8 votes to catch up the leader—Stewart. If these 8 votes out of the remaining 25 go to Daly, we are left with 17 votes. Assuming that these votes are distributed evenly between the two candidates—Daly and Stewart—we see that Daly needs at least half of the rest to ensure a win, that is, 9. Summing 8 and 9, we get the required number of votes that will bring Daly a win.

Since the difference between the leader and Howarth is 13, we give these votes to Howarth to make a tie and are left with 12 votes out of the remaining 25. To ensure a win, Howarth needs more than a half of these, namely, $12/2 = 6$ plus 1, so she needs $7 + 13 = 20$ votes. We summarize the results in the following table:

Candidate	Votes	Needed Votes
Daly	17	**17**
Howarth	12	**20**
Russo	21	**15**
Stewart	25	**11**

Example 1.2.5. Minimum Votes to Win
If 202 votes are cast, what is the smallest number of votes a winning candidate can have in a 4-candidate race that is to be decided by plurality? What is it if there are 203 or 204 votes?

Solution Assuming that all voters are spread uniformly between these four candidates, we need to divide 202 by 4:

$$\frac{202}{4} = \frac{200 + 2}{4} = \frac{200}{4} + \frac{2}{4} = 50 + \frac{2}{4}.$$

Note that this number is not divisible by 4 (in whole numbers). If every candidate receives 50 votes, we are left with 2 to be distributed among 4 candidates:

$$\boxed{50}^{\;2} \quad \boxed{50} \quad \boxed{50} \quad \boxed{50}$$

Since we cannot split 2 votes between two candidates—it will lead to a tie—we have to give these 2 votes to one of the alternatives. So the answer is 52 votes.

Now assume that there are 203 votes. Again dividing this number by 4, we are faced with a dilemma of how to distribute 3 votes among four candidate so that one will be a winner with the least number of votes. This can be achieved only if one candidate gets 2 votes and another one grabs 1 vote:

$$\boxed{50}^{\;2} \quad \boxed{50}^{\;1} \quad \boxed{50} \quad \boxed{50}$$

Therefore, the smallest number of votes to win is again 52. If there are 204 votes, we encounter the same answer: 52 votes, as the following distribution shows:

$$\boxed{50}^{\;2} \quad \boxed{50}^{\;1} \quad \boxed{50}^{\;1} \quad \boxed{50}$$

When there are 205 votes, the minimum number of votes to win, 52, can be achieved as follows:

$$\boxed{51}^{\;1} \quad \boxed{51} \quad \boxed{51} \quad \boxed{51}$$

1.2.1 Runoff Method

The two-round system is known as **runoff voting**, where the second round is used to ensure that the winner is elected by a majority. Runoff voting is sometimes used as a generic term to describe any system involving any number of rounds in voting, with eliminations after each round. However, we will use this term only for a two-round system subject to a strict rule of elimination on the first round. Needless to say that there are many other variants of the two-round system with different rules for eliminating candidates, which allow more than two candidates to proceed to the second round. Under these systems it is sufficient for a candidate to receive a plurality of votes (i.e., more votes than anyone else) to be elected in the second round. In elections for the French National Assembly any candidate with less than 12.5% of the total vote is eliminated in the first round, and all remaining candidates are permitted to stand in the second round, in which a plurality is sufficient to be elected. Under some variants of runoff voting there is no formal rule for eliminating candidates, but, rather, candidates who receive few votes in the first round are expected to withdraw voluntarily.

A runoff method enjoys simplicity inherited from plurality because a voter simply marks her or his choice. Top-two runoff voting is the second most common method used in elections. It consists of two rounds, each of them is conducted using plurality voting. If no candidate has an absolute majority of votes (i.e., 50% plus one) in the first round, then the two candidates with the most votes proceed to a second round, from which all others are excluded. Since there are only two competitors in the second stage, the majority rule defines the winner.

Runoff voting is widely used around the world for the election of legislative bodies and directly elected presidents. Typically, the runoff method is implemented in two elections separated by two weeks. For example, it is used in French presidential, legislative, and cantonal elections, and also to elect the presidents of Argentina, Austria, Brazil, Bulgaria, Chile, Croatia, Cyprus, Dominican Republic, Finland, Ghana, Guatemala, Indonesia, Poland, Portugal, Romania, Russia, Serbia, Zimbabwe, and many others.

There is one major disadvantage in this method—it is expensive in both money and time. Al elections are expensive for a government—it pays for organization, control, and making it public; competitors also spend a lot of time and resources before an election; the electorate spends time watching debates and participating in an election. However, the runoff method is much harder to manipulate than the plurality method. It is intended to reduce the potential for eliminating "wasted" votes by tactical voting. Therefore, many countries decide to make the investment into a runoff election by paying twice so that the winner gets the majority support of the population.

Below is an example of a ballot used in runoff voting. As a result of counting all ballots, only two candidates survive (at most) to the second round. An example of a second round ballot is presented below.

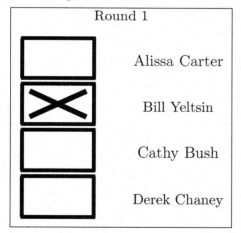

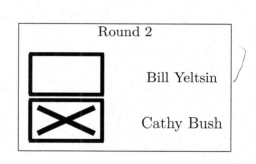

A much more efficient way to implement the runoff method without conducting two separate elections is to use preference ballots. For any pair of competitors, such ballots tell us exactly whom the voters prefer. That is, it identifies which candidate is preferred compared to another one. For instance, if a voter puts candidate A into the third place, and candidate B into the fifth place, we know that the voter prefers A to B.

Example 1.2.6. Runoff Method

The principal at a school has just retired and the school board must hire a replacement. The four nominees for this position are Mr. Berrio, Mrs. Cuna, Mr. Ferrari, and Mrs. Songin (B, C, F, and S, respectively). After interviewing the four finalists, each of the 9 school board members gets to order the candidates by means of preference ballots as shown in the tables. We present two equivalent tables, one with ordering candidates and another one with write-in ballots.

<table>
<tr><td rowspan="2"></td><td colspan="5">Number of Votes (9)</td><td rowspan="9" style="text-align:center">or</td><td rowspan="2">**Ranking**</td><td colspan="5">Number of Votes</td></tr>
<tr><td>1</td><td>1</td><td>2</td><td>2</td><td>3</td><td>1</td><td>1</td><td>2</td><td>2</td><td>3</td></tr>
<tr><td>Berrio (B)</td><td>4</td><td>1</td><td>4</td><td>3</td><td>2</td><td>First choice</td><td>F</td><td>B</td><td>S</td><td>F</td><td>C</td></tr>
<tr><td>Cuna (C)</td><td>2</td><td>4</td><td>3</td><td>2</td><td>1</td><td>Second choice</td><td>C</td><td>S</td><td>F</td><td>C</td><td>B</td></tr>
<tr><td>Ferrari (F)</td><td>1</td><td>3</td><td>2</td><td>1</td><td>4</td><td>Third choice</td><td>S</td><td>F</td><td>C</td><td>B</td><td>S</td></tr>
<tr><td>Songin (S)</td><td>3</td><td>2</td><td>1</td><td>4</td><td>3</td><td>Fourth choice</td><td>B</td><td>C</td><td>B</td><td>S</td><td>F</td></tr>
</table>

The latter table shows us that the plurality method is inconclusive in our case because there is a tie between two candidates—Mrs. Cuna and Mr. Ferrari—both got 3 first-place votes. No candidate has an absolute majority of first-place votes (in this election that would be 5 because the total number of votes is 9). Applying the runoff method, we eliminate the losing candidates, B and S, so the two other candidates with the most first-place votes, C and F, proceed to a second round. The supporters of B and S must now vote for one of the two remaining candidates because their favorite candidates have been eliminated. The second round leads to the following tally: C gets 5 first-place votes whereas F gets 4 first-place votes, so C wins the election based on the runoff method.

<table>
<tr><td rowspan="2">**Ranking**</td><td colspan="5">Number of Votes</td><td rowspan="7" style="text-align:center">⟹</td><td rowspan="2">**Ranking**</td><td colspan="5">Number of Votes</td></tr>
<tr><td>1</td><td>1</td><td>2</td><td>2</td><td>3</td><td>1</td><td>1</td><td>2</td><td>2</td><td>3</td></tr>
<tr><td>First choice</td><td>F</td><td>☐</td><td>☐</td><td>F</td><td>C</td><td>First choice</td><td>F</td><td>F</td><td>C</td><td>F</td><td>C</td></tr>
<tr><td>Second choice</td><td>C</td><td>☐</td><td>F</td><td>C</td><td>☐</td><td>Second choice</td><td>C</td><td>C</td><td>F</td><td>C</td><td>F</td></tr>
<tr><td>Third choice</td><td>☐</td><td>F</td><td>C</td><td>☐</td><td>☐</td><td></td><td></td><td></td><td></td><td></td><td></td></tr>
<tr><td>Fourth choice</td><td>☐</td><td>C</td><td>☐</td><td>☐</td><td>F</td><td></td><td></td><td></td><td></td><td></td><td></td></tr>
</table>

Example 1.2.7. Two Parties

One candidate from the right party receives 15% of the vote, and nine candidates from the left party receive 6%, 7%, 8%, 9%, 10%, 10%, 11%, 12%, and 13% of the vote. The right winger will win a plurality election, despite the fact that 85% of the voters preferred a leftist candidate.

In a runoff, a second round would be held between the right wing candidate with 15% and the leftist candidate with 13% of the initial vote. Since an 85% majority of the voters preferred leftist candidates, the remaining leftist would likely win with ease.

Of course, it is not certain that the candidate with 13% is the best representative of the leftists. This illustrates the basic limitation of the two-round runoff's effectiveness.

Example 1.2.8. French Presidential Election, 2002

In the 2002 French presidential election, the two contenders described by the media as having the possibility to win were Jacques Chirac and Lionel Jospin. However, a third contender, Jean-Marie Le Pen, unexpectedly obtained slightly more than Jospin in the first round of elections:

- Jacques Chirac 19.88%

- Jean-Marie Le Pen 16.86%

- Lionel Jospin 16.18%

Since none of the candidates obtained an absolute majority, a second round was organized with the first two candidates from the list. Jean-Marie Le Pen is, however, a very controversial politician, and in the second round a vast majority of the voters rejected him:

- Jacques Chirac 82.21%

- Jean-Marie Le Pen 17.79%

This example demonstrates how the two candidates from the first round might not be the favorite candidates of the population. □

1.2.2 Electoral College

The Electoral College was established by the founding fathers as a compromise between election of the president by Congress and election by popular vote. It was the Committee of Eleven, formed to work out various details including the mode of election of the president of the United States, that recommended the election to be conducted by a group of people apportioned among the states in the same numbers as their representatives in Congress. Later, the Convention approved the Committee's proposal, with minor modifications, on September 6, 1787. Although the United States Constitution refers to "Electors" or "electors," the name "Electoral College" was not used until the early 1800s. Since that time the name "Electoral College" came into general usage as the collective designation for the electors selected to cast votes for president and vice president. It was first written into federal law in 1845.

Presidential electors are selected on a state-by-state basis, as determined by the laws of each state. Each state currently uses its statewide popular vote on Election Day to appoint electors. Although ballots list the names of the presidential candidates, voters within the 50 states and Washington, D.C. actually choose electors for their state when they vote for president and vice president. These presidential electors in turn cast electoral votes for those two offices. Even though the aggregate national popular vote is calculated by state officials and media organizations, the national popular vote is not the basis for electing a president or vice president.

A candidate must receive an absolute majority of electoral votes (currently 270) to win the presidency. If no candidate receives a majority in the election for president, or vice president, that election is determined via a contingency procedure. The Office of the Federal Register coordinates the functions of the Electoral College on behalf of the Archivist of the United States, the states, the Congress, and the American people. The Office of the Federal Register operates as an intermediary between the governors and secretaries of state of the various States and the Congress. It also acts as a trusted agent of the Congress in the sense that it is responsible for reviewing the legal sufficiency of the certificates before the House and Senate accept them as evidence of official state action.

Despite that electors pledge to stick with results of elections in their state, sometimes they break the pledge and vote for another candidate.

The president of the United States is not elected by a plurality of votes nationwide, but by 538 members (since 1964) of the Electoral College, called electors. This number consists of 435 seats in the House of Representatives, 100 senators (two from each state), and 3 members from the District of Columbia, in total 538. Each state, except for Maine and Nebraska, elects its members of the Electoral College by the winner-take-all method, namely, the winner of the plurality vote in a state is entitled to all the electors from the state. These two states give an elector to the winner of the plurality of votes in each congressional district and in addition give two electors to the winner of the plurality of the statewide vote. Hence, the electors in Maine and Nebraska may be split among several candidates.

The following table shows the number of electoral votes (EV) to which each state and the District of Columbia was entitled during the 2004 and 2008 presidential elections (see §2.5):

State	EV	State	EV	State	EV	State	EV
Alabama	9	Indiana	11	Nebraska	5**	S. Carolina	8
Alaska	3	Iowa	7	Nevada	5	S. Dakota	3
Arizona	10	Kansas	6	New Hampshire	4	Tennessee	11
Arkansas	6	Kentucky	8	New Jersey	15	Texas	34
California	55	Louisiana	9	New Mexico	5	Utah	5
Colorado	9	Maine	4**	New York	31	Vermont	3
Connecticut	7	Maryland	10	North Carolina	15	Virginia	13
Delaware	3	Massachusetts	12	North Dakota	3	Washington	11
Florida	27	Michigan	17	Ohio	20	W. Virginia	5
Georgia	15	Minnesota	10	Oklahoma	7	Wisconsin	10
Hawaii	4	Mississippi	6	Oregon	7	Wyoming	3
Idaho	4	Missouri	11	Pennsylvania	21	D.C.*	3
Illinois	21	Montana	3	Rhode Island	4	Total electors	538

* Washington, D.C., although not a state, is granted three electoral votes by the Twenty-third Amendment.

** Maine and Nebraska electors distributed by way of the Congressional District Method.

Let us look at some historical examples. In the 1980 election, Ronald Reagan beat president

Jimmy Carter by only 50.7% to 41.0% in the national popular vote, but in the electoral vote the split was 489 to 49, with Reagan taking 92.6% of all votes.

In the 2000 election, George W. Bush narrowly won the November 7 election, with 271 electoral votes to Albert Gore's 266 (with one elector abstaining in the official tally). In this election, Bush got 50,456,002 votes or 47.87%, whereas Gore, had 50,999,897 popular votes or 48.38%. The third candidate, Ralph Nader, collected 2,882,955 votes or 2.74% popular votes. There is no doubt that Gore won this election according to plurality, but ultimately the outcome was decided in the Supreme Court. This was the third case in U.S. elections in which the President-elect failed to win the popular vote. The first had been in the 1876 election, and the second was in the 1888 election.

The U.S. presidential election of 2000 was one of the closest elections in U.S. history, decided by only 527 votes in the swing state of Florida. On election night, the media prematurely declared a winner twice based on exit polls before finally deciding that the Florida race was too close to call. It would turn out to be a month before the election was finally certified after numerous court challenges and recounts. Republican candidate George W. Bush won Florida's 25 electoral votes by a razor-thin margin of the popular votes (2,912,790 votes against 2,912,253).

1.2.3 The Single Transferable Vote

The term **single transferable vote** actually incorporates two different voting systems: the Hare method and the Coombs method. Whichever of the methods is applied, voters must rank the names of alternatives on the ballot according to their preference. Number 1 refers to the voter's first choice, number 2 to the second, and so on. There is no universal agreement on whether to force the voter to rank all candidates, or some of them, or only pick up at least one choice. In contrast to the two-stage runoff method, which reduces voter's power to some extent, the single transferable voting takes into account all candidates. The latter may also require only one election if preference ballots are used. However, the single transferable voting system includes multiple iterations by identifying at each step the best alternative and the worst alternative; if the best one wins the majority, then the iteration stops and this option is declared the winner, otherwise the worst alternative is eliminated. The votes of the eliminated alternative are transferred to the next choice candidate, and iteration proceeds. Such transformation of votes is similar to redistribution of medals in Olympic Games when a medal holder does not pass a drug test followed by the holder's elimination. The distinction between the Hare method (1861) and the Coombs method (1964) is based on the way of defining the worst candidate.

The concept of transferable voting was first proposed by Thomas Wright Hill (1763 – 1851) in 1821. The **Hare** system or **instant runoff voting** (IRV) or the **plurality with sequential eliminations** (which is the most descriptive of the three names) is a multiple-round voting method. It was introduced by the British lawyer, Thomas Hare (1806 – 1891). In every round, alternatives are compared based on the plurality method—only the first-place votes are counted. If one of the alternatives (candidates) collects more than 50% first-place votes, then that alternative is the winner. If the majority rule cannot determine a winner, the alternative (or alternatives if there is a tie) with the *fewest* first-place votes is eliminated, and a new round of calculations is held. The eliminated candidate is called the loser. The votes of the loser are passed on to the candidate that the voters placed as second choice on the ballots, and these votes are recounted. These rounds are repeated until one candidate (if any) obtains more than half of the votes. The final round may lead to one of the following three cases.

1. If the final round of election results in elimination of all candidates (when there is a tie between them) but one, the leader is announced to be a winner.

2. If two alternatives participate in the final round, the alternative is determined by the majority rule; otherwise it is a tie.

3. If three or more candidates are left in the final round, and they share the same number of first places, the Hare method is inconclusive because of the tie.

In comparison to other voting schemes in practical use, the single transferable vote (STV) is somehow complicated from voter's point of view because it requires ranking of all candidates. However, despite its relative complexity, the STV or its variants are used in the presidential elections in Ireland and Sri Lanka, in elections for the parliaments of Ireland, Tasmania, Malta, Papua New Guinea, and the Fijian House of Representatives; for the senates of Northern Ireland, Australia, and South Africa; for local authorities in Australia, Canada, the United Kingdom, New Zealand, and the USA. The Hare system is currently used, in different forms, to elect the President of the European Parliament, the speakers of the Canadian and British Houses of Commons, the various party nominees for president of the United States, and the host city of the Olympic Games.

The one-round elections have advantages in avoiding voters having to go back to the polls, and saving the government money on elections. The single transferable vote allows the voter to fully express her or his preferences between all listed alternatives, it is not biased by influence of the first-round results (as in a runoff), and therefore, by the opinions of other voters. This electoral system greatly reduces voter's incentives to vote insincerely. However, the system does not completely eliminate the possibility of manipulation. The main drawback of the single transferable vote is its complexity, in particular, the impossibility to count votes in local offices—the national result aggregates results of all local offices. The counting process is greatly facilitated through the use of information technology, as the French electoral experiment shows (see Example 1.2.17, page 29).

Example 1.2.9. The IRV Method

Candidates	Number of Votes (13)			
	5	4	3	1
Alex	1	4	3	3
Boris	2	2	4	1
Cardona	3	1	2	4
Dennis	4	3	1	2

or

Ranking	Number of Votes			
	5	4	3	1
First choice	A	C	D	B
Second choice	B	B	C	D
Third choice	C	D	A	A
Fourth choice	B	A	B	C

A 13-member board used the instant runoff voting to elect a chair. The four candidates were Alex (A for short), Boris (B), Cardona (C), and Dennis (D). The preference rankings on the 13 ballots are listed above.

Boris is eliminated after the first round, and we are left with the rankings:

	Number of Votes (13)			
Ranking	5	4	3	1
First choice	A	C	D	D
Second choice	C	D	C	A
Third choice	D	A	A	C

Since Cardona and Dennis have fewer first-place votes than Alex, the latter one wins the election. Note that Alex, who is also the least wanted candidate, will also be the winner according to the plurality. So Alex is the plurality winner and the plurality loser. Using the runoff method, we eliminate Boris and Dennis, leaving two best finishers:

	Number of Votes			
Ranking	5	4	3	1
First choice	A	C	C	A
Second choice	C	A	A	C

The table shows that Alex gets six first-place votes whereas Cardona has seven; therefore, she is the winner in the runoff election.

Now assume that instead of a 13-member board, the company has 14 members. Adding one ballot, we supposedly have

Candidates	Number of Votes (14)			
	5	4	3	2
Alex	1	4	3	3
Boris	2	2	4	1
Cardona	3	1	2	4
Dennis	4	3	1	2

or

Ranking	Number of Votes			
	5	4	3	2
First choice	A	C	D	B
Second choice	B	B	C	D
Third choice	C	D	A	A
Fourth choice	B	A	B	C

The first round eliminates Boris since he has two first-place votes. This yields

	Number of Votes (14)			
Ranking	5	4	3	2
First choice	A	C	D	D
Second choice	C	D	C	A
Third choice	D	A	A	C

After the second round, C is eliminated and the tallies become

	Number of Votes			
Ranking	5	4	3	2
First choice	A	D	D	D
Second choice	D	A	A	A

Obviously, Dennis will win this election. However, the runoff method will yield the following rankings after the first stage:

	Number of Votes (14)			
Ranking	5	4	3	2
First choice	A	C	C	A
Second choice	C	A	A	C

So the runoff method is inconclusive for 14-member board elections because there is a tie between Alex and Cardona.

Example 1.2.10. Hare System with Ties

Consider the following sequence of the preference list ballots:

Number of Votes (16)							
1	2	3	2	2	3	1	2
A	A	C	C	B	B	D	D
B	C	B	D	A	D	A	C
C	B	D	A	D	A	C	B
D	D	A	B	C	C	B	A

So A has 3 first-place votes, B has 5 first-place votes, C has 5 first-place votes, and D has 3 first-place votes. Eliminating A and D, we get

Number of Votes (16)							
1	2	3	2	2	3	1	2
B	C	C	C	B	B	C	B
C	B	B	B	C	C	B	C

Since both candidates, B and C, have the same number of first-place votes, the Hare system is inconclusive for this election.

Let us consider another sequence of the four preference list ballots:

Number of Votes (18)							
2	3	2	2	3	3	1	2
A	A	C	C	B	B	D	D
B	C	B	D	A	D	A	C
C	B	D	A	D	A	C	B
D	D	A	B	C	C	B	A

Eliminating D with only 3 first-place votes, we obtain the following results:

Number of Votes (18)							
2	3	2	2	3	3	1	2
A	A	C	C	B	B	A	C
B	C	B	A	A	A	C	B
C	B	A	B	C	C	B	A

The IRV method is inconclusive in our case because all three candidates, A, B, and C, get the same number of first-place votes (6).

Example 1.2.11. IRV versus Runoff

In order to promote a wider viewpoint, a 13-member board of directors of a company has decided to increase its size by adding one to four members. The preference schedule regarding the number of members to add is as follows:

Options	Number of Votes (13)					
	2	2	2	2	3	2
One	1	1	4	4	3	4
Two	2	4	1	1	4	3
Three	3	2	3	2	1	2
Four	4	3	2	3	2	1

Applying plurality, we see that options *One* and *Two* both have four first-place votes, whereas option *Three* has three first-place votes, and option *Four* has only two first-place votes. In the runoff between option *One* and option *Two*, the ballots are as follows:

Options	Number of Votes (13)					
	2	2	2	2	3	2
One	1	1	4	4	3	4
Two	2	4	1	1	4	3

Option *One* defeats option *Two* by a score of 7 to 6. Hence Option *One* is the runoff winner.

With the Hare system, we first eliminate option *Four* as having the least number of first-place votes. This leads to the following preference rankings:

Options	Number of Votes (13)					
	2	2	2	2	3	2
One	1	1	3	3	2	3
Two	2	3	1	1	3	2
Three	3	2	2	2	1	1

Options *One* and *Two* now have only four first-place votes compared to the five first-place votes that option *Three* has. Therefore, these two options, *One* and *Two*, are now deleted, leaving option *Three* as the winner under the IRV system. This example shows that different voting methods may choose distinct winners.

Example 1.2.12. Olympic Host City Election, 2016

Rio de Janeiro, Brazil, was chosen as the host of the 2016 Summer Olympics. The selection was made on November 2, 2009, by the 121-member International Olympic committee in Copenhagen, Denmark. The cities in contention were Rio de Janeiro (Brazil), Madrid (Spain), Tokyo (Japan), and Chicago (USA). All four countries pressed hard and organized a strong campaign to lobby the Olympic committee. For instance, the president of the United States, Barack Obama, visited Copenhagen to express the country's support for Chicago.

The voting process used by the International Olympic committee is based on the plurality with sequential eliminations method. In the first round of voting the tallies were as follows:

First Round

Rio de Janeiro, Brazil	26
Madrid, Spain	28
Tokyo, Japan	22
Chicago, USA	18

Thus, Chicago was eliminated from the competition. The second round of votes yielded the following results:

Second Round

Rio de Janeiro, Brazil	46
Madrid, Spain	29
Tokyo, Japan	20

So we see that Tokyo lost two supporters in the second stage. The final round of votes gave the following standing:

Third Round

Rio de Janeiro, Brazil	66
Madrid, Spain	32

Note that the number of votes increased with each round. Initially, only 94 members of the committee participated in the election. The next round was determined by 95 votes, and in the final stage 98 members cast their ballots.

This example explicitly shows that if we believe that the Hare method decides things based on the relative strength of ranking, we would have to believe that in the first round voters are only concerned about their first choice and view all other options as equal. Then, as candidates are eliminated, new preferences are revealed, and given full weight, as if they were first preferences. This means that the choices that are initially very low on a voter's ballot can end up weighted just as strongly as first preferences.   □

The previous example shows that, unlike plurality, the lower choices may dominate in later stages of elections when some alternatives are eliminated. The Hare method is a commonly proposed replacement for plurality. Often opponents of the instant runoff voting make the claim that it is capricious. Let us see why by considering some examples.

Example 1.2.13. Consider the following election between three alternatives (we label them with A, B, and C):

Ranking	Number of Votes (23)								Number of Votes					
	8	2	4	3	1	5			8	2	4	3	1	5
First choice	A	A	B	B	C	C	or	A	1	1	2	3	2	3
Second choice	B	C	A	C	A	B		B	2	3	1	1	3	2
Third choice	C	B	C	A	B	A		C	3	2	3	2	1	1

The first round gives the following results:
A – 10 first-place votes, B – 7 first-place votes, and C – 6 first-place votes.
The IRV proceeds by dropping C, and A gets an additional vote, but B receives 5 votes.
Round 2: A – 11; B – 12 first-place votes. So B becomes the IRV winner.

Note the plurality winner is alternative A, with 10 first-place votes.

Now, let us imagine a similar election. All that will be changed is that two people will rank alternative A lower on their ballots. No one will change their ballots in any other way. Suppose that we have the following preference rankings:

Ranking	Number of Votes (23)								Number of Votes					
	8	2	4	3	1	5			8	2	4	3	1	5
First choice	A	C	B	B	C	C	or	A	1	2	2	3	2	3
Second choice	B	A	A	C	A	B		B	2	3	1	1	3	2
Third choice	C	B	C	A	B	A		C	3	1	3	2	1	1

However, A becomes the winner according to the Hare system. This is despite the fact that the only difference between these two elections is that two people who preferred A to C in the first, prefer C to A in the second. This is not reasonable evidence that A should win. Let us see why.

Round 1: A – 8 first-place votes, B – 7 first-place votes, and C – 8 first-place votes.
Therefore, B is eliminated, and A gets 4 additional votes, but C receives 3 additional votes.
Round 2: A – 12; C – 11 first-place votes. So A becomes the winner. □

Now we turn our attention to the determination of the worst candidate/alternative, called the **loser**. When preference rankings are provided, the loser under any of the plurality methods is determined as the winner of the reverse rankings (last place becomes the first one, and so on).

Example 1.2.14. Worst Candidate
We consider another example, but now instead of looking for the best candidate, we seek

the worst one—called the loser. Obviously, if voters are sincere, their rankings would all
be exactly reversed. We could then use a plurality method to find the answer to the new
question, that is, "who is the worst candidate?" Suppose we have the following ranking
table:

Ranking	Number of Votes (30)					
	9	1	2	7	9	2
First choice	A	A	B	B	C	C
Second choice	B	C	A	C	A	B
Third choice	C	B	C	A	B	A

Since A has 10 first-place votes, B gets 9 first-place votes, and C has 11 first-place votes,
we see that C is the plurality winner. Let us see who is the worst candidate, by reversing
the order in the ranking table:

Reverse Ranking	Number of Votes (30)					
	9	1	2	7	9	2
First choice	C	B	C	A	B	A
Second choice	B	C	A	C	A	B
Third choice	A	A	B	B	C	C

Calculating the first-place votes in the reverse table, we get that A has 9 first-place votes,
B gets 10 first-place votes, and C has 11 first-place votes in the reverse rankings. So C is
the plurality loser and the plurality winner.

Suppose we want to apply the runoff method. Eliminating B from the competition, we see
that candidate C defeats candidate A with score 18:12. So C is the runoff winner. In the
reverse table, we eliminate candidate A and again find that C beats B with the same score
18:12. Therefore, the runoff method chooses the same candidate for the winner and for the
loser.

Now we consider another election between three candidates (A, B, and C), where 9 voters
cast their ballots as follows:

Ranking	Number of Votes					Number of Votes			
	4	2	3				4	2	3
First choice	B	C	A	or	A	3	2	1
Second choice	C	A	B		B	1	3	2
Third choice	A	B	C		C	2	1	3

The Hare method proceeds by dropping C, then A wins versus B. Now we reverse the
ranking over to obtain

Reverse Ranking	Number of Votes					Number of Votes		
	4	2	3			4	2	3
First choice	A	B	C	or	A	1	2	3
Second choice	C	A	B		B	3	1	2
Third choice	B	C	A		C	2	3	1

In the IRV, B is dropped, and A wins against C. So, if the voters use plurality with sequential eliminations, they can choose the same candidate both for best and worst. Hence, each of three plurality methods—the plurality, the runoff, and the Hare method—contradicts itself! □

At the end of this section, we consider another method, which is very similar to the Hare system, created by Clyde Hamilton Coombs (1912 – 1988), an American psychologist specializing in the field of mathematical psychology. This method now bears his name, also called the **Coombs rule**. Similar to the IRV, it consists of several stages. It is an alternative to the Hare method. According to the Coombs rule, the worst result is defined by the *level of rejection*, whereas the IRV considers the worst result as the *level of support*.

The **Coombs method** requires voters to rank all of the candidates on their ballot. It starts with no alternatives eliminated. If at any time one candidate is ranked first (among non-eliminated candidates) by an absolute majority of the voters, then this is the winner. As long as this is not the case, the candidate that has more last-place votes (again among non-eliminated candidates) is eliminated. Repeat the process until an alternative has a majority in first-place votes among non-eliminated candidates. So far no examples exist showing a practical application of the Coombs method.

Example 1.2.15. Coombs Worst Alternative

The members of a local theater organization must decide which new play they would like to put on. The preference rankings of the members are as follows:

Play	Number of Votes (13)							
	1	2	1	3	1	2	1	2
Cats	1	1	2	2	3	3	4	4
Chicago	2	3	1	4	1	4	2	3
The Producers	3	4	3	1	4	2	1	2
Stomp	4	2	4	3	2	1	3	1

Suppose that the members want first to determine the worst alternative in order to eliminate it from future considerations. To apply the Coombs rule, we first reverse the preference table and determine the winner based on the Coombs method. The winner in this reversed election would be the loser for the original one.

Play	Number of Votes (13)							
	1	2	1	3	1	2	1	2
Cats	4	4	3	3	2	2	1	1
Chicago	3	2	4	1	4	1	3	2
The Producers	2	1	2	4	1	3	4	3
Stomp	1	3	1	2	3	4	2	4

Counting the number of first-place votes to check whether there is a majority winner, we see that *Cats* and *The Producers* get 3 first-place votes, each, *Chicago* has 5 first-place votes, *Stomp* gets 2 first-place votes. So there is no majority winner. Next we calculate the number of last-place votes: *Cats* has 3 fourth-place votes, *Chicago* gets 2, *The Producers* and *Stomp* get 4 last-place votes. In the first round, we eliminate the worst options that have 4 last-place votes. This leads to the competition between only two options:

Play	Number of Votes (13)							
	1	2	1	3	1	2	1	2
Cats	4	4	3	3	2	2	1	1
Chicago	3	2	4	1	4	1	3	2

or

Play	Number of Votes (13)							
	1	2	1	3	1	2	1	2
Cats	2	2	1	2	1	2	1	1
Chicago	1	1	2	1	2	1	2	2

Therefore, *Chicago* defeats *Cats* with a score 8:5, which leads *Chicago* to be the Coombs loser.

Example 1.2.16. IRV versus Coombs

Consider a fiercely competitive election between five candidates, A, B, C, D, and E. The preference schedule is presented here.

Ranking	Percentage of Votes				
	17%	20%	23%	21%	19%
First choice	A	B	C	D	E
Second choice	B	A	D	E	D
Third choice	C	E	A	B	C
Fourth choice	D	C	E	A	B
Fifth choice	E	D	B	C	A

According to the Hare method, candidate A with 17% of first-place votes is dropped in the first round, and her votes are transferred to candidate B. The second round eliminates candidate E. In the third round, candidate C is eliminated leaving B and D in the final competition, which yields the winner—candidate D. Note that candidate C was the plurality winner, dropped in the third round.

The Coombs method starts with dropping the least wanted candidate, which has more last-place votes. We look into the last row of the table and see that the method eliminates candidate B in the first round because he has 23% of last-place votes—more than anybody. The second round eliminates candidate E, with 40% of last-place votes. The third round yields the following percentage of last-place votes: candidate A has 42%, C has 21%, and D has 37%. Therefore, A is dropped, and C becomes the winner since he defeats D with a score 60:40. So we see that the Coombs rule confirms the plurality winner.

Example 1.2.17. Presidential Elections in France, 2007

The French presidential election uses a two-round electoral system—runoff method. During the 2007 campaign, a voting experiment was conducted in two of the eleven voting districts with the aim to check whether the single transferable vote procedure can offer a credible alternative to the expensive two-round system, in terms of simplicity from a voter's point of view. The following tables show the results of elections, but they do not present the detailed preference schedules.

Vote count according to the Hare method

Names	Iterations (% of votes received)										
	1	2	3	4	5	6	7	8	9	10	11
Bayrou	21.1	21.1	21.2	21.6	21.9	22.4	22.7	22.8	23.2	**27.4**	
Besancenot	6.49	6.61	6.72	7.17	7.17	7.74	8.98	10.5	**12.4**		
Bové	1.168	1.68	**1.68**								
Buffet	2.13	2.13	2.24	2.24	2.24	2.81	**3.27**				
De Villiers	1.90	1.90	1.90	**2.02**							
Laguiller	2.24	2.24	2.24	2.47	2.58	**2.58**					
Le Pen	6.72	6.72	6.83	6.84	7.51	7.52	7.52	**7.77**			
Nihous	0.45	**0.45**									
Royal	22.2	22.2	22.2	22.5	22.5	23.1	23.2	24.4	25.1	32	**45.85**
Sarkozy	32.8	32.8	32.8	32.8	33.7	33.7	34.1	34.3	39.1	40.5	**54.15**
Schivardi	**0.11**										
Voynet	2.02	2.02	2.02	2.24	**2.24**						

Vote count according to the Coombs rule

Candidate	Iterations (% of votes received)										
	1	2	3	4	5	6	7	8	9	10	11
Bayrou	21.1	21.4	21.4	21.5	21.8	22.2	22.7	22.7	23.2	27.4	**51.97**
Besancenot	6.49	7.62	7.74	7.85	7.96	8.53	8.89	11.3	**12.4**		
Bové	1.68	1.68	1.68	1.68	**1.68**						
Buffet	2.13	2.13	2.13	2.24	2.24	2.24	**2.81**				
De Villiers	1.90	2.13	2.13	**2.13**							
Laguiller	2.24	2.58	2.58	2.58	2.8	**3.14**					
Le Pen	**6.72**										
Nihous	0.45	0.56	**0.56**								
Royal	22.2	22.7	22.7	22.8	22.8	23.1	23.3	24.2	25.1	**32**	
Sarkozy	32.8	37	37	37.1	38.4	38.5	38.8	39.0	39.1	40.5	**48.03**
Schivardi	0.11	**0.11**									
Voynet	2.02	2.02	2.02	2.02	2.02	2.24	2.36	**2.59**			

This experiment shows that voters understood preference ranking pretty well, with blank or spoil ballots of 6.98%. In Australia, a country where the Hare method has been used in legislative elections for many years and where voting is obligatory, blank and spoil votes accounted for 3.95% of ballots in the 2007 election.

In the two-round electoral system—runoff—used in France, the 10 worst candidates were eliminated in the first round during the 2007 campaign. In contrast, the single transferable vote drops one candidate at a time. Using the preference schedule presented here, we see that the first candidate to be eliminated is Gérard Schivardi who received 0.11% of votes cast. The table shows sequential eliminations according to the Hare method. The percentages of the first-place ballots of the candidate that is dropped are given in bold font. The final round gave results very similar to the national elections where Nicolas Sarkozy defeated Ségolène Royal with 53.06% versus 46.94% of votes.

The next table shows sequential eliminations according to the Coombs rule. The first candidate to be eliminated was Le Pen, with 6.72% first-place votes because he had more last-place votes than any other competitor. Moreover, the Coombs rule produced a different winner: François Bayrou, but not the runoff winner.

Example 1.2.18. Not All Stages are Needed

The Eastview Condominium Association must determine what color to paint the wood trim of three buildings. The four choices are yellow, light blue, gray, and green. The preference rankings of the association members are summarized in the following table:

Colors	Number of Votes (21)					
	1	4	5	4	2	5
Yellow	1	1	3	3	3	4
Light Blue	2	3	4	1	4	3
Gray	3	4	1	4	2	2
Green	4	2	2	2	1	1

First, we apply the Hare method by calculating first-place votes for every alternative. Since option *Light Blue* has the least number of first-place votes (4), this alternative is eliminated, and its four votes are transferred to the next option—*Green*. In the second round, we have the following preference schedule:

Colors	Number of Votes (21)					
	1	4	5	4	2	5
Yellow	1	1	3	2	3	3
Gray	2	3	1	3	2	2
Green	3	2	2	1	1	1

Counting the first-place votes, we see that option *Green* has 11 first-place votes—the majority out of 21 votes. Therefore, we terminate counting and declare this option to be the Hare winner.

Colors	Number of Votes (21)					
	1	4	5	4	2	5
Yellow	1	1	2	3	2	3
Light Blue	2	3	3	1	3	2
Green	3	2	1	2	1	1

Let us apply the Coombs rule to the same preference schedule. The worst scenario would getting eliminated in the first round. So we count the number of last-place votes: *Yellow* has 5, *Light Blue* has 7, *Gray* has 8, and *Green* has 1. Hence option *Gray* is dropped yielding the previous table.

Again, the Coombs winner is *Green* because this option has 12 first-place votes—the majority. This example shows that multi-round counting as in the IRV or Coombs method may be terminated before arriving to the final stage when one of the alternatives has the majority of first-place votes.

Problems.

1. The management of a company has decided to treat their office staff to dinner. The votes on what type of food to have are the following: Chinese, Italian, French, Mexican, Italian, Chinese, Mexican, French, Italian, Italian, Mexican, French. Which type of food will they eat if the decision is by plurality?

2. A club is holding an election to choose its president. There are four candidates: Boris (B, for short), Alex (A), Peter (P), and Samantha (S). The result of voting is as follows: B, A, A, S, P, P, S, A, B, S, B, S. Who will be the president of the club if the election is determined by plurality?

3. Suppose that 76 votes are cast in an election among 3 candidates: Arriaza, Barrett, and Coenen. After the first 56 votes are counted, the tallies are

   Arriaza – 18;   Barrett – 21;   Coenen – 17.

   What is the minimal number of the remaining votes each of the candidates can receive and be assured of a win?

4. Suppose that 90 votes are cast in an election among 3 candidates: Jones, Kontos, and Liebman. After the first 55 votes are counted, the tallies are

   Jones – 17;   Kontos – 18;   Liebman – 20.

   What is the minimal number of the remaining votes each of the candidates can receive and be assured of a win?

5. Suppose that 130 votes are cast in an election among 4 candidates: Damato, Erickson, Foray, and Gamba. After the first 104 votes are counted, the tallies are

   Damato 27,   Erickson 21,   Foray 25,   Gamba 31.

   What is the minimal number of the remaining votes each of the candidates can receive and be assured of a win?

6. Suppose that 100 votes are cast in an election among 4 candidates: A, B, C, and D. After the first 75 votes are counted, the tallies are

   A — 17,   B — 12,   C — 21,   D — 25.

   What is the minimal number of the remaining votes each of the candidates can receive and be assured of a win?

7. Suppose that 300 votes are cast in an election among 4 candidates: Lopes, Mateo, Renzi, and Songin. After the first 280 votes are counted, the tallies are

Lopes	71
Mateo	66
Renzi	78
Songin	65

   What is the minimal number of the remaining votes each of the candidates can receive and be assured of a win?

8. Suppose that 140 votes are cast in an election among 5 candidates: Adewale, Borcuk, Duarte, Ford, and Healey. After the first 100 votes are counted, the tallies are

   Adewale   13
   Borcuk   15
   Duarte   21
   Ford   24
   Healey   27

   What is the minimal number of the remaining votes each of the candidates can receive and be assured of a win?

9. Suppose that 130 votes are cast in an election among 5 candidates: Andraka, Beagen, Frias, Klotz, and Romano. After the first 105 votes are counted, the tallies are

   Andraka   25
   Beagen   26
   Frias   20
   Klotz   19
   Romano   15

   What is the minimal number of the remaining votes each of the candidates can receive and be assured of a win?

10. If 202 votes are cast, what is the smallest number of votes a winning candidate can have in a 4-candidate race that is to be decided by plurality? What if there are 203 or 204 votes?

11. If 303 votes are cast, what is the smallest number of votes a winning candidate can have in a 4-candidate race that is to decided by plurality?

12. If 205 votes are cast, what is the smallest number of votes a winning candidate can have in a 4-candidate race that is to be decided by plurality?

13. If 202 votes are cast, what is the smallest number of votes a winning candidate can have in a 5-candidate race that is to be decided by plurality? What is if there are 203 or 204 votes?

14. If 205 votes are cast, what is the smallest number of votes a winning candidate can have in a 5-candidate race that is to be decided by plurality?

In Problems 15 – 30, determine the winner and the loser using

(a) the plurality method;

(b) the plurality method followed by a runoff between the top two finishers;

(c) the Hare method;

(d) the Coombs method.

15. Suppose three candidates—Bertoncini, Faiola, and Lombardi—are running for mayor, and the preference rankings of the voters are broken down into percentages in the following table. Determine the winner and the loser for each of plurality methods.

	Percentage of Votes					
	19%	14%	18%	21%	16%	12%
Bertoncini	1	1	2	3	3	2
Faiola	2	3	1	1	2	3
Lombardi	3	2	3	2	1	1

16. Eleven members of a committee must decide what kind of flooring to install in a community room. The preference rankings of the committee members are listed here. Determine the winner and the loser for each of plurality methods.

	Number of Votes (11)					
	3	1	1	2	2	2
Carpet	1	1	2	2	3	3
Ceramic tile	2	3	1	3	1	2
Wood	3	2	3	1	2	1

17. Three candidates are running for president of a sport community center. The preference rankings of the voters are presented here. Determine the winner and the loser for each of plurality methods.

Candidate	Number of Votes (47)					
	11	5	8	7	10	6
Reis	1	1	2	3	3	2
Sherman	3	2	1	1	2	3
Waldman	2	3	3	2	1	1

18. Fifteen soccer players must decide which of three fields they are going to use for practice. The preference rankings of the committee members are listed here. Determine the winning field and the losing field for each of plurality methods.

Field	Number of Votes (15)					
	2	3	3	2	1	4
North field	2	3	3	2	1	1
Reserve field	1	1	2	3	2	3
South field	3	2	1	1	3	2

19. Fifteen players from a soccer team want to choose the type of food to be served at a team party. Each of the players has an order of preference for the different types of food. The preference rankings of the players are listed in the following table. Which type of food would win and lose using each of four plurality methods?

	Number of Votes (15)					
	4	1	2	3	3	2
Fish	3	3	3	1	2	4
Pasta	2	1	1	4	3	2
Chicken	4	2	4	3	1	1
Steak	1	4	2	2	4	3

20. The 195 parents of children in a middle school were asked to vote on having mandatory school uniforms, optional school uniforms, badges, or no school uniforms. The preference rankings of the parents are as follows:

Option	Number of Votes (195)							
	34	16	31	18	29	19	25	23
Mandatory uniform	1	1	4	3	4	2	3	2
Optional uniform	2	4	1	1	2	3	2	4
Badge	3	2	3	4	1	1	4	3
No uniform	4	3	2	2	3	4	1	1

21. Sixteen players from a rugby team want to choose the type of food to be served at a team party. Each of the players has an order of preference for the different types of food. The

preference rankings of the players are listed in the following table. Which type of food would win using each of the four plurality methods?

Number of Votes:	4	2	2	3	3	2
First choice	Steak	Pasta	Pasta	Fish	Chicken	Steak
Second choice	Pasta	Chicken	Steak	Steak	Fish	Pasta
Third choice	Fish	Fish	Fish	Chicken	Pasta	Chicken
Fourth choice	Chicken	Steak	Chicken	Pasta	Steak	Fish

22. A poll of members of a community asked people about their opinions on a curfew time for young people under the age of 16. Their preference rankings broke down into the following percentages.

	Percentage of Voters							
Option	19%	3%	18%	7%	16%	10%	12%	15%
9 pm	1	1	2	2	3	3	4	4
12 midnight	2	4	3	1	2	4	1	3
1 am	3	2	4	3	4	1	2	1
No curfew	4	3	1	4	1	2	3	2

23. Thirteen people at a picnic want to select an activity for the afternoon. Their preference rankings of the possibilities are listed below. Which sport activity will win? Which sport activity will lose?

	Number of Votes (13)							
Activity	3	2	2	1	2	1	1	1
Soccer	1	4	2	4	3	1	4	3
Volleyball	2	2	1	1	4	4	3	2
Basketball	3	1	3	2	1	2	2	4
Tennis	4	3	4	3	2	3	1	1

24. Determine the winner and the loser using four plurality methods in the following election:

Number of votes:	12	10	8	6	3
First choice	A	C	D	B	C
Second choice	B	A	B	D	D
Third choice	C	B	C	C	B
Fourth choice	D	D	A	A	A

25. A math class is asked by the instructor to vote among four possibilities to determine the due date of their homework assignment: M (Monday), T (Tuesday), W (Wednesday), and F (Friday). Determine the winner and the loser using four plurality methods in the following election:

	Number of Votes (41)								
2	3	8	7	4	2	4	3	8
M	M	M	T	T	T	W	W	F
T	T	W	W	M	W	F	M	W
W	F	T	F	W	M	T	F	T
F	W	F	M	F	F	M	T	M

26. Determine the winner and the loser using four plurality methods in the following election:

Number of votes:	2	4	8	6	3
First choice	A	B	C	D	C
Second choice	B	D	B	A	D
Third choice	C	A	D	B	A
Fourth choice	D	C	A	C	B

27. Determine the winner and the loser using four plurality methods in the following election:

Alternative	Number of Votes (7)				
	2	1	2	1	1
A	1	1	4	3	4
B	3	4	1	2	3
C	2	3	2	4	1
D	4	2	3	1	2

28. Determine the winner and the loser using four plurality methods in the following election:

Ranking	Number of Votes (11)				
	2	2	1	3	3
First choice	A	B	A	C	D
Second choice	D	C	B	A	B
Third choice	C	D	D	B	C
Fourth choice	B	A	C	D	A

29. Suppose an election is held at the local department of education. The candidates are Professors Baglama, Eaton, Finizio, Grove, and Ladas (B, E, F, G, and L). Determine the winner and the loser using four plurality methods given the following preference rankings:

Number of Votes (23)					
5	3	5	4	3	3
B	E	F	G	E	F
E	G	L	F	B	L
F	B	G	E	G	B
G	L	B	L	L	E
L	F	E	B	F	G

30. Determine the winner and the loser using four plurality methods in the following election:

Number of votes	1	1	1	1	1	1
First choice	A	B	C	D	E	A
Second choice	B	C	B	A	D	B
Third choice	C	A	E	C	B	C
Fourth choice	D	D	D	E	C	E
Fifth choice	E	E	A	B	A	D

In Problems 31 – 34, consider elections between five candidates, labeled by A, B, C, D, and E. Based on provided preference rankings, determine the winner and the loser (if any) using four plurality methods.

31.

	Number of Votes (23)									
Candidate	2	3	4	2	1	1	4	3	2	1
A	1	1	2	2	3	3	5	3	5	5
B	2	3	1	5	4	5	1	4	2	4
C	3	2	5	4	1	1	4	2	4	3
D	4	5	4	1	2	2	3	5	1	2
E	5	4	3	3	5	4	2	1	3	1

32.

Number of votes:	2	1	1	2	1	1
First choice	A	E	A	B	C	D
Second choice	B	B	D	E	B	E
Third choice	C	D	C	D	E	B
Fourth choice	D	C	E	C	A	A
Fifth choice	E	A	B	A	D	C

33.

	Number of Votes (23)									
Candidate	2	3	2	2	1	3	4	3	2	1
A	1	1	2	2	3	3	4	4	5	4
B	3	3	1	5	4	5	1	2	2	5
C	2	2	5	4	1	1	5	3	4	3
D	4	5	4	1	2	2	3	5	1	2
E	5	4	3	3	5	4	2	1	3	1

34.

Number of votes:	2	2	1	1	1
First choice	C	B	C	D	A
Second choice	E	A	A	B	E
Third choice	D	E	D	E	B
Fourth choice	A	D	E	A	C
Fifth choice	B	C	B	C	D

In Problems 35 – 42, show that the elections based on the plurality method, the plurality method followed by a runoff between the top two finishers, the Hare method, and the Coombs method provide four different winners.

35.

	Number of Votes (15)				
Candidate	5	4	3	2	1
A	1	5	5	5	5
B	2	1	4	4	4
C	3	4	1	3	2
D	4	2	3	1	3
E	5	3	2	2	1

36.

Number of votes:	15	7	13	5	2
First choice	T	A	C	P	A
Second choice	P	P	B	A	B
Third choice	A	C	P	B	C
Fourth choice	B	B	A	C	P
Fifth choice	C	T	T	T	T

37.

Candidate	Number of Votes (17)					
	2	4	5	3	2	1
A	1	3	2	2	3	3
B	4	1	3	4	5	4
C	3	5	1	5	4	2
D	2	2	5	1	2	5
E	5	4	4	3	1	1

38.

Ranking	Number of Votes (23)									
	4	3	3	3	3	2	2	1	1	1
First choice	A	A	B	B	C	C	D	D	E	E
Second choice	D	E	C	E	D	E	E	B	D	C
Third choice	C	B	E	D	E	B	C	E	C	B
Fourth choice	B	D	D	C	B	A	A	A	A	D
Fifth choice	E	C	A	A	A	D	B	C	B	A

39.

Candidate	Number of Votes (27)									
	2	3	4	5	1	3	4	2	2	1
A	1	1	2	3	3	3	3	4	5	5
B	3	3	1	5	4	5	1	2	2	4
C	2	2	5	4	1	1	5	3	4	3
D	4	5	4	1	5	4	4	1	1	2
E	5	4	3	2	2	2	2	5	3	1

40.

Number of votes:	5	2	2	3	1	1	3
First choice	A	B	B	C	D	D	E
Second choice	D	E	D	E	E	E	C
Third choice	C	D	E	D	B	C	B
Fourth choice	B	A	C	A	A	B	A
Fifth choice	E	C	A	B	C	A	D

41.

Candidate	Number of Votes (38)				
	8	9	5	4	2
A	2	5	2	1	1
B	3	3	1	3	3
C	1	4	5	4	2
D	4	2	4	2	4
E	5	1	3	5	5

42.

Number of votes:	5	2	2	2	1	1	1	1	2
First choice	A	B	B	C	C	D	D	E	E
Second choice	D	E	D	E	D	E	E	D	C
Third choice	C	C	E	D	E	A	C	C	B
Fourth choice	B	D	C	A	B	B	B	B	D
Fifth choice	E	A	A	B	A	C	A	A	A

In Exercises 43 – 44, show that the plurality method contradicts itself, that is, it chooses the same candidate for both the winner and the loser. The preference rankings are presented here.

43.

	Number of Votes (15)			
Alternative	6	4	3	2
A	1	3	2	2
B	3	1	1	3
C	2	2	3	1

44.

Number of votes:	6	5	2
First choice	A	B	C
Second choice	B	C	B
Third choice	C	A	A

In Exercises 45 – 46, show that the plurality method followed by a runoff between the two best finishers contradicts itself, that is, it chooses the same candidate both for the winner and for the loser. The preference rankings are presented here.

45.

	Number of Votes (17)			
Alternative	7	6	2	2
A	1	2	3	3
B	3	1	1	2
C	2	3	2	1

46.

Number of votes:	6	2	6	5
First choice	C	C	A	B
Second choice	B	A	C	A
Third choice	A	B	B	C

In Exercises 47 – 52, show that each of three plurality methods—the plurality, the plurality method followed by a runoff between the two top finishers, and the Hare method—contradicts itself, that is, each of these methods chooses the same candidate both for the winner and for the loser. The preference rankings are presented here.

47.

	Number of Votes (13)					
Alternative	1	2	3	2	1	4
A	1	3	2	3	4	2
B	4	1	1	4	2	3
C	3	4	4	1	3	1
D	2	2	3	2	1	4

48.

Number of votes:	3	8	4	2	1	9
First choice	A	B	C	A	C	D
Second choice	B	A	A	C	D	B
Third choice	D	C	D	B	A	C
Fourth choice	C	D	B	D	B	A

49.

	Number of Votes (11)				
Candidate	1	2	3	2	3
A	1	2	3	1	4
B	2	1	4	3	3
C	4	4	1	4	1
D	3	3	2	2	2

50.

Ranking	Number of Votes (23)					
	2	5	2	4	5	5
First choice	A	A	B	C	C	D
Second choice	B	D	A	B	D	C
Third choice	C	B	D	A	B	A
Fourth choice	D	C	C	D	A	B

51.

Candidate	Number of Votes (25)									
	5	1	1	3	1	3	4	2	2	3
A	1	3	5	2	5	3	4	3	2	5
B	3	5	1	5	4	5	1	2	1	2
C	2	2	2	4	1	1	5	4	5	4
D	4	1	4	1	2	2	3	5	4	3
E	5	4	3	3	3	4	2	1	3	1

52.

Number of votes:	5	2	2	2	1	1	2	2
First choice	A	B	B	C	C	D	D	E
Second choice	D	E	D	E	D	E	E	A
Third choice	C	A	E	D	A	A	C	B
Fourth choice	B	D	C	B	E	B	B	C
Fifth choice	E	C	A	A	B	C	A	D

In Exercises 53 – 54, show that the Coombs method contradicts itself, that is, it chooses the same candidate both for the winner and for the loser. The preference schedules are presented here.

53. Show that in the following election between five competitors the Coombs method contradicts itself.

Candidate	Number of Votes (33)					
	7	5	8	5	2	6
A	1	3	4	3	3	5
B	2	5	3	4	2	1
C	5	4	1	2	5	3
D	3	2	5	1	1	4
E	4	1	2	5	4	2

54. Show that in the following election between five competitors (A, B, E, K, and S) the Coombs method contradicts itself, that is, it chooses the same candidate both for the winner and for the loser. The preference rankings are presented here.

Ranking	Number of Voters (41)							
	2	3	6	5	7	3	8	7
First choice	A	A	B	B	E	E	K	S
Second choice	B	S	A	E	A	A	E	K
Third choice	E	B	S	A	K	B	S	E
Fourth choice	K	E	K	S	S	K	A	B
Fifth choice	S	K	E	K	B	S	B	A

1.3   Point Distribution Systems

Not all elections are determined by simply counting who gets the most first-place votes and declaring that person to be the winner (plurality methods). It often occurs that we need not only to know who wins the election, but also who comes in second, third, and so on. For instance, an Olympic soccer tournament results in the distribution of three kinds of medals—gold, silver, and bronze. A condominium association needs to elect a president, a vice president, and a treasurer. There are many other similar elections when it is required to know who finishes first, second, and so on (see §1.7). So we need a voting method that can provide us a distribution of preferences. Obviously, plurality methods are useless in this case simply because they are based on counting only first-place votes. Instead, **point allocation** methods allow us to get a full distribution of voters' preferences.

In France during the latter half of the eighteenth century, two major contributions in voting theory were made by Jean-Charles de Borda (1733 – 1799) and the Marquis de Condorcet (1743 – 1794), both members of l'Académie Royale des Science (France). From these two mathematicians sprang two streams of approaches on the problem of designing voting rules. In 1770, de Borda [3] read a paper before the Academy (Paris) in which he proposed the voting method which now bears his name.

The Borda count was developed independently several times, but is named for Jean-Charles de Borda, who devised the system in 1770. Jean-Charles de Borda (1733 – 1799) was a French mathematician, physicist, political scientist, and sailor. The French Academy of Sciences used Borda's method to elect its members for about two decades until it was quashed by Napoleon Bonaparte who insisted that his own method be used after he became president of the Académie in 1801. As a contemporary of the Marquis de Condorcet, he was engaged in many scholarly debates regarding the merits of their respective voting systems.

Borda's idea was to fix the main drawback of plurality methods—they do not take into account the voter's relative preferences for all candidates. Since only the first place is counted, the plurality methods do not reflect the voter's opinion about other alternatives besides the first choice; therefore, an election based on plurality does not entirely reflect the population's choice as a whole. A ballot with preference ranking gives complete information about the voter's will, but the problem is how to aggregate such ballots to come up with the best social choice. Some modifications of plurality (see §1.2.3) try to utilize the simplicity of its counting procedure with preference rankings. Another idea, presented in this section, is to transfer a preference ranking into a sequence of weights, and then use corresponding scores to determine a winner.

The **Borda count**, known also as weighted voting, is a form of preferential voting where the rankings are converted into points, and the candidate who receives the most points is declared the winner. Each ballot contains an ordering of the n candidates from best to worst. (Voters are not allowed to omit a candidate they know nothing about, and are not allowed to regard two candidates as equal.) The Borda method is a point count system, where a first-choice vote is worth a fixed number of points, a second-choice vote is worth a fixed number of points, and so on. The winner is the candidate with the most total points. In this book, we use a popular formula for Borda count that allocates the last place on a ballot to be worth zero points, the second to last place to be worth one point, and so on until the first place, which is worth one less than the number of candidates running. If there are n alternatives, then the first place is rewarded with $n - 1$ points, second place is worth $n - 2$ points, and so on, the last position is not rewarded at all. This means that every voter distributes $0 + 1 + \cdots + n - 1 = \dfrac{n(n - 1)}{2}$ points in an election with n alternatives. If there are m ballots cast, the total number of points that all candidates can earn in an election using the Borda count method is $m\,n(n - 1)/2$.

Depending on where the method is used, the exact allocation of points varies. It does not matter how many points are assigned to the last place; but it is essential that the difference in points between each sequential ranks is 1. Some authors give 1 point to the last place, 2 points to the next place, and so on, the first place earns n points. The winner will be exactly the same as if you assign $(n - 1)$ points to the first place—you just operate with larger numbers in your count: the total number of points each voter distributes in the election becomes $1 + 2 + \cdots + n = \dfrac{n(n + 1)}{2} = \dfrac{n(n - 1)}{2} + n$. So you just add n points into circulation without changing the outcomes.

Since a distribution of points between candidates may lead to a tie (or ties), the system might be inconclusive in this case, and an additional rule to break ties is required. Assuming that there is no tie, every participant in an election is associated with an integer—called the **Borda score**. This allows us to rank all candidates/alternatives from best to worst. The option with the least number of points is naturally called the **Borda loser**. Therefore, there is no need to reverse preference rankings on the ballots to determine the loser as we did in the previous section. Moreover, in contrast to plurality methods, point distribution systems do not contradict themselves in the sense that the winner and the loser are always two different options.

The Borda system is currently used for the election of two ethnic minority members of the National Assembly of Slovenia, and, in modified forms, to elect members of the Parliament of Nauru (formerly known as Pleasant Island). The Borda count is popular in determining awards for sports in the United States. It is also used in determining the Most Valuable Player in Major League Baseball and in the NBA, by the Associated Press and United Press International to rank players in NCAA sports, and other contests. The Eurovision Song Contest also uses a positional voting method similar to the Borda count, with a different distribution of points. It is used for wine trophy judging by the Australian Society of Viticulture and Oenology. It is also used throughout the world by various private organizations and competitions.

Along with the advantages Borda system shares with other preferential voting systems, the point distribution method tends to elect broadly acceptable candidates, rather than those supported by the majority. This is sometimes used as an argument against the Borda method. Unlike plurality voting systems, in the Borda count it is possible for a candidate who is the first preference of an absolute majority of voters to fail in an election; this is because the Borda count affords greater importance to a voter's lower preferences than most other methods. Some of its supporters see the Borda count tendency to favor candidates supported by a broad consensus among voters, rather than the candidate who is necessarily the favorite of a majority, as a method that promotes consensus and avoids the "tyranny of the majority." When a majority candidate is strongly opposed by a large minority of the electorate, the Borda winner may have higher overall utility than the majority winner. Advocates favor the use of Borda's system in cases where the electorate is divided into two unequal parts (due to their religion, for example). The candidates for such an approach could be Northern Ireland, Israel, the Balkans, and Kashmir.

The Borda method is a highly inelegant system that has little merit for use in public elections. For one thing, the strength of a given person's vote is highly variable as it affects the competitions between different candidates. There are two principal objections to the Borda count. One of them is **strategic nomination**. The biggest problem with Borda's system is that it reacts badly to "strategic voting." Borda's system encourages a strategic approach by parties to nominations. An extra candidate increases the cardinal number of points in the system and alters the distribution of points between other candidates. Whereas a minority faction may increase its chances of prevailing over a majority by offering more candidates,. Another drawback is **tactical voting**. Borda's system supports tactical voting to an even greater extent than the plurality method. Lower preferences can harm the first choice.

Example 1.3.1. Borda System versus Plurality

A small company decides to use the Borda system to determine what day of the week to have the company picnic. The four choices are Tuesday, Thursday, Friday, and Saturday. Employees ranked their choices as shown here.

	Number of Votes (13)					
	3	**2**	**2**	**2**	**2**	**2**
Tuesday	1	2	3	2	4	3
Thursday	2	1	4	3	3	4
Friday	3	4	1	4	2	1
Saturday	4	3	2	1	1	2

The numbers at the top of each column in the table show how many voters submitted the particular preference ballot. For example, the **3** means that three voters had preference ballots choosing Tuesday first, Thursday second, Friday third, and Saturday fourth (in list notation, Tuesday $\succ$ Thursday $\succ$ Friday $\succ$ Saturday). Notice that Tuesday has 3 first-place votes, Thursday has 2, Friday and Saturday have 4 first-place votes each.

Since the first place earns 3 points, the second place is rewarded with 2 points, the third place is given 1 point, and the last place worth nothing, we get the following Borda scores for each alternative:

Alternatives	Number of Points				
	1st-Place Votes×3 Points	2nd-Place Votes×2 Points	3rd-Place Votes×1 Points	4th-Place Votes×0 Points	Total Number of Points
Tuesday	$3 \times 3 = 9$	$4 \times 2 = 8$	$4 \times 1 = 4$	$2 \times 0 = 0$	21
Thursday	$2 \times 3 = 6$	$3 \times 2 = 6$	$4 \times 1 = 4$	$4 \times 0 = 0$	16
Friday	$4 \times 3 = 12$	$2 \times 2 = 4$	$3 \times 1 = 3$	$4 \times 0 = 0$	19
Saturday	$4 \times 3 = 12$	$4 \times 2 = 8$	$2 \times 1 = 2$	$3 \times 0 = 0$	22

To check our calculations, we add the numbers of points each voter distributed: $21 + 16 + 19 + 22 = 78$. Since every voter distributed 6 points and there are 13 voters, the total is $6 \times 13 = 78$, so our calculations are correct.

Therefore, Saturday will be declared the winner based on Borda's count. However, some workers suggested using another voting method. Let us see which alternative wins according to one of the plurality methods. Counting first-place votes, we find that Tuesday gets 3 first-place votes, Thursday has only 2, and two other alternatives, Friday and Saturday, share the same number of first-place votes—4. Therefore, the plurality method ends up in a tie between Friday and Saturday. The runoff method eliminates Tuesday and Thursday as the least wanted options leaving Friday and Saturday in the final pool:

	Number of Votes (13)					
	3	2	2	2	2	2
Friday	3	4	1	4	2	1
Saturday	4	3	2	1	1	2

or

	Number of Votes (13)					
	3	2	2	2	2	2
	F	S	F	S	S	F
	S	F	S	F	F	S

So we see that Friday is the winner according to the runoff method because it has one more vote than Saturday.

The Hare method would suggest to drop the option with the least number of first-place votes, which is Thursday. Lifting all other candidates that are below Thursday by one place, we get

	Number of Votes (13)					
	3	2	2	2	2	2
Tuesday	1	1	3	2	3	3
Friday	2	3	1	3	2	1
Saturday	3	2	2	1	1	2

Friday and Saturday have only 4 first-place votes, while Tuesday has 5. Hence, Tuesday is the winner based on the Hare method.

The Coombs rule eliminates Thursday and Friday in the first round because they both get 4 last-place votes whereas Tuesday has 2 last-place votes and Saturday has 3. The second round gives Saturday a preference compared to Tuesday with the score 8:5 (meaning that eight of the voters prefer Saturday to Tuesday, and five of the voters prefer Tuesday to Saturday). Therefore, the Coombs method chooses Tuesday as a winner.

Example 1.3.2. Borda System Conflicts with Plurality Methods

Let us show with examples that the Borda system conflicts with other methods of aggregation. Suppose that the preference ranking based on write-in ballots is given here.

Ranking	Number of Votes (13)			
	5	4	3	1
First choice	A	B	C	C
Second choice	B	A	D	D
Third choice	C	C	B	A
Fourth choice	D	D	A	B

Borda's scores of each alternative are as follows:

A: $5 \times 3 + 4 \times 2 + 1 \times 1 = 24$,
B: $4 \times 3 + 5 \times 2 + 3 \times 1 = 25$,
C: $4 \times 3 + 9 \times 1 = 21$,
D: $4 \times 2 = 8$.

To check our calculations, we add all points in circulation to obtain $24 + 25 + 21 + 8 = 13 \times 6 = 78$, correct! In this election, there is no clear winner because no one gains the majority of votes. The plurality, runoff, and Hare methods nominate A as the winner. However, the Borda count (as well as the Coombs system) defines B to be a winner.

Example 1.3.3. Eliminating a Weak Candidate

Consider an election involving 4 candidates and 7 voters:

	Number of Votes (7)		
	3	2	2
A	1	4	3
B	2	1	4
C	3	2	1
D	4	3	2

Borda scores of candidates are as follows:

A: $3 \times 3 + 2 \times 1 = 11$,
B: $2 \times 3 + 3 \times 2 = 12$,
C: $2 \times 3 + 3 \times 2 + 3 \times 1 = 13$,
D: $2 \times 2 + 2 \times 1 = 6$.

Here, C is the Borda winner, and D finishes last. Moreover, suppose it is discovered that the far-last-place loser D, top-ranked by nobody, was a criminal and a non-citizen and hence was not eligible to be a candidate. We eliminate D from the election, and do everything over again with the same orderings of A, B, C in all votes.

Surely we will get the same winner as before, right? Wrong! The result of the Borda election now is completely reversed! This follows from the preference table followed by calculations.

	Number of Votes		
	3	2	2
A	1	3	2
B	2	1	3
C	3	2	1

A: $3 \times 2 + 2 \times 1 = 8,$
B: $2 \times 2 + 3 \times 1 = 7,$
C: $2 \times 2 + 2 \times 1 = 6.$

The winner becomes A, who scored only third in the 4-run election. So the Borda method violates the rule that if a weak candidate is taken away from the rankings, the results of the rankings would not change.

We see that candidate D is the Borda loser since he has the lowest Borda score. To support our claim, we reverse the order in the preference schedule to obtain

	Number of Votes		
	3	2	2
A	4	1	2
B	3	4	1
C	2	3	4
D	1	2	3

The Borda points for each candidate are as follows:
A: $2 \times 3 + 2 \times 2 = 10,$
B: $2 \times 3 + 3 \times 1 = 9,$
C: $3 \times 2 + 2 \times 1 = 8,$
D: $3 \times 3 + 2 \times 2 + 2 \times 1 = 15.$

Hence, candidate D is indeed the Borda loser. So there is no need to reverse the order in the preference schedule to determine the loser—Borda scores allow us to identify the winner and the loser right away.

Example 1.3.4. The Borda System Goes Against the Majority Rule
The preference rankings

Ranking	Number of Votes (10)			
	7	1	1	1
First choice	A	C	B	B
Second choice	B	B	C	D
Third choice	C	D	D	C
Fourth choice	D	A	A	A

show that A is the clear majority winner because of support by 70% of the voters. However, the Borda count shows B to be the winner instead because A is ranked last by the remaining 3 voters. This could be due to tactical voting by the remaining 3 voters, which is another flaw in the Borda count system. On the other hand, A is the winner using each of the four plurality methods.

Example 1.3.5. Strategic Voting
In the real world, the Borda count is less likely to be implemented unless the number of voters is relatively small. National elections based on this system may have disastrous effects, from political sabotage to illegal coercion. Therefore, this system should be used with caution in order to prevent any false winner from being produced. The following example shows the vulnerability of the Borda system to strategic voting.

Once upon a time at a major university in Europe, the economics department was hiring a new colleague. There were 4 applicants: One was a world-class macroeconomist, one was a world-class microeconomist. The other two were mediocre, one clearly better than the other. When it came time to select one by voting using Borda, about half the department preferred the macroeconomist, with the microeconomist as their second favorite. The rest of the department preferred the microeconomist, with the macroeconomist as their second favorite. They each had a pretty good understanding of each other's sincere preferences, and thus expected the election would be close. You can guess what happened: When they voted, each voter raised the two mediocre candidates over his/her second favorite, hoping to manipulate the outcome if the Borda count was close. As a result, one of the mediocre candidates was voted everyone's second choice and had the largest Borda count.

In short, strategy has an enormous impact in the real world, and over 90% of real voters act strategically and not honestly, given the chance. Another, different sort of example where strategic voting is known to have had a huge impact: a 1950 Gallup poll showed 38% of British voters wanted to vote for the Liberal Party but only 9% did. That is exactly why third parties always die out and the USA retains a 2-party domination.

The Pacific Island Republic of Kiribati (pop. 60,000; formerly the Gilbert Islands)—a nation located in the central tropical Pacific Ocean—was the only country in the world to adopt the Borda system for the selection of presidential candidates from among members of parliament. (Nauru uses its own, different, weighted positional voting system, which seems to work better than Borda.) It failed immediately as a consequence of massive "strategic voting" and suffered from exactly this kind of problem as the two "most popular" candidates were eliminated and two dark horses who did not campaign and were not recognized as serious contenders were nearly elected. Consequently, Kiribati soon abandoned the Borda system for elections and went back to the (often superior for strategic voters) plurality system and now has 2-party domination.

1.3.1 Some Other Systems

Some modifications of the Borda system are known to compensate for the following serious flaws:

- It can fail to elect a candidate who is the first choice of a strict majority.

- It is extremely vulnerable to tactical voting.

There are other formulas for assigning points (or weights) for each ranking of a candidate. In Nauru, a distinctive formula is used based on increasingly small fractions of points. Under the system, a candidate receives 1 point for first place, $\frac{1}{2}$ a point for second place, $\frac{1}{3}$ for third place, and so on. This method is far more favorable to candidates with many first preferences than the conventional Borda count; it also substantially reduces the impact of electors indicating late preferences at random because they have to complete the full ballot.

Example 1.3.6. Fractional Count
We reconsider the ballot table from Example 1.3.2, page 44. The new score (we may call it the Nauru score) becomes:

A: $6 \times 1 + 5 \times \frac{1}{2} + 3 \times \frac{1}{3} + 4 \times \frac{1}{4} = 10\frac{1}{2} = 10.5$,

B: $5 \times 1 + 6 \times \frac{1}{2} + 4 \times \frac{1}{3} + 3 \times \frac{1}{4} = 10\frac{1}{12} \approx 10.0833$,

C: $7 \times 1 + 11 \times \frac{1}{3} = 10\frac{2}{3} \approx 10.6666$,

D: $7 \times \frac{1}{2} + 11 \times \frac{1}{4} = 6\frac{1}{4} = 6.25$.

Hence we see that alternative C is the winner according to the Nauru count, whereas the Borda system nominates alternative B. □

Some other variations of the Borda system are known, which can effect the result of a given election, but the principle remains the same. For instance, some systems reward the first place with more than 1 point to emphasize the role of the first place. If there are n candidates in an election, its simplest modification assigns n points to the first place in a run between n alternatives. The second place is rewarded with only $n-2$ points, the third place is given $n-3$ points, and so on, exactly as in a Borda count (by subtracting 1 point). We may call it the **first-place promotion** score. Therefore, the first-place promotion uses the weights $(n, n-2, n-3, \ldots, 0)$ whereas the Borda system is based on the weights $(n-1, n-2, n-3, \ldots, 0)$.

Example 1.3.7. First-place Promotion
Consider the preference ballot table

	Number of Votes (17)			
Ranking	7	5	4	1
First choice	A	B	C	D
Second choice	B	C	D	B
Third choice	C	A	A	C
Fourth choice	D	D	B	A

The Borda count yields

A: $7 \times 3 + 9 \times 1 = 30$,

B: $5 \times 3 + 8 \times 2 = 31,$
C: $4 \times 3 + 5 \times 2 + 8 \times 1 = 30,$
D: $1 \times 3 + 4 \times 2 + 2 = 11.$

Therefore, B is the Borda winner and D is the Borda loser. On the other hand, the first-place promotion yields
A: $7 \times 4 + 9 \times 1 = 37,$
B: $5 \times 4 + 8 \times 2 = 36,$
C: $4 \times 4 + 5 \times 2 + 8 \times 1 = 34,$
D: $1 \times 4 + 4 \times 2 + 2 = 12.$
Now A becomes the winner.

Example 1.3.8. National Hockey League Trophies and Awards
All sport leagues in every country reward players when the regular season is over. As an example, we consider the National Hockey League (NHL) that has 28 trophies and awards. There are some trophies that are awarded to players based on their statistics during the regular season; they include, among others, the Art Ross Trophy for the league scoring champion (goals and assists), the Maurice "Rocket" Richard Trophy for the goal-scoring leader, and the William M. Jennings Trophy for the goalkeeper(s) for the team with the fewest goals against them. The other player trophies are voted on by the Professional Hockey Writers' Association or the team general managers.

It is interesting to note that members of the Professional Hockey Writers' Association select five candidates using a specific Borda system. Each individual voter ranks her or his top five candidates on a 10-7-5-3-1 points system (that is, the first place is rewarded with 10 points, second place—with 7 points, and so on). This allows the league to name three finalists and the trophy is awarded at the NHL Awards ceremony after the playoffs. In the case of a tie, it is broken based on the plurality method—the player who has more first-place votes is the winner.

For example, the closest the voting for the Hart Trophy was in the 2001–02 season, when Jose Theodore and Jarome Iginla tied in the total voting. The tiebreaker for choosing the Hart Trophy winner in such a case is number of first-place votes: Theodore, who had 86 first-place votes to Iginla's 82, claimed it.

Example 1.3.9. NHL Count versus Borda Count

	Number of Votes (39)					
	4	3	5	8	13	6
Burrito	1	1	4	2	5	4
Chile Rellenos	3	3	1	3	3	5
Enchilada	4	2	2	1	4	3
Fajita	2	4	5	5	2	1
Tamales	5	5	3	4	1	2

In order to decide which new item to add to its menu, a local Mexican restaurant did a market survey in which customers were asked to rank their preferences for Burrito, Chile Rellenos, Enchilada, Fajita, and Tamales. The results of counting the ballots are shown here.

To facilitate counting points for each alternative, we summarize the number of places in the following table.

	1st place	2nd place	3rd place	4th place	5th place
Burrito	7	8	0	11	13
Chile Rellenos	5	0	28	0	6
Enchilada	8	8	6	17	0
Fajita	6	17	0	3	13
Tamales	13	6	3	8	7

Our counting procedure shows that the Borda scores (4-3-2-1-0 point system) are as follows:
Burrito: $7 \times 4 + 8 \times 3 + 11 \times 1 = 63$;
Chile Rellenos: $5 \times 4 + 28 \times 2 = 76$;
Enchilada: $8 \times 4 + 8 \times 3 + 6 \times 2 + 17 \times 1 = 85$;
Fajita: $6 \times 4 + 17 \times 3 + 3 \times 1 = 78$;
Tamales: $13 \times 4 + 6 \times 3 + 3 \times 2 + 8 \times 1 = 84$.

So Enchilada is the Borda winner and Burrito is the Borda loser. Now using NHL weight distribution (based on 10-7-5-3-1 point distribution), we obtain:
Burrito: $7 \times 10 + 8 \times 7 + 11 \times 3 + 13 \times 1 = 172$;
Chile Rellenos: $5 \times 10 + 28 \times 5 + 6 \times 1 = 196$;
Enchilada: $8 \times 10 + 8 \times 7 + 6 \times 5 + 17 \times 3 = 217$;
Fajita: $6 \times 10 + 17 \times 7 + 3 \times 3 + 13 \times 1 = 201$;
Tamales: $13 \times 10 + 6 \times 7 + 3 \times 5 + 8 \times 3 + 7 \times 1 = 218$.

Therefore Tamales becomes the winner based on NHL point distribution. □

The biggest problem in a practical application of a voting system is the way to handle ballots that are not fully filled in—called truncated ballots—in which a voter has not expressed a total list of preferences. Borda's method suggests several approaches.

- The simplest method is to allow voters to rank as many or as few candidates as they wish, but simply give every unranked candidate the minimum number of points. For example, if there are 10 candidates, and a voter votes for candidate A first and candidate B second, leaving everyone else unranked, candidate A receives 9 points, candidate B receives 8, and all other candidates receive zero. However, this method allows strategic voting in the form of bullet voting: voting for only one candidate and leaving every other candidate unranked. This variant makes a bullet vote more effective than a fully-ranked ballot.

- Voters can simply be obliged to rank all candidates. This is the method used in Nauru.

- Voters can be permitted to rank a subset of the total number of candidates but obliged to rank all of those, with all unranked candidates being given zero points. This system was used in Kiribati.

- Use the modified Borda count.

In a **modified Borda count** (MBC) on a ballot of n options/candidates, if a voter casts preferences for only m options (where $m < n$), a first preference gets m points, a second preference $m-1$ points, and so on. This method effectively penalizes voters who do not rank a full ballot, by diminishing the number of points their vote distributes among candidates.

In a five-option election, a person who votes for only one option thus gives her favorite alternative just 1 point, and 0 to the rest. A voter who casts the ballot with two options gives her/his first preference 2 points, her/his second preference 1 point, and other three options 0 points. To ensure that your favorite gets the maximum 4 points, therefore, you should cast either all five preferences or four. Then your favorite gets 4 points, your second preference gets 3 points, and so on, just like in a Borda count. The modified Borda count encourages voters to submit a fully marked ballot (or drop the last choice).

The modified Borda count has been used by the Irish Green Party to elect its chairperson.

Example 1.3.10. Modified Borda Count

The Heisman Memorial Trophy Award (usually known colloquially as the Heisman Trophy or the Heisman), named after the former Brown University college football player and coach John Heisman, is awarded annually by the Heisman Trophy Trust to the most outstanding player in collegiate football. Balloting is open for all football players in all divisions of college football.

The 2009 Heisman finalists were Toby Gerhart, Stanford University; Mark Ingram, University of Alabama; Colt McCoy, University of Texas; Ndamukong Suh, University of Nebraska; and Tim Tebow, University of Florida. The voting for the 2009 Heisman Trophy, which was awarded to Mark Ingram, is summarized in the following table. The winner is determined based on 3-2-1 point distribution (modified Borda count).

	First	Second	Third	Tally
Mark Ingram	227	236	151	1304
Toby Gerhart	222	225	160	1276
Colt McCoy	203	188	160	1145
Ndamukong Suh	161	105	122	815
Tim Tebow	43	70	121	390

The tally for Mark Ingram is calculated as follows: $227 \times 3 + 236 \times 2 + 151 \times 1 = 681 + 472 + 151 = 1304$. □

1.3.2   Nanson's and Baldwin's Methods

The Borda count can be combined with an instant runoff procedure to create hybrid election methods that are called Nanson's and Baldwin's methods.

The **Nanson method** is based on the original work of the Australian mathematician Edward J. Nanson (1850 – 1936). This method consists of several rounds and starts with the Borda count. After that, Nanson's method eliminates those choices from a Borda count tally that are at or below the average Borda count score, then the ballots are re-tallied as if the remaining candidates were exclusively on the ballot. This process is repeated if necessary until a single winner remains or there is a tie.

Another method was developed by Joseph Mason Baldwin (1878 – 1945), an Australian government astronomer. A professor of mathematics at the University of Melbourne, he became acquainted with the work of Edward J. Nanson, and wrote an article describing an elimination method for the Borda count based on Nanson's research. This method is referred to as **Baldwin's method**.

According to him, candidates are voted for on ranked ballots as in the Borda count. Then, the points are tallied in a series of rounds. In each round, the candidate with the fewest points is eliminated, and the points are re-tallied as if that candidate were not on the ballot.

Example 1.3.11.   Nanson's Method
A small employee-owned company is voting on a merger with one of its competitors. After 42 share holders cast their ballots, we get the following preference table.

	Number of Ballots (42)				
Ranking	15	7	13	5	2
First choice	E	A	C	D	A
Second choice	D	D	B	C	B
Third choice	A	C	A	B	C
Fourth choice	B	B	D	A	D
Fifth choice	C	E	E	E	E

Calculating the number of places every alternative gets, we obtain

	1st places	2nd places	3rd places	4th places	5th places
A	9	0	28	5	0
B	0	15	5	22	0
C	13	5	9	0	15
D	5	22	0	15	0
E	15	0	0	0	27

Our counting procedure shows that the Borda scores are as follows:
A: $9 \times 4 + 0 \times 3 + 28 \times 2 + 5 \times 1 = 97;$

B: $0 \times 4 + 15 \times 3 + 5 \times 2 + 22 \times 1 = 77;$
C: $13 \times 4 + 5 \times 3 + 9 \times 2 + 0 \times 1 = 85;$
D: $5 \times 4 + 22 \times 3 + 0 \times 2 + 15 \times 1 = 101;$
E: $15 \times 4 + 0 \times 3 + 0 \times 2 + 0 \times 1 = 60.$

The total number of points is $97 + 77 + 85 + 101 + 60 = 420$, which corresponds to 42×10 because every voter has 10 points to allocate between five alternatives. Since the average is $420/5 = 84$, companies B and E are eliminated because their scores are less than the average. This yields the new preference rankings:

	Number of Ballots (42)				
	15	7	13	5	2
First choice	D	A	C	D	A
Second choice	A	D	A	C	C
Third choice	C	C	D	A	D

Since there are only three alternatives, the first place is rewarded with 2 points, the second place gives 1 point, and the last place contributes nothing. So the Borda scores become:

A: $9 \times 2 + 28 = 46;$
C: $13 \times 2 + 7 = 33;$
D: $20 \times 2 + 7 = 47;$

with total of $126 = 42 \times 3$. The average is $126/3 = 42$, so we eliminate alternative C to obtain the final pool of two options: A and D. Since A defeats D with score 24:18, we see that alternative A is Nanson's winner whereas alternative D is Borda's winner.

Example 1.3.12. Baldwin's Method

Before a conference, a panel of participants voted on the topic for the keynote address. The choices were Global Warming, Information Technologies, Health Insurance, Education, and Finance. The votes are summarized in the following preference table.

Alternatives	Number of Votes (41)				
	7	11	9	10	4
Education	4	4	5	2	3
Finance	2	1	3	3	4
Health Insurance	1	2	1	4	1
Information Technologies	3	3	2	5	2
Global Warming	5	5	4	1	5

To facilitate counting points for each alternative, we summarize the number of places in the following table.

	1st place	2nd place	3rd place	4th place	5th place
Education	0	10	4	18	9
Finance	11	7	19	4	0
Health Insurance	20	0	11	10	0
Information Technologies	0	24	7	0	10
Global Warming	10	0	0	9	22

Our counting procedure shows that the Borda scores are as follows:

Education: $0 \times 4 + 10 \times 3 + 4 \times 2 + 18 \times 1 = 56$

Finance: $11 \times 4 + 7 \times 3 + 19 \times 2 + 4 \times 1 = 107$;

Health Insurance: $20 \times 4 + 11 \times 2 + 10 \times 1 = 112$;

Information Technologies: $24 \times 3 + 7 \times 2 = 86$;

Global Warming: $10 \times 4 + 9 \times 1 = 49$.

Hence we see that "Health Insurance" is the Borda winner and option "Global Warming" is the Borda loser. Eliminating the loser, we get the following preference rankings:

	Number of Votes (41)				
	7	11	9	10	4
Education	4	4	4	1	3
Finance	2	1	3	2	4
Health Insurance	1	2	1	3	1
Information Technologies	3	3	2	4	2

Since there are only four alternatives left, the first place is rewarded with 3 points, the second one gets 2 points, and the third earns 1 point. When we tally the points,

Education gets $10 \times 3 + 4 \times 1 = 34$;

Finance gets $11 \times 3 + 17 \times 2 + 9 \times 1 = 76$;

Health Insurance gets $20 \times 3 + 11 \times 2 + 10 \times 1 = 92$;

Information Technologies gets $13 \times 2 + 18 = 44$.

The total number of points is 246, which is consistent with 41 ballots times 6 points each voter can allocate. The next round eliminates the loser—Education—which yields the following preference table:

	Number of Ballots (41)				
	7	11	9	10	4
Finance	2	1	3	1	3
Health Insurance	1	2	1	2	1
Information Technologies	3	3	2	3	2

Ranking in the fourth column can be united with the second one, and the last column can be combined with the third one:

	7	21	13
Finance	2	1	3
Health Insurance	1	2	1
Information Technologies	3	3	2

The Borda count for each alternative becomes
Finance: $21 \times 2 + 7 = 49$;
Health Insurance: $20 \times 2 + 21 = 61$;
Information Technologies: $13 \times 1 = 13$.

Eliminating the loser—Information Technologies—we arrive at a competition between two options. Applying the majority rule, we see that "Finance" beats "Health Insurance" with a score of 21:20. So Baldwin's method rewards the option "Finance," while the Borda count identifies "'Health Insurance" as the winner.

Problems.

1. What is the total number of points that all candidates can earn in an election using the Borda method if there are five candidates and 23 voters?

2. What is the total number of points that all candidates can earn in an election using the Borda method if there are four candidates and 35 voters?

3. What is the maximum number of points that a candidate can earn in an election using the Borda count if there are 12 candidates and 17 voters?

4. What is the minimum number of points that a candidate can earn in an election using the Borda count if there are 7 candidates and 87 voters?

5. An election is held among 4 candidates (A, B, C, and D). Each column in the following rankings shows the percentage of voters voting that way.

	Percentage of Votes							
Ranking	3%	5%	8%	10%	13%	15%	20%	26%
First choice	A	C	B	D	D	B	A	C
Second choice	B	D	C	B	A	C	D	A
Third choice	C	A	D	A	B	D	C	B
Fourth choice	D	B	A	C	C	A	B	D

Assume that there are 500 votes; find the number of votes for each column in the given preference schedule and then find the winner of the election under the Borda count method.

6. An election is held among 4 candidates (A, B, C, and D). Each column in the following rankings shows the percentage of voters voting that way.

	Percentage of Votes				
Ranking	30%	24%	18%	16%	12%
First choice	A	C	B	B	D
Second choice	D	B	D	A	C
Third choice	B	D	A	D	A
Fourth choice	C	A	C	C	B

Assume that there are 150 votes; find the number of votes for each column in the given preference schedule and then find the winner of the election under the Borda count method.

7. An election is to be decided using the Borda system. There are five candidates (A, B, C, D, and E) in this election.

   (a) How many points are given out by one ballot?

   (b) If there are 175 votes in the election, what is the total number of points given out to the candidates?

   (c) If candidate A gets 333 points, candidate B gets 254 points, candidate C gets 174 points, and candidate D has 112 points, how many points did candidate E get?

8. An election is to be decided using the Borda system. There are four candidates (A, B, C, and D) and 13 voters in this election. If candidate A gets 22 points, candidate B gets 19 points, candidate C gets 18 points, and candidate D gets 16 points, how many voters rank candidate A the best?

9. The most outstanding player in collegiate football is rewarded with the Heisman Memorial Trophy. The winner is determined by the modified Borda count when the first place receives 3 points, second place receives 2 points, and the third place receives 1 point. All other places in a preference ballot contribute zero points.
   The 2008 Heisman Trophy voting breakdown is presented in the following table.

	First	Second	Third	Total
Sam Bradford, Oklahoma	300	315	196	
Colt McCoy, Texas		288	230	1640
Tim Tebow, Florida	309		234	1575
Graham Harrell, Texas Tech	13	44		213
Michael Crabtree, Texas Tech	3	27	116	
Shonn Greene, Iowa		9	32	65

Find the missing entries in the table.

10. The Cy Young Award was first introduced in 1956 by Commissioner of Baseball Ford Frick in honor of Hall of Fame pitcher Cy Young, who died in 1955. The Cy Young Award is an honor given annually in baseball to the best pitchers in Major League Baseball (MLB), one each for the American League (AL) and National League (NL). The winner is determined by the modified Borda count when the first place gives 3 points, second place gives 2 points, and the third place gives 1 point. All other places in a preference ballot contribute zero points. The 2009 Cy Young Award voting results in the AL are presented here.

	First	Second	Third	Total
Zack Greinke, Kansas City	25	3		134
Felix Hernandez, Seattle	2	23		80
Justin Verlander, Detroit		0	9	14
CC Sabathia, New York	0	2	7	
Roy Halladay, Toronto			11	11

Find the missing entries in the table.

11. Three cities, Atlanta, Beijing, and de Cannes are competing to be the host city for a song festival. The final decision is made by a secret vote of 31 members of the executive committee.

A day before the actual election is to be held, a straw poll is conducted by the executive committee just to see how the public react on the competition. The results of the straw poll are shown in the table. Determine a winner based on Borda's count.

	Number of Votes (31)			
City	10	8	7	6
Atlanta	1	2	3	1
Beijing	2	3	1	3
de Cannes	3	1	2	2

12. Suppose that, in a survey, people were asked to rank the ice cream flavors chocolate, strawberry, and vanilla in order from their first to last choice. Determine the most favorite ice cream based on Borda's count.

	Percentage of Voters					
Flavor	11%	32%	4%	19%	8%	26%
Chocolate	1	1	2	2	3	3
Strawberry	2	3	1	3	1	2
Vanilla	3	2	3	1	2	1

13. Suppose that a university is considering whether to expand, decrease, or maintain the current level of computer support to students. Such projected change in service will affect service's fees, with more service resulting in higher fees. The preference ranking of the student body regarding the level of service are presented here.

    (a) Which option would the students choose using the Borda method?

    (b) Which option would the students choose using the Borda method with the first-place promotion?

	Percentage of Voters					
Option	27%	7%	6%	19%	26%	15%
Expand	1	1	2	2	3	3
Decrease	2	3	1	3	1	2
Maintain	3	2	3	1	2	1

14. Three candidates—Bergeron, Savard, and Wideman—ran for a seat on city council. Polls indicate that the preference rankings of the voters broke down into the following percentages.

	Percentage of Voters					
Candidate	13%	7%	26%	22%	10%	22%
Bergeron	1	1	2	2	3	3
Savard	2	3	1	3	1	2
Wideman	3	2	3	1	2	1

    (a) Who would be the winner of the election if the Borda method had been used?

    (b) Who would be the winner of the election if the Borda method with the first-place promotion had been used?

15. Fifteen players from a soccer team want to choose a type of Japanese appetizer. Each of the players has an order of preference for the different types of appetizers. The preference ranking of the players are listed in the following table. Which type of food would win using the Borda count?

	Number of Votes (15)					
	4	1	2	3	3	2
Edamame	3	3	3	1	2	4
Gyoza	2	1	1	4	3	1
Shumai	4	2	4	3	1	2
Tempura	1	4	2	2	4	3

16. A family wants to open an Austrian restaurant. On opening day, they offered (among many others) four famous Viennese dishes, and asked customers to rank each food favorite to least favorite. The choices were Apfelstrudel, Gulasch, Tafelspitz, and Wiener Schnitzel. The results of their voting are summarized in the following table. Using the Borda count method, determine the most favorite dish and the least favorite dish.

	Percentage of Votes							
	8%	17%	12%	14%	10%	16%	12%	11%
Apfelstrudel	2	2	3	3	1	1	4	4
Gulasch	3	1	4	1	4	3	3	2
Tafelspitz	1	4	2	4	2	2	1	3
Wiener Schnitzel	4	3	1	2	3	4	2	1

17. Sixteen players from a rugby team want to choose a type of snack. Each of the players has an order of preference for the different types of snacks. The preference rankings of the players are listed in the following table. Which type of food would win using the Borda count?

Rank	Number of Votes (total 16)				
	4	2	2	3	5
1st choice	Taco	Pizza	Pizza	Hamburgers	Hot Dogs
2nd choice	Pizza	Hot Dogs	Taco	Taco	Hamburgers
3d choice	Hamburgers	Hamburgers	Hamburgers	Hot Dogs	Pizza
4th choice	Hot Dogs	Taco	Hot Dogs	Pizza	Taco

18. Thirty members of the student council must decide what would be the opening act for the yearly school talent show. The preference rankings of the committee members are listed here. Determine the winner and the loser using the Borda count.

	Number of Votes (30)						
	2	7	7	6	3	4	1
Dancing	1	3	2	4	1	1	4
Poetry	4	2	1	3	2	2	1
Singing	2	4	4	1	3	4	2
Stepping	3	1	3	2	4	3	3

19. In an election for homecoming queen, the candidates are Ashley, Beatrice, Clara, and Donna. Determine the winner and the loser based on the Borda count if the preference schedule for the election is as follows:

	Number of Votes (55)								
	3	4	9	9	2	5	8	3	12
Ashley	1	1	1	4	2	3	4	2	3
Beatrice	2	2	3	1	1	1	3	4	4
Clara	4	3	4	3	4	4	2	3	1
Donna	3	4	2	2	3	2	1	1	2

20. There are four candidates running for president of a sorority: Ashley, Casey, Melissa, and Julia. Who wins the election based on the Borda count? Who is the loser?

	Number of Votes (30)							
	2	6	3	3	3	4	4	5
Ashley	1	1	3	4	2	3	4	2
Casey	2	4	1	1	3	4	2	3
Melissa	3	2	4	2	1	1	3	4
Julia	4	3	2	3	4	2	1	1

21. An election is held among four candidates: Abramovich (A), Berezovsky (B), Lesin (L), and Potanin (P). Each column in the following preference schedule shows the percentage of voters voting that way. Using Borda count, determine the winner.

Ranking	Percentage of Voters							
	18%	12%	13%	12%	8%	17%	9%	11%
First choice	A	A	B	B	L	L	P	P
Second choice	P	L	P	A	B	P	L	B
Third choice	B	P	A	L	A	A	A	L
Fourth choice	L	B	L	P	P	B	B	A

22. An instructor has a class of 34 students. She asked her students to pick four favorite colors and rank them. The results are summarized in the following table. Determine the winner and the loser using the Borda count method.

Option	Number of Votes (34)				
	9	7	8	3	7
Pink	1	4	2	3	3
Blue	2	3	1	2	4
Yellow	3	1	4	1	2
Green	4	2	3	4	1

In Problems 23 – 34, determine the winner using

(a) the Borda count;

(b) the Borda method with the first-place promotion;

(c) the modified Borda method with 3-2-1 weights;

(d) the Nauru count;

(e) Nanson's method;

(f) Baldwin's method.

23. Ninety-three members of a fraternity voted to determine who of four candidates would be the president for the next two years. The results of voting are presented here.

Ranking	Number of Ballots (79)					
	3	17	21	16	9	13
First choice	A	A	B	C	C	D
Second choice	C	D	A	D	B	C
Third choice	B	C	D	B	A	A
Fourth choice	D	B	C	A	D	B

24. A poll of members of a community asked people about their opinions on a curfew time for young teenagers under the age of 16. Their preference rankings broke down into the following percentages.

	Percentage of Voters							
Option	3%	7%	16%	12%	15%	19%	18%	10%
9 pm	1	1	2	1	3	2	4	4
12 midnight	2	4	3	2	2	3	1	3
1 am	3	2	4	3	4	1	2	1
No curfew	4	3	1	4	1	4	3	2

25. Four candidates (A, B, C, and D) are running for mayor in a small town. Being at the cutting edge of electoral reform, the city would like to test six different scoring voting methods. A straw poll is conducted by the committee to see the public reaction on their reform. The preference schedule is shown here.

	Number of Ballots (172)					
Ranking	17	23	42	33	9	48
First choice	A	A	B	C	C	D
Second choice	B	D	C	D	A	A
Third choice	C	B	A	B	B	C
Fourth choice	D	C	D	A	D	B

26. Members of a hockey team are asked to select the most valuable defender on the team among four fellow hockey players: Chara, Ference, Stuart, and Wideman. The preference schedule is shown here.

	Number of Voters (22)								
Defenceman	5	2	3	3	2	1	3	2	1
Chara	1	1	2	2	3	3	4	4	4
Ference	2	4	1	1	4	4	2	3	3
Stuart	4	3	3	4	1	2	3	1	2
Wideman	3	2	4	3	2	1	1	2	1

27. According to the school's bylaws, its principal should be chosen out of four appointed candidates. The preference schedule is shown here.

	Number of Ballots (38)					
Ranking	1	6	11	3	7	10
First choice	A	A	B	C	C	D
Second choice	C	B	A	A	D	C
Third choice	D	C	D	B	B	A
Fourth choice	B	D	C	D	A	B

28. A hockey team will be choosing the number-one pick in the upcoming draft. After narrowing the list of candidates to five players (Steven Stamkos, Kyle Okposo, Kyle Turris, Drew Doughty, and Nikita Filatov), the coaches and team executives meet to discuss the candidates and eventually choose the team's first pick on the draft. The voting results are presented here.

Rookie	Number of Voters (19)									
	2	3	1	2	2	1	2	2	3	1
Stamkos	1	1	2	2	3	3	4	4	5	5
Okposo	2	5	4	1	5	2	1	3	4	3
Turris	3	4	3	3	1	5	2	5	2	4
Doughty	4	2	1	5	2	4	5	1	3	1
Filatov	5	3	5	4	4	1	3	2	1	2

29. The editorial board of a local magazine is having an election to choose the "Auto Repair Shop of the Year." The candidates are Blackstone, Intensive Care, Rebello, Tellstone, and Wayney. The preference schedule for the election is given in the following table.

Auto Repair	Number of Voters (15)									
	1	2	1	2	1	2	1	2	1	2
Blackstone	1	1	2	2	3	3	4	4	5	5
Intensive Care	2	5	1	3	4	5	2	3	1	4
Rebello	3	4	5	4	1	2	1	2	3	2
Tellstone	4	2	4	5	2	1	5	1	4	3
Wayney	5	3	3	1	5	4	3	5	2	1

30. Determine the winner in the following election between five candidates: A, B, C, D, and E.

Ranking	Number of Ballots (47)						
	3	7	9	5	9	8	6
First choice	A	D	E	C	B	D	E
Second choice	B	A	B	D	A	A	C
Third choice	C	C	D	B	E	C	B
Fourth choice	E	E	C	A	C	E	A
Fifth choice	D	B	A	E	D	B	D

31. Determine the winner in the following election between five candidates: A, B, C, D, and E.

	Number of Voters (27)									
	3	2	1	4	3	2	1	5	4	2
A	1	1	2	3	4	4	5	2	5	3
B	2	5	1	1	3	3	3	4	4	2
C	3	3	3	5	1	1	2	5	2	4
D	4	2	4	4	2	5	1	1	3	5
E	5	4	5	2	5	2	4	3	1	1

32. Determine the winner in the following election between five candidates: A, B, C, D, and E.

Ranking	Number of Ballots (31)					
	3	4	5	3	9	7
First choice	A	D	E	C	B	D
Second choice	B	E	C	D	A	A
Third choice	C	A	D	B	E	B
Fourth choice	E	C	B	A	C	E
Fifth choice	D	B	A	E	D	C

33. Determine the winner in the following election between five candidates: A, B, C, D, and E.

	Number of Voters (31)									
	2	3	4	5	1	2	3	4	5	2
A	1	1	2	2	3	3	4	4	5	5
B	5	3	4	1	5	2	1	2	4	3
C	4	2	5	4	2	1	3	5	3	1
D	3	4	3	5	4	4	5	1	1	2
E	2	5	1	3	1	5	2	3	2	4

34. Determine the winner in the following election between five candidates: A, B, C, D, and E.

Ranking	Number of Ballots (29)					
	1	7	9	3	5	4
First choice	D	A	E	C	B	D
Second choice	B	C	B	D	A	C
Third choice	C	E	D	A	C	B
Fourth choice	A	D	A	B	D	A
Fifth choice	E	B	C	E	E	E

In Problems 35 – 44, determine a winner using

(a) the plurality method;

(b) the plurality method followed by a runoff between the top two finishers;

(c) the Hare system;

(d) the Coombs method;

(e) the Borda count;

(f) the Nauru system;

(g) Nanson's method;

(h) Baldwin's method.

35. Thirty-three people were registered to attend a conference. The person in charge of planning the conference decided to conduct participant voting on a group recreational activity. The four choices were dancing, golf, movies, and theater. The preference rankings of the people are given below. Which activity won?

	Number of Votes (27)							
	4	2	5	2	7	3	1	3
Dancing	1	1	2	2	3	4	3	4
Golf	2	4	1	1	4	3	4	2
Movies	4	3	3	4	1	2	1	3
Theater	3	2	4	3	2	1	2	1

36. Eleven members of a committee must decide what kind of flooring to install in a conference room. The preference rankings of the committee members are listed here. Which type of flooring would be chosen?

Votes:	1	2	3	1	2	2
First choice	Ceramic	Wood	Carpet	Carpet	Ceramic	Granite
Second choice	Granite	Ceramic	Wood	Ceramic	Granite	Wood
Third choice	Carpet	Granite	Granite	Wood	Carpet	Ceramic
Fourth choice	Wood	Carpet	Ceramic	Granite	Wood	Carpet

37. Find the winner in the election based on the following preference rankings.

Number of Voters (27)

	3	1	3	2	3	4	2	3	3	3
A	1	1	2	2	3	3	4	4	5	5
B	5	4	4	5	1	4	1	2	1	3
C	3	5	5	1	2	5	2	3	4	1
D	4	2	3	3	4	1	5	1	3	2
E	2	3	1	4	3	3	3	5	2	4

38. Find the winner in the election based on the following preference schedule.

Votes:	3	6	3	1	5	2	4
First choice	E	A	B	C	D	E	C
Second choice	C	C	E	D	C	D	D
Third choice	B	D	D	A	E	B	A
Fourth choice	A	E	A	B	B	A	E
Fifth choice	D	B	C	E	A	C	B

39. Find the winner in the election based on the following preference rankings.

Number of Voters (21)

	3	1	4	3	2	1	2	1	3	1
A	1	1	2	2	3	3	4	4	5	5
B	2	5	5	1	4	2	1	3	4	3
C	5	4	1	5	1	5	2	5	2	4
D	4	2	3	3	2	4	5	1	3	1
E	3	3	4	4	5	1	3	2	1	2

40. Find the winner in the election based on the following preference schedule.

Votes:	4	9	7	3	6	4	2
First choice	E	A	B	C	D	E	C
Second choice	D	C	C	D	B	A	E
Third choice	A	B	D	A	C	D	A
Fourth choice	B	D	E	B	E	C	B
Fifth choice	C	E	A	E	A	B	D

41. Find the winner in the election based on the following preference schedule.

Number of Voters (20)

	4	2	3	2	3	1	1	2	1	1
A	1	1	3	5	5	3	5	4	4	3
B	5	3	1	1	4	5	4	3	2	5
C	3	5	5	4	1	1	2	5	3	4
D	4	2	2	2	3	2	1	1	5	2
E	2	4	4	3	2	4	3	2	1	1

42. Find the winner in the election based on the following preference schedule.

Votes:	7	5	6	7	5	4	1
First choice	A	B	C	D	E	B	D
Second choice	E	C	B	E	C	A	B
Third choice	C	E	E	C	D	E	A
Fourth choice	D	A	D	A	A	C	E
Fifth choice	B	D	A	B	B	D	C

43. Find the winner in the election based on the following preference schedule.

	Number of Voters (25)									
	5	3	3	2	1	1	2	1	4	3
A	1	1	2	4	3	5	4	3	5	5
B	4	3	1	1	4	2	2	5	2	4
C	2	5	4	2	1	1	3	2	3	2
D	5	2	3	5	2	3	1	1	4	3
E	3	4	5	3	5	4	5	4	1	1

44. Find the winner in the election based on the following preference schedule.

Votes:	1	2	4	3	4	4	3
First choice	A	B	C	D	E	A	D
Second choice	E	A	D	C	B	B	C
Third choice	D	E	B	E	A	C	B
Fourth choice	B	C	E	A	C	D	E
Fifth choice	C	D	A	B	D	E	A

The following two problems require application of the Borda count that has been specially modified based on 5-3-1 distribution of points.

45. In the NBA (the national basketball league), the best rookie of the year is determined by a special modification of the Borda count: in ranking ballot, the first place is assigned 5 points, the second place is given 3 points, and the third place is rewarded with 1 point. All other places are ignored. Twenty-five ESPN experts cast their ballots at the end of the 2009 season, which are summarized in the following table. Who was rewarded with the title?

	Number of Voters (25)						
Rookies	4	4	5	3	4	3	2
Derrick Rose	1	2	5	3	4	4	5
Brook Lopez	2	1	4	5	3	5	2
Kevin Love	3	5	1	4	2	3	1
Eric Gordon	4	3	2	1	5	1	4
Michael Beasley	5	4	3	2	1	2	3

46. In the NBA (the national basketball league), the best rookie of the year is determined by a special modification of the Borda count: in ranking ballot, the first place is assigned 5 points, the second place is given 3 points, and the third place is rewarded with 1 point. All other places are just ignored. Forty one ESPN experts cased their ballots at the end of 2010 season, which are summarized in the following table. Who was rewarded with the title?

Rookies	Number of Voters (41)								
	4	5	4	6	9	4	3	3	3
Tyreke Evans	1	2	4	3	2	4	5	3	1
Brandon Jennings	3	1	2	4	4	3	2	1	3
Omri Casspi	2	3	1	2	3	2	1	5	4
Jonny Flynn	5	4	5	1	5	1	3	4	2
James Harden	4	5	3	5	1	5	4	2	5

In Problems 47 – 50, find a winner using the modified Borda count based on a 3-2-1 distribution of points. Also determine the number of points each of the three best finishers get in this election.

47. The 2009 Cy Young Award for the National Baseball League voting results are presented here.

	First	Second	Third	Total
Tim Lincecum, San Francisco	11	12	9	
Chris Carpenter, St. Louis	9	14	7	
Adam Wainwright, St. Louis	12	5	15	
Javier Vazquez, Atlanta	0	1	0	
Dan Haren, Arizona	0	0	1	1

48. Find the winner in the election based on the following preference schedule.

Option	Number of Ballots Cast (33)									
	1	4	5	6	4	2	1	2	5	3
A	1	1	2	2	3	3	4	4	5	5
B	5	3	4	1	5	2	2	3	4	1
C	3	2	5	3	1	4	3	5	1	2
D	2	4	3	5	4	1	5	1	2	3
E	4	5	1	4	2	5	1	2	3	4

49. In December 2009, fifteen ESPN experts (Brian Bennett, Beano Cook, Rece Davis, David Duffey, Heather Dinich, Brad Edwards, Pat Forde, Tim Griffin, Chris Low, Ivan Maisel, Todd McShay, Ted Miller, Adam Rittenberg, Mark Schlabach, and Graham Watson) cast their ballots in order to determine a prospective Heisman Trophy winner. The results are presented here.

Player	First	Second	Third	Forth	Fifth	Total
Toby Gerhart	7	2	2	4	0	
Mark Ingram	4	8	2	1	0	
Ndamukong Suh	3	5	6	0	1	
Colt McCoy	1	0	3	6	5	
Others	0	0	2	4	9	

50. The best rookie of the year in 2009 for the National Baseball League was chosen from 11 candidates. The results of preference rankings are presented here.

Player	First	Second	Third	Total Points
Chris Coghlan	17	6	2	
J.A. Happ	10	11	11	
Tommy Hanson	2	6	9	
Andrew McCutchen	2	5	0	
Casey McGehee	1	3	4	
Others	0	6	9	

For elections given in Problems 51 – 56, determine which of the four plurality methods supports the Borda winner and the Borda loser.

51. Find the winner and the loser in the election between four competitors (A, B, C, and D) based on the following preference schedule.

	Number of Voters (9)						
Ranking	3	1	1	1	1	1	1
First choice	A	A	B	B	C	C	D
Second choice	D	B	C	C	B	D	C
Third choice	B	C	D	A	D	B	B
Fourth choice	C	D	A	D	A	A	A

52. Find the winner and the loser in the election between four competitors (A, B, C, and D) based on the following preference schedule.

	Number of Voters (11)							
Candidate	3	1	1	2	1	1	1	1
A	1	1	4	3	4	3	4	3
B	2	4	1	2	3	2	2	1
C	3	2	3	1	2	1	3	4
D	4	3	2	4	1	4	1	2

53. A local university has to decide which of its five schools—Liberal Arts, Business, Engineering, Law, or Medicine—should increase enrollment. The preference schedule of the board voting is presented here. Determine the winner and the loser in this election.

	Number of Voters (39)						
School	7	3	9	6	7	3	4
Liberal Arts	5	3	4	1	4	2	1
Business	4	1	5	4	2	1	2
Engineering	3	2	3	2	1	5	4
Law	2	5	1	3	5	4	3
Medicine	1	4	2	5	3	3	5

54. Five candidates—Awesomeman (A), Bestman (B), Easyman (E), Kindman (K), and Smartman (S) are running for the office. Use the preference table to determine the winner and the loser.

	Number of Voters (35)							
Ranking	2	3	3	3	4	3	8	9
First choice	A	A	B	B	E	E	K	S
Second choice	E	K	A	E	A	S	A	B
Third choice	K	E	S	S	K	B	S	A
Fourth choice	S	B	K	A	B	A	B	C
Fifth choice	B	S	E	K	S	K	E	D

55. Find the winner and the loser in the election between five competitors (A, B, C, D, and E) based on the following preference schedule.

Ranking	Number of Voters (23)									
	2	3	1	4	1	3	2	2	2	3
First choice	A	A	B	D	C	C	B	E	B	E
Second choice	C	B	E	C	A	D	E	C	D	C
Third choice	D	E	D	B	E	B	A	D	C	D
Fourth choice	B	C	A	E	D	A	D	B	A	B
Fifth choice	E	D	C	A	B	E	C	A	E	A

56. Find the winner and the loser in the election between five competitors (A, B, C, D, and E) based on the following preference schedule.

Candidate	Number of Voters (25)									
	2	3	1	4	1	3	4	2	2	3
A	1	5	2	3	1	5	4	3	1	2
B	2	4	1	5	4	3	2	5	4	1
C	3	3	5	1	2	4	1	4	5	4
D	4	2	3	4	5	1	3	2	3	5
E	5	1	4	2	3	2	5	1	2	3

57. Twenty five math students are taking a survey to determine the class' favorite NBA basketball team. Each student has their preference on which team is their favorite. The options include the Boston Celtics (B), Miami Heats (M), Los Angeles Lakers (L), New York Knicks (N), and Orlando Magic (O).

Choices	Number of Votes (25)							
	5	3	2	4	3	3	3	2
First choice	B	M	L	N	O	L	O	M
Second choice	M	L	O	B	N	O	L	B
Third choice	L	O	M	L	M	B	M	N
Fourth choice	N	B	N	O	B	M	N	L
Fifth choice	O	N	B	M	L	N	B	O

The last three exercises discuss equivalent versions of the Borda count method.

58. **Equivalent Borda count (Variation 1)** Consider the following variation of the Borda count described in §1.3: the last place is rewarded with k points, next to the last place is given $k + 1$, and so on, the first place is worth $n + k - 1$ points, where n is the number of alternatives in the election and k is a nonnegative integer. Explain why this variation is equivalent to the original Borda count described in the chapter (where $k = 0$).

59. **Equivalent Borda count (Variation 2)** Another commonly used variation of the Borda system is as follows. A first place is worth 0 points, second place is rewarded with 1 point, and so forth, the last place is given $n - 1$ points, where n is the number of alternatives in the election. The option with the fewest points is the winner. Explain why this variation is equivalent to the original Borda count described in the chapter.

60. **Equivalent Borda count (Variation 3)** The average ranking of an option is obtained by taking the numbered place of the alternative on each of the ballots, adding these numbers, and dividing by the number of ballots. Explain why the option with the best average ranking is the Borda winner.

1.4 Head-to-Head Comparisons

Competition is a natural human behavior that we observe throughout the centuries. The participants may be people, teams, organizations, parties, countries, or even ideas, propositions, or options. Probably the most famous and oldest human competition is a sport activity. During a long history, people develop several fair methods to determine a single winner. Since sport activities are well-known, we will use sport terminology and examples in our exposition of methods presented in this section.

Our next voting methods to discuss are based on head-to-head comparisons of candidates or options. Almost all sport competitions use one of these methods to determine a winner. It is not surprising that the voting theory of the pairwise comparison methods has roots in sport competitions and inherits some rules from them. A head-to-head contest is like a round robin tournament in which every team or competitor is matched against every other team/competitor. Election results are interpreted as being a set of matchups between any pair of candidates. Thus, the act of ranking a candidate above another (or ranking one, but not the other) is interpreted as one vote for the higher candidate in a matchup between the two. The method[1] was proposed in the eighteenth century by French mathematician and philosopher Marie Jean Antoine Nicolas Caritat, the Marquis de Condorcet (1743 – 1794). This single-winner election method is defined as follows.

In a head-to-head comparison, a candidate or option that defeats all other competitors is called the **Condorcet winner**. A candidate or alternative that is not defeated in pairwise contests with other competitors is called a **weak Condorcet winner**. Any voting system that chooses the Condorcet winner when it exists is known as a Condorcet method. We will also say in this case that the voting system satisfies **Condorcet's principle** or **Condorcet's criterion** for selecting a winner. A Condorcet loser is a candidate who is less preferred than all other candidates in pairwise matchups. A weak Condorcet loser is a candidate who is defeated by or ties every other candidate in a head-to-head comparison.

Condorcet's method is one of several pairwise methods that is used in single-seat elections (president, governor, mayor, champion, etc.). Unlike most methods which make you choose the lesser of two evils, Condorcet's method and other pairwise methods let you rank the candidates in the order in which you would see them elected. A Condorcet ballot asks the voter to rank the candidates from top to bottom. There are some variations in different methods; some allow the voter to rank candidates equally, and often the voter need not rank all candidates, but only their top choices. Condorcet voting is arguably the fairest voting method, because if there is a candidate who would beat every other candidate in head-to-head contests, that candidate will always be chosen as the winner. Unlike plurality voting, Condorcet's method rarely provides a strong incentive to voters to vote insincerely. Thus, Condorcet's method allows alternative parties to grow and provide real choices to voters.

---

[1]Condorcet rediscovered and popularized the method that was known centuries before, at least to Ramon Llull in the thirteenth century.

An election based on head-to-head comparisons may or may not have a Condorcet winner. But when one does exist, it seems difficult to dispute that this choice ought to be the election winner. The probability that there is a Condorcet winner decreases as the number of options or voters increases. If a Condorcet winner exists, it is unique, but there could be many weak Condorcet winners (if any). Similarly, there can be more than one weak Condorcet loser. Any Condorcet method must be augmented by another rule to decide elections in the case when the method is inconclusive.

If candidate A defeats candidate B, we abbreviate it as $A \succ B$ or $B \prec A$; if there is a tie between two competitors, we abbreviate it as $A \sim B$. As usual, candidate A beats candidate B if more voters rank A over B. When preference ballots are used, A beats B ($A \succ B$) if more voters have ranked A over B. The number of voters ranking the alternative over its competitor is the measure of how much it lost by. This number is usually referred to as the score of competition, and denoted by the ratio (for example, if team A beats team B three times, and is defeated only once, we say that team A defeats team B with the score 3:1). The set of all weak Condorcet winners is called the **Schwartz set** for a particular election[2]. Note that this set might be empty.

The results of pairwise comparisons are typically displayed in a matrix form. Recall that a matrix is a rectangular table with some entries. Suppose, for instance, that there are four alternatives to choose from: A, B, C, and D. A typical ballot may look as $\begin{bmatrix} C \\ A \\ D \\ B \end{bmatrix}$ or $\langle C,\ A,\ D,\ B \rangle$ in vector-row notation or $C \succ A \succ D \succ B$ in list notation. A voter who wants to express this order between these candidates should fill out the table, called the election table, as follows.

	A	B	C	D
A	–	1	0	1
B	0	–	0	0
C	1	1	–	1
D	0	1	0	–

Every entry in the table indicates the result in comparison of one alternative in the row with the other alternative in the column. We put "1" when the row-competitor defeats the column-alternative, and "0" otherwise. For example, the first row in this table tells us that alternative A beats alternative B, is defeated by C, and prevails over D. Matrices of this kind are useful because they can be easily added together to give the overall results of an election. The sum of all ballots in an election is called the sum matrix. For instance, adding the following comparison matrices filled out by three voters

---

[2]The Schwartz set is named for political scientist Thomas Schwartz from UCLA.

	A	B	C	D
A	–	1	0	1
B	0	–	0	0
C	1	1	–	1
D	0	1	0	–

	A	B	C	D
A	–	1	1	1
B	0	–	0	0
C	0	1	–	1
D	0	1	0	–

	A	B	C	D
A	–	0	0	0
B	1	–	1	0
C	1	0	–	0
D	1	1	1	–

,

we get

	A	B	C	D
A	–	2	1	2
B	1	–	1	0
C	2	2	–	2
D	1	3	1	–

or

$$A{:}B = 2{:}1 \qquad B{:}C = 1{:}2$$
$$A{:}C = 1{:}2 \qquad B{:}D = 0{:}3$$
$$A{:}D = 2{:}1 \qquad C{:}D = 2{:}1$$

So we see that C is the Condorcet winner because it wins a majority in all of the pairwise comparisons. These three ballots can be tallied into the following preference table:

	Number of Votes (3)	
	2	1
A	2	3
B	4	2
C	1	4
D	3	1

or

Votes	2	1
First choice	C	D
Second choice	A	B
Third choice	D	A
Fourth choice	B	C

Example 1.4.1. The mayor of Smalltown is being chosen in an election using the head-to-head comparison method. The four candidates are Paul (the former town mayor), Alex (the head of the town council), Sarah (director of the finance department), and Boris (former district attorney). The results of 500 registered voters are summarized in the preference schedule below.

	Number of Votes			
	125	135	117	123
Alex	4	1	1	3
Boris	2	2	2	2
Paul	1	4	3	4
Sarah	3	3	4	1

or

Votes	125	135	117	123
First choice	Paul	Alex	Alex	Sarah
Second choice	Boris	Boris	Boris	Boris
Third choice	Sarah	Sarah	Paul	Alex
Fourth choice	Alex	Paul	Sarah	Paul

The Condorcet method is actually a round robin tournament so, to begin, we need an organized way of displaying all possible head-to-head matchups. Perhaps the simplest way to do this is with a table giving each candidate a row and a column.

	Alex	Boris	Paul	Sarah
Alex				
Boris				
Paul				
Sarah				

This table allows us to determine the winner of each pairwise comparison. In each of the four columns in the preference table, we compare two candidates to see who is ranked above. For instance, if we want to determine who is the winner in the head-to-head matchup between Alex and Boris, we read every column in the preference table. The first column shows that 125 voters put Boris in the second place, but Alex gets only the last place. Therefore these 125 voters prefer Boris to Alex. The next two columns show that Alex beats Boris since there are 252 (135 + 117) voters who prefer Alex to everyone else. The last column claims that with the second place against the third place Boris defeats Alex attracting 123 votes. Hence in this election Alex beats Boris with the score 252 to 248. We record this score in the two cells that are intersections of columns and rows with their names. To see the winner of the head-to-head matchup between Alex and Boris, just go to the Alex row and move over to the Boris column or go to the Boris row and look over to the Alex column.

	Alex	Boris	Paul	Sarah
Alex		252:248		
Boris	248:252			
Paul				
Sarah				

Another way to obtain this score between Alex and Boris is to eliminate all other candidates in the preference table (or isolate these two candidates, Alex and Boris, on each ballot) and count which candidate is preferred on a majority of ballots:

Number of Votes				
	125	135	117	123
Alex	4	1	1	3
Boris	2	2	2	2

From this truncated table, we count the number of voters that rank Alex above Boris (or vice versa). In a similar way, we fill out all cells in the tables to obtain

	Alex	Boris	Paul	Sarah
Alex		252:248	375:125	252:248
Boris	248:252		375:125	377:123
Paul	125:375	125:375		242:258
Sarah	248:252	123:377	258:242	

Of course, the "diagonal" cells in the table are not filled. Also, some cells are not needed. For example, the winner of the head-to-head matchup between Paul and Alex could also have been recorded by going to the Alex row and moving over to the Paul column. Therefore, we can fill only half of the total cells—either above the diagonal:

	Alex	Boris	Paul	Sarah
Alex		252:248	375:125	252:248
Boris			375:125	377:123
Paul				242:258
Sarah				

or below the diagonal:

	Alex	Boris	Paul	Sarah
Alex				
Boris	248:252			
Paul	125:375	125:375		
Sarah	248:252	123:377	258:242	

Instead of typing the score, we can just print either the name of the winner:

	Alex	Boris	Paul	Sarah
Alex				
Boris	Alex			
Paul	Alex	Boris		
Sarah	Alex	Boris	Sarah	

or the symbol $\prec/\succ$ according to our abbreviation:

	Alex	Boris	Paul	Sarah
Alex				
Boris	$\prec$			
Paul	$\prec$	$\prec$		
Sarah	$\prec$	$\prec$	$\succ$	

This table may contain the symbol $\sim$ if there is a tie between two candidates. Note that the score tables are used in all sport activities when the results of games should be recorded and stored. In tournament competitions such as soccer, basketball, hockey, and many others, teams play against each other's twice (it could be more than twice)—one time in its own city, the next time in the competitor's location. In this case, home results are filled in the upper part of the table (above the main diagonal), away results are stored in the lower part of the table.

Now we can return to the preference table and claim that Alex is the Condorcet winner because he defeats Boris with score 252:248, Paul with score 375:125, and Sarah with score 252:248. On the other hand, Paul is the Condorcet loser because he was beaten by all competitors. □

With four candidates, we see that we need to make 6 comparisons to fill out the table. In general, if there are n alternatives, there are n^2 total number of cells in the table, from which we need to subtract n diagonal elements (a competition with itself does not mean a real matchup). So we are left with $n^2 - n = n(n-1)$ cells. Since we need to fill only half of the table cells, we get $\boxed{n(n-1)/2}$ number of comparisons. Note that in sport activities, when n teams play against each other twice, there are $n(n-1)$ matchups.

Actually, when a Condorcet winner is supposed to be determined, we do not need to make all $n(n-1)/2$ of these comparisons. In our previous example, since we find that Alex beats Boris, then the latter one is eliminated from the competition because Boris cannot be a Condorcet winner. However, Boris may affect the results of the election if, for instance, he beats Paul, eliminating him from being a Condorcet winner.

Example 1.4.2. An Election with No Condorcet Winner

Now suppose that in the previous example Alex drops out of the election (we do not discuss the reasons for such a move). This leads to a new distribution in the preference table (because voters that previously supported Alex split their ballots between the other three candidates):

Number of Votes (500)

	125	132	117	126
Boris	2	1	1	3
Paul	1	3	2	2
Sarah	3	2	3	1

or

Votes	125	132	117	126
First choice	Paul	Boris	Boris	Sarah
Second choice	Boris	Sarah	Paul	Paul
Third choice	Sarah	Paul	Sarah	Boris

Making a table of winners, we get

	Boris	Paul	Sarah
Boris		Paul	Boris
Paul	Paul		Sarah
Sarah	Boris	Sarah	

Since there is neither row nor column with a single candidate, we conclude that this election does not reveal a winner based on the Condorcet method. Moreover, we observe that Boris defeats Sarah (B $\succ$ S) but lost to Paul (B $\prec$ P); however, Sarah beats Paul (S $\succ$ P). Hence we see that the binary relation $\succ/\prec$ is not transitive. That is, B $\succ$ S, S $\succ$ P, but B $\prec$ P.

Example 1.4.3. A Weak Condorcet Winner and Loser

Now suppose that in the previous example one voter changed his or her mind (from 3,2,1 to 1,3,2), which leads to the following results:

Number of Votes (500)

	125	133	117	125
Boris	2	1	1	3
Paul	1	3	2	2
Sarah	3	2	3	1

or

Votes	125	133	117	125
First choice	Paul	Boris	Boris	Sarah
Second choice	Boris	Sarah	Paul	Paul
Third choice	Sarah	Paul	Sarah	Boris

From the preference table, it follows that Boris is tied with Paul (250 : 250), and defeats Sarah (375 : 125). On the other hand, Sarah beats Paul (258 : 242), which makes Boris the weak Condorcet winner. On the other hand, Paul is the weak Condorcet loser because he did not win any head-to-head matchup. □

Example 1.4.4. Condorcet and Plurality Winners and Losers

In the race for one seat in the Senate between three candidates (call them A, B, and C), the exit poll suggests the following set of preference links:

29%	4%	28%	5%	27%	7%
A	A	B	B	C	C
B	C	A	C	A	B
C	B	C	A	B	A

Is there a Condorcet winner? A Condorcet loser?
Who won using plurality voting? Who was the plurality loser?

To answer these questions, we first calculate the number of first-place votes: A — 33%, B — 33%, and C — 34%. So candidate C is the plurality winner and the plurality loser because he has 57% of last-place votes—more than any other candidates.

Comparing A and B, we see that A defeats B with the score A:B = 60:40, and A beats C with 61:39. So A is the Condorcet winner. Since B beats C with the score 62:38, candidate C is the Condorcet loser. Therefore, we observe that the plurality method and pairwise comparison disagree on determination of the winner, but they choose the same loser.

A similar situation may occur when the runoff method is used. For instance, the runoff winner—Nicolas Sarkozy—in the 2007 French presidential elections (see Example 1.2.17 on page 29) was predicted to be defeated by the Condorcet winner—François Bayrou.

Example 1.4.5. A Weak Condorcet Winner and Loser at the Same Time

When the number of votes is even, it may happen that head-to-head comparisons may choose the same alternative to be the winner and the loser. This means that determination of the weak Condorcet winner may contradict itself!

Again, consider an election between four candidates: Alex, Boris, Chris, and Donna. Suppose that the exit poll has the following set of preference links:

Candidate	Percentage							
	27%	18%	9%	14%	9%	14%	4.5%	4.5%
Alex	1	4	2	4	1	3	3	3
Boris	2	1	4	2	3	1	2	4
Chris	3	2	1	3	4	4	1	2
Donna	4	3	3	1	2	2	4	1

Making the pairwise comparisons, we see that Alex tied with all other candidates, so he is a weak Condorcet winner because he was not defeated. Another candidate, Boris, defeats the others: Boris : Chris = 82:18 and Boris : Donna = 63.5:36.5. Hence Boris is also the weak Condorcet winner.

Since Donna is defeated by Chris with the score of 41.5:58.5, we claim that Donna is the weak Condorcet loser, sharing with Alex, who did not win in all matchups. Therefore, Alex is the weak Condorcet winner and the loser at the same time.   □

One of the advantages of the Condorcet system is its ability to eliminate the least wanted alternative without abandoning more wanted alternatives. It seems that Condorcet's method is by far the best of the methods studied so far. At least it is the fairest method, at first glance. However, let us analyze the Condorcet method with more scrutiny—we will discuss the conditions of fairness in §1.6.

The main drawback of the Condorcet system is its inability to determine a winner in many practical cases; in particular, when the level of competitors is close. It is most likely that Condorcet's method will be inconclusive in such a tournament. Hence, the Condorcet system should be accompanied with another rule when a winner is not determined by the system. In real sport contests, participants' strengths are close to each other and the results of their matchups are unpredictable (otherwise an audience would not enjoy watching such competitions), so in round robin tournaments every competitor is usually defeated by someone—no Condorcet winner.

Unlike the plurality methods (see §1.2), pairwise comparison methods give the same weight to the winner independently of how many candidates fall between two competitors on the ballot. However, the ranked pair clearly does give the greatest preference to higher choices, in the sense that all lower choices are less preferred. Yet head-to-head comparison methods do not place more weight on a voter's choice of first over second candidate than seventh over eighth. This emphasizes that Condorcet's method should be adjusted if we want to use it in practical situations.

When the number of competitors, n, is large enough, the number of comparisons, $n(n-1)/2$, grows very fast because the Condorcet method requires a round robin tournament (every alternative should be compared with all others). For example, if there are 20 teams, the champion is determined after $20 \cdot 19/2 = 190$ matchups. So every team must play (in one round) 19 games. This is effective with team competitions such as soccer or hockey. However, some tournaments cannot afford round robin competitions. For instance, in boxing or tennis contests, a player may not be able to complete the tournament after losing a match due to physical or emotional conditions—as players need time for recovery.

Another shortcoming could be caused by the impossibility of comparing alternatives at once—as in almost all sport activities. While it is possible to conduct a political or business comparison during, say one day, sport activities require either a lengthy competition or elimination of competitors that lose their matchups. It is true that lengthy competitions attract audiences for some activities such as sports, dancing, and singing, but it may be an obstacle in others. To speed up the determination of a single winner, there is another well-known method, which will require a new definition.

By an **agenda** we understand an ordering (or listing in some order) of competitors. For example, in a tennis competition, all participants are placed in some order so that high-ranking players would not compete against each other in early stages. The same principle is maintained for all tournaments of teams' competitions in soccer, basketball, baseball, hockey, and other sport activities—the best teams would meet in the late stages. So in sports, the agenda is the schedule of the competition.

Sequential pairwise voting compares pairs of candidates according to an agenda to determine which one would defeat the other. After such comparison, a loser is eliminated, and the winner moves on to confront the next candidate on the list. This process continues throughout the entire agenda, and the one remaining at the end is declared the winner.

An agenda may be the result of a collective decision made by the entire group or it may be created by one or more members. Obviously, the agenda may be manipulated to try to control the electoral outcome. In certain situations, knowledge of other people's preferences can enable one voter to manipulate the outcome. To apply this method, it is assumed that there is either an odd number of voters, or a tie resolution could be broken in some way or other. Therefore, a sequential pairwise voting assumes that in any head-to-head matchup, one of the competitors defeats the other (no ties).

As we see in some sport competitions (tennis), the winner is usually determined by sequential pairwise competitions according to a preassigned schedule (agenda). The objective of an agenda is to reduce the number of matchups in determination of a Condorcet winner. Strictly speaking, the sequential pairwise winner is not necessarily the Condorcet winner just because it is not a round robin tournament. However, in real-life competitions, there are often some strong candidates and some weak candidates, so the agenda utilizes the observation that a strong candidates will always defeat weak ones (which is not always true). Hence, the agenda can greatly affect the outcome of the competition. Needless to say, a sequential pairwise voting arises not only in sport or music activities, but also in legislative processes.

Rankings		
Voter #1	Voter #2	Voter #3
A	C	B
B	A	D
D	B	C
C	D	A

Table 75: Example 1.4.6.

Example 1.4.6. Sequential Pairwise Voting Depends on Agenda

The example below shows that in the absence of a Condorcet winner, the election outcome is highly dependent on the competitor's agenda. The sequential pairwise voting method uses all of the information from the preference table, but not all at once. It clearly meets the Condorcet criteria, but the later an alternative is introduced, the better its chances of winning. World and Olympic competitions clearly show the importance of the order of the comparisons: they "seed" competitors according to a special ranking so the best players meet each other as late as possible. Clearly it is unreasonable to attempt to rig a political election in this fashion.

Suppose we have a set of three voter preferences as follows:

It is possible to devise agendas that result in each of the alternatives winning:

Agenda 1	A vs B $(2:1)$	$\Longrightarrow$	A vs C $(1:2)$	$\Longrightarrow$	C vs D $(1:2)$		D wins
Agenda 2	B vs C $(2:1)$	$\Longrightarrow$	A vs B $(2:1)$	$\Longrightarrow$	A vs D $(2:1)$		A wins
Agenda 3	A vs C $(1:2)$	$\Longrightarrow$	C vs B $(1:2)$	$\Longrightarrow$	B vs D $(3:0)$		B wins
Agenda 4	A vs B $(2:1)$	$\Longrightarrow$	A vs D $(2:1)$	$\Longrightarrow$	A vs C $(1:2)$		C wins

Several amendments of the Condorcet method are widely used. These voting systems use preferential ballots to make pairwise comparison counts of n candidates, and determine sequence scores to identify the most popular choice.

One of them is **Copeland's method** in which the winner is determined by finding the candidate with the most pairwise victories. This method has its roots in sport activities where the teams with the greatest number of victories in regular season matchups make it to the playoffs. To make it close to a chess contest, A. H. Copeland [13] suggested to use a score function. More precisely, in a pairwise competition between two alternatives A_i and A_j $(i \neq j$ and $i, j = 1, 2, \ldots, n)$, the comparison score would be

$$s_{i,j} = \begin{cases} 1, & \text{if } A_i \text{ beats } A_j, \\ \frac{1}{2}, & \text{if there is a tie between } A_i \text{ and } A_j, \\ 0, & \text{if } A_j \text{ defeats } A_i. \end{cases} \tag{1.4.1}$$

Other equivalent weights of wins, ties, and losses are known, for example: $\left(\frac{1}{3}, \frac{1}{6}, 0\right)$ or $(1, 0, -1)$. The latter weights are used in ice hockey to evaluate a player's impact on the ice: a player is given $+1$ rating if his team scores while he is on the ice, and -1 if the other team scores. The **Copeland score** for each alternative A_i $(i = 1, 2, \ldots, n)$ is defined as

$$C(A_i) \equiv C(i) = \sum_{k \neq i} s_{i,k}. \tag{1.4.2}$$

Traditionally the next step is to define a Condorcet's winner as a candidate with Copeland score equal to $n - 1$ (as an alternative that defeats all others). When a Condorcet winner does not exist, we define a Copeland winner as the alternative with the highest score. In other words, the winner is the alternative for which the number of alternatives it beats, minus the number of alternatives that beat it, is the greatest. If there are two or more candidates with the same Copeland score, the system is inconclusive and a tie-breaker is needed.

The Copeland score is used to rank the candidates when they need to be ordered. It identifies each candidate with a point on the number line; therefore, transitivity of the election ranking is inherited from the transitivity of points on the line. Recall that a Condorcet system is not transitive as Example 1.4.2 shows.

Since the problem of choosing a voting system can be interpreted as a process of pooling information to resolve conflicts, the Copeland scoring can help groups clarify the nature and strength of "consensus" by deliberating about which type of collective outcome best fits their judgment. For instance, voters might consider whether the intervals between the candidates are best approximated by Copeland scoring (or neither). A decision system for conflict resolution may be limited by factors such as group size, decision tasks, and so on. Critics argue that the Copeland method puts too much emphasis on the quantity of pairwise victories rather than on the magnitude of victories (or conversely, of the defeats). Since it leads to ties and inconclusion similar to the Condorcet system, let us consider one of the approaches to break ties.

Originally, the Marquis de Condorcet suggested using the notation $n_{i,j}$ for every pair of alternatives, A_i and A_j, $i, j = 1, 2, \ldots, n$, where $n_{i,j}$ is the number of voters ranking alternative i over alternative j minus the number of voters ranking j over i. Therefore, $n_{i,j}$ is the margin by which alternative A_i defeats alternative A_j. A Condorcet winner (loser) is the alternative i for which all $n_{i,j}$, $j \neq i$, are positive (negative). Adding all these values, we obtain the **Condorcet score** for every alternative:

$$N(A_i) \equiv N(i) = \sum_{k \neq i} n_{i,k}. \tag{1.4.3}$$

Since Copeland's method may result in a tie, there must be a tie-breaker method specified with it. Since a universal rule to break ties is not known, every election or competition chooses its own method. When the Copeland system is inconclusive, the winner could be determined with either the highest Condorcet score or their pairwise matchup. A combination of these tie-breakers is used in sport competitions such as hockey, soccer, and handball. If two or more teams share the same number of points in a tournament, then the winner is the team that defeats others in pairwise matchups, followed by a difference between goals scored against goals allowed. In sport competitions, such a difference is used to calculate the Condorcet score, while the outcome (a win, tie, or loss) of their pairwise matchup is considered the Copeland score using awards (1 point, $\frac{1}{2}$, or 0).

Example 1.4.7. Copeland's Method
Consider the following preference rankings between three candidates—Alex, Boris, and Chris (we abbreviate them as A, B, and C, respectively):

Number of votes (28)					
6	5	3	3	2	9
A	A	B	B	C	C
B	C	C	A	A	B
C	B	A	C	B	A

From the table above, we see that Boris defeats Alex with a score 15:13, which means that fifteen voters prefer Boris to Alex, and thirteen voters prefer Alex to Boris. Therefore, Boris gets 1 point in their pairwise matchup. Chris beats Boris with a score 16:12, but Alex makes a tie with Chris, 14:14. Therefore, there is no Condorcet winner. However, Chris is a weak Condorcet winner because he is not defeated. Their Copeland scores

$$C(\text{Alex}): = \frac{1}{2}, \qquad C(\text{Boris}): = 1, \qquad C(\text{Chris}): = 1\frac{1}{2}$$

identify the winner uniquely—it is Chris.

Example 1.4.8. Condorcet's Scores
The preference rankings of 13 players from a volleyball team for their favorite fruit are as follows.

Number of Votes (13)

	1	2	3	4	1	2
Apple	1	2	3	4	1	1
Mango	2	3	4	1	2	4
Orange	3	4	1	2	4	3
Pear	4	1	2	3	3	2

The score table for these four alternative is as follows (the margin is written in parentheses):

	Apple	Mango	Orange	Pear	Margin loss	Margin win	Condorcet score
Apple		9:4 (+5)	6:7 (−1)	4:9 (−5)	−6	+5	−1
Mango	4:9 (−5)		8:5 (+3)	6:7 (−1)	−6	+3	−3
Orange	7:6 (+1)	5:8 (−3)		8:5 (+3)	−3	+4	+1
Pear	9:4 (+5)	7:6 (+1)	5:8 (−3)		−3	+6	+3

Since every alternative is defeated, there is no Condorcet winner. The Copeland method is also inconclusive because two options (Orange and Pear) have two wins and the other two options have two losses. To break ties, we can use either a head-to-head comparison or the Condorcet scores. Since Orange prevails over the Pear option in a paired contest, the former is the winner. However, the Condorcet scores

N(Apple) = −1,   N(Mango) = −3,   N(Orange) = +1,   N(Pear) = +3.

identify another winner—Pear.

Problems.

1. Suppose that 24 soccer teams participate in a round-robin tournament. How many matches must be scheduled for the entire tournament?

2. In an election with four candidates, what is the maximum number of columns possible in the preference schedule?

3. In the race for one seat in the Senate between three candidates (call them A, B, and C), the exit poll suggests the following set of preference links:

21%	12%	20%	13%	19%	15%
A	A	B	B	C	C
B	C	A	C	A	B
C	B	C	A	B	A

Is there a Condorcet winner? A Condorcet loser?
Who won using plurality voting? Who was plurality loser?

4. In the race for one seat in the Senate between three candidates (call them A, B, and C), the exit poll suggests the following set of preference links:

24%	10%	25 %	7%	23%	11%
A	A	B	B	C	C
B	C	A	C	A	B
C	B	C	A	B	A

Is there a Condorcet winner? A Condorcet loser?
Who won using the IRV voting? Who was the IRV loser?

5. In order to decide which new soup to add to its menu, a local Russian restaurant conducted a market survey in which customers were asked to rank their preferences for Borscht, Okroshka, and Schi. The results of counting the questionnaires are shown in the table here. Use the pairwise comparison method to determine which soup is the Condorcet winner (if any). Which one is the Condorcet loser? Does the plurality method support the Condorcet decision?

	Number of Ballots Cast (823)					
	210	86	115	146	158	108
Borscht	3	2	3	2	1	1
Okroshka	2	3	1	1	3	2
Schi	1	1	2	3	2	3

6. There are three nominees for the opening position: Dan, Kevin, and Justin. The preference schedule is presented below. Who is the Condorcet winner (if any)? Who is the Condorcet loser? Does the plurality method support the Condorcet decision?

	Number of Ballots Cast (627)					
	98	86	115	102	118	108
Dan	3	2	3	2	1	1
Kevin	2	3	1	1	3	2
Justin	1	1	2	3	2	3

7. The returning members of a soccer team decide to select a team captain for the next season using the pairwise comparison method. The preference rankings are listed below. Who is the Condorcet winner? Who is the Condorcet loser?

	Number of Voters (11)					
	1	1	3	2	2	2
Drogba	3	2	3	2	1	1
Beckham	2	3	1	1	3	2
Cahill	1	1	2	3	2	3

8. A small company decided to use the pairwise comparison method to determine on what day of the week to have a company picnic. The three choices were Thursday, Friday, and Saturday. Employees ranked their preferences as shown below. Which option is the Condorcet winner? Which option is the least desired?

	Number of Voters (27)					
	5	6	3	4	4	5
Thursday	1	1	2	2	3	3
Friday	2	3	1	3	1	2
Saturday	3	2	3	1	2	1

9. A group of student leaders at a local university were asked to vote on the aspects of college life to target improvement over the next year. The choices were Athletic facilities, Campus security, Dining facilities, and Transportation service. The results of the vote were summarized in the following preference table. Use the pairwise comparison method to determine the Condorcet winner and the Condorcet loser (if any). Does any one of the four plurality methods (plurality, runoff, Hare, or Coombs) support the Condorcet choices?

	Number of Voters (93)					
	17	28	20	15	8	5
Athletic facilities	1	4	1	4	3	3
Campus security	2	3	4	1	2	4
Dining facilities	4	2	3	2	1	1
Transportation service	3	1	2	3	4	2

10. The members of a local theater are voting for the type of play they will perform next season. The choices are a Comedy, Drama, Mystery, or Greek tragedy. The results of voting are summarized in the following preference table. Use the pairwise comparison method to determine the Condorcet winner and the Condorcet loser (if any). Check whether the election has a weak Condorcet winner because the number of ballots is even. Does any one of the four plurality methods support the Condorcet choices?

	Number of Voters (44)						
	7	8	5	10	6	3	5
Comedy	1	2	3	4	4	3	1
Drama	4	1	2	3	2	1	4
Mystery	2	4	4	1	3	4	2
Greek tragedy	3	3	1	2	1	2	3

11. Imagine an election to choose which food to eat for dessert. There are 21 people having dessert and four options: Apple Pie, Fruit, Ice Cream, and Nuts. The preference schedule is presented below. Use the pairwise comparison method to determine the Condorcet winner and the Condorcet loser (if any).

Option	Number of Voters (21)							
	1	2	3	3	2	2	1	7
Apple Pie	1	1	1	2	4	3	2	3
Fruit	2	3	4	1	1	4	3	2
Ice Cream	3	4	2	3	2	1	1	4
Nuts	4	2	3	4	3	2	4	1

12. The members of a hockey team are asked to select the most valuable forward in the team from four candidates: Krejci, Lucic, Ryder, and Sobotka. The preference schedule is shown below. Who is the Condorcet winner (if any)?

	Number of Voters (22)							
	6	2	4	3	2	1	3	1
Krejci	1	1	4	3	2	3	4	3
Lucic	2	3	1	1	4	2	2	4
Ryder	3	4	2	4	1	1	3	2
Sobotka	4	2	3	2	3	4	1	1

13. Find the winner of the election using the head-to-head comparison method. Does the Condorcet winner exist? Who is the Condorcet loser?

Ranking	Number of Voters (37)								
	3	2	5	5	1	3	1	7	10
First choice	A	A	A	B	B	B	C	C	D
Second choice	B	B	C	C	A	C	D	A	C
Third choice	C	D	B	D	C	A	B	D	A
Fourth choice	D	C	D	A	D	D	A	B	B

14. Using the preference table below, find the winner of the election using the head-to-head comparison method. Does the Condorcet winner/loser exist?

Ranking	Number of Voters (34)						
	6	7	8	3	2	2	6
First choice	A	B	C	D	D	E	E
Second choice	D	E	D	E	A	C	A
Third choice	C	D	B	B	C	B	C
Fourth choice	B	A	E	A	E	A	B
Fifth choice	E	C	A	C	B	D	D

In Problems 15 – 24, the preference schedule of an election has been presented. Determine the Copeland winner and Copeland loser (if any) in the given election. Use the Condorcet scores to break ties, if needed.

15.

Option	Number of Voters (19)							
	1	2	3	3	2	2	1	5
A	1	1	3	2	5	4	3	2
B	2	5	1	5	3	2	1	3
C	3	4	5	1	2	3	5	4
D	4	3	2	4	4	5	4	1
E	5	2	4	3	1	1	2	5

16.

Ranking	Number of Ballots Cast (28)							
	3	2	4	1	6	5	2	5
First choice	A	A	B	E	C	D	D	E
Second choice	B	E	C	B	A	B	B	B
Third choice	C	D	A	C	E	A	E	C
Fourth choice	D	C	E	D	B	E	C	A
Fifth choice	E	B	D	A	D	C	A	D

17.

Option	Number of Voters (16)							
	1	1	2	4	3	1	2	2
A	2	3	4	1	3	4	2	3
B	1	1	1	2	4	2	3	4
C	4	2	3	3	1	1	4	2
D	3	4	2	4	2	3	1	1

18.

Ranking	Number of Voters (20)								
	3	2	1	2	2	4	1	2	3
First choice	A	A	B	B	B	C	C	D	D
Second choice	B	D	A	D	C	A	B	A	B
Third choice	C	B	D	C	A	D	A	B	C
Fourth choice	D	C	C	A	D	B	D	C	A

19.

Option	Number of Voters (24)								
	3	2	1	3	3	4	2	5	1
A	4	2	3	1	1	3	4	2	3
B	1	1	1	2	3	4	2	3	4
C	3	4	2	4	2	1	1	4	2
D	2	3	4	3	4	2	3	1	1

20.

Ranking	Number of Voters (28)								
	3	4	5	2	3	2	2	6	1
First choice	A	A	B	B	C	C	C	D	D
Second choice	B	C	D	C	D	A	D	B	A
Third choice	C	D	A	D	A	B	B	C	B
Fourth choice	D	B	C	A	B	D	A	A	C

21.

Option	Number of Voters (18)							
	1	2	3	1	2	3	2	4
A	1	5	2	4	2	2	1	5
B	2	4	1	2	5	3	5	4
C	3	2	4	3	1	4	3	3
D	4	1	3	5	3	1	4	2
E	4	3	5	1	4	5	2	1

22.

Ranking	Number of Voters (38)							
	5	3	3	5	4	4	7	7
First choice	A	A	B	B	C	C	D	E
Second choice	C	B	C	D	E	B	B	A
Third choice	D	E	D	E	D	A	E	C
Fourth choice	E	D	E	A	A	E	C	B
Fifth choice	B	C	A	C	B	D	A	D

23.

	Number of Voters (44)							
Ranking	5	4	3	5	4	6	9	8
First choice	A	A	B	B	C	C	D	E
Second choice	D	E	C	A	E	B	E	B
Third choice	B	D	E	C	D	A	B	D
Fourth choice	C	B	D	E	A	E	A	C
Fifth choice	E	C	A	D	B	D	C	A

24.

	Number of Voters (23)						
Ranking	5	3	3	5	3	2	2
First choice	A	A	B	C	D	D	E
Second choice	B	D	E	E	C	C	B
Third choice	C	B	A	D	B	B	A
Fourth choice	D	C	C	A	E	A	C
Fifth choice	E	E	D	B	A	E	D

In Problems 25 – 32, the preference schedule of an election has been presented. Determine the winner using the sequential pairwise comparison method according to a specified agenda.

25. There are three candidates—Loureiro, Pacheco, and Ramos—running for a seat on the city council. Polls indicate that the preference rankings of the voters broke down into the following percentages.

	Percentage of Votes					
Candidate	16%	18%	19%	13%	12%	22%
Loureiro	1	1	2	2	3	3
Pacheco	2	3	3	1	2	1
Ramos	3	2	1	3	1	2

Who would have been the winner of the election if sequential pairwise voting with the agenda Ramos, Loureiro, Pacheco had been used? Who is the winner under another agenda: Pacheco, Ramos, Loureiro?

26. Candidates Arshavin, Fabregas, and Ramsey are running for president of a professional society. The election is to be decided using sequential pairwise voting, and the preference rankings of the voters break down into the following percentages.

	Percentage of Votes					
Candidate	24%	11%	19%	14%	19%	13%
Arshavin	1	1	2	2	3	3
Fabregas	2	3	3	1	2	1
Ramsey	3	2	1	3	1	2

Who is the winner with the agenda Arshavin, Fabregas, Ramsey? Who will win the election under another agenda Ramsey, Fabregas, Arshavin?

27. Consider the following set of preference lists:

	Number of Voters (17)							
	2	1	2	1	3	2	3	3
A	1	1	1	2	3	3	4	4
B	2	3	4	1	2	4	3	1
C	3	4	3	4	1	2	2	2
D	4	2	2	3	4	1	1	3

Calculate the winner using sequential pairwise voting with the agenda A, B, C, D. Who will win the election under another agenda: D, B, A, C?

28. Consider the following set of preference lists:

	Number of Voters (19)							
	1	2	3	2	1	4	3	3
A	1	4	1	3	2	3	2	4
B	2	3	2	4	1	4	3	1
C	3	2	4	1	3	1	4	2
D	4	1	3	2	4	2	1	3

Calculate the winner using sequential pairwise voting with the agenda B, A, C, D. Who will win the election under another agenda: D, C, B, A?

29. Consider the following set of preference lists:

Ranking	Number of Voters (13)				
	3	4	3	2	1
First choice	A	B	C	D	D
Second choice	D	A	B	C	A
Third choice	C	D	A	B	C
Fourth choice	B	C	D	A	B

Calculate the winner using sequential pairwise voting with the agenda A, B, C, D. Who will win the election under another agenda: A, C, D, B?

30. Consider the following set of preference lists:

Ranking	Number of Voters (17)					
	3	1	4	5	2	2
First choice	A	B	C	D	A	B
Second choice	B	C	D	B	D	C
Third choice	C	D	A	A	C	D
Fourth choice	D	A	B	C	B	A

Calculate the winner using sequential pairwise voting with the agenda B, A, D, C. Who will win the election under another agenda: C, D, A, B?

31. Consider the following set of preference lists:

	Number of Voters (11)							
	1	2	2	1	2	1	1	1
A	1	1	2	3	4	3	3	4
B	2	4	3	4	1	1	4	2
C	4	3	1	1	2	4	2	3
D	3	2	4	2	3	2	1	1

Calculate the winner using sequential pairwise voting with the agenda D, C, B, A. Who will win the election under another agenda: A, B, C, D?

32. Consider the following set of preference lists:

Ranking	Number of Voters (7)				
	2	1	2	1	1
First choice	A	C	D	C	B
Second choice	B	D	C	A	D
Third choice	D	B	A	B	A
Fourth choice	C	A	B	D	C

Calculate the winner using sequential pairwise voting with the agenda A, B, C, D. Who will win the election under another agenda: D, C, B, A?

In Problems 33 – 36, the preference rankings of a competition between four candidates, A, B, C, and D, have been presented. Calculate the winner using

(a) plurality method,

(b) the Borda count method,

(c) the Hare system,

(d) sequential pairwise voting with the agenda: A, B, C, D.

Show that in these elections every voting method produces a different winner (everyone wins).

33.

	Number of Voters (16)					
	5	2	2	4	2	1
A	1	4	2	4	4	4
B	3	1	1	2	3	2
C	4	2	3	1	2	3
D	2	3	4	3	1	1

34.

Ranking	Number of Voters (7)				
	2	2	1	1	1
First choice	B	C	B	D	A
Second choice	A	A	C	C	D
Third choice	D	B	A	A	C
Fourth choice	C	D	D	B	B

35.

	Number of Voters (17)				
	3	4	2	4	4
A	2	1	4	2	2
B	4	2	2	1	3
C	1	3	3	4	1
D	3	4	1	3	4

36.

Ranking	Number of Voters (7)					
	1	2	1	1	1	1
First choice	A	A	B	B	C	D
Second choice	B	C	D	C	D	C
Third choice	C	D	A	A	B	B
Fourth choice	D	B	C	D	A	A

86   *Chapter 1. Social Choice*

In Problems 37 – 40, the preference rankings of a competition between five candidates, A, B, C, D, and E, have been presented. Calculate the winner using

(a) plurality method,

(b) the Borda count method,

(c) the Hare system,

(d) sequential pairwise voting with the agenda: A, B, C, D, E.

Show that in these elections every voting method produces a different winner (everyone wins).

37.

	Number of Voters (9)						
	2	1	2	1	1	1	1
A	2	3	4	1	4	2	2
B	3	4	1	2	2	4	5
C	1	1	3	5	5	5	1
D	5	5	2	4	3	1	3
E	4	2	5	3	1	3	4

38.

	Number of Voters (9)					
Ranking	3	1	2	1	1	1
First choice	C	C	B	A	E	D
Second choice	A	D	A	B	B	B
Third choice	B	E	D	E	D	C
Fourth choice	D	A	E	D	C	A
Fifth choice	E	B	C	C	A	E

39.

	Number of Voters (13)					
	4	3	2	1	1	2
A	1	3	5	4	5	5
B	4	1	4	3	2	4
C	5	2	1	5	4	3
D	3	5	3	2	3	1
E	2	4	2	1	1	2

40.

	Number of Voters (8)							
Ranking	1	1	1	1	1	1	1	1
First choice	E	A	E	B	C	A	E	D
Second choice	D	E	C	A	A	C	D	C
Third choice	C	D	A	D	D	D	C	A
Fourth choice	B	C	B	C	E	B	B	B
Fifth choice	A	B	D	E	B	E	A	E

In Exercises 41 – 50, the preference schedule of an election has been presented. Determine which type of plurality method (plurality, runoff between the top two finishers, Hare, and Coombs) chooses the Condorcet winner or the Copeland winner. By reversing the orders of rankings determine which type of plurality method (plurality, runoff between the top two finishers, Hare, and Coombs) chooses the Condorcet loser or the Copeland loser.

41. Ten friends wanted to go out for lunch. They decided to vote on where to go. The options were Taco Bell, Subway, McDonald's, and KFC. The preference schedule is presented here.

Option	Number of Voters (10)					
	3	1	2	1	2	1
Taco Bell	1	2	4	3	4	3
Subway	4	3	1	2	3	4
McDonald's	3	1	3	4	1	2
KFC	2	4	2	1	2	1

42. Fifty-five people were on the guest list to attend Celia's party. Once all of them had arrived, Celia decided to have each person vote on which music they preferred to hear. The results of their rankings are presented below.

Option	Number of Voters (55)							
	4	5	8	7	12	9	6	4
Taylor Swift	4	3	2	3	1	2	2	4
Lil Wayne	3	4	3	1	2	4	1	3
Lady Gaga	2	1	4	2	4	1	3	2
The Fray	1	2	1	3	3	3	4	1

43. Find the winner and the loser (if any) in the following election between four alternatives:

	Number of Voters (27)						
	2	3	5	1	2	6	8
A	1	1	4	2	4	3	2
B	4	3	1	1	2	4	4
C	2	4	2	3	1	1	3
D	3	2	3	4	3	2	1

44. Find the winner and the loser (if any) in the following election between four alternatives:

Ranking	Number of Voters (21)						
	3	2	1	4	2	3	6
First choice	A	A	B	B	C	C	D
Second choice	B	C	D	A	D	B	C
Third choice	D	B	C	D	A	A	B
Fourth choice	C	D	A	C	B	D	A

45. A classroom of 30 students were surveyed about their favorite hobbies. The results of their preference rankings are summarized in the following table.

Hobby	Number of Ballots Cast (30)							
	5	3	4	2	1	3	4	8
Drawing	1	1	2	4	3	2	3	4
Singing	4	3	1	1	1	4	2	2
Shopping	2	4	3	2	4	1	1	3
Traveling	3	2	4	3	2	3	4	1

46. The professional videogame organization Major League Gaming (MLG) is planning to organize a tournament between twenty five teams. The participants must vote on four possible games to be played. The four choices are Call of Duty (C), Halo 3 (H), Gears of War (G), and Project Gotham Racing (P). The preference rankings of the teams are summarized in the following table.

Ranking	Number of Ballots Cast (25)						
	5	3	6	2	2	4	3
First choice	C	C	H	G	G	P	P
Second choice	H	P	G	H	C	H	G
Third choice	G	H	P	P	H	G	H
Fourth choice	P	G	C	C	P	C	C

47. A group of 44 students were asked to take a poll regarding their favorite musical instrument. Their preference rankings are as follows:

Instrument	Number of Ballots Cast (44)							
	5	6	4	7	3	8	2	9
Guitar	1	1	3	4	2	3	4	2
Drums	3	2	1	1	4	2	3	4
Flute	2	4	4	2	1	1	2	3
Tambourine	4	3	2	3	3	4	1	1

48. College band directors across the northeastern part of the United States are trying to set up a showcase to promote college marching bands to high school students. Four universities expressed their willingness to host it next year. They are the University of Massachusetts at Amherst, the University of Rhode Island, the University of New Hampshire, and the University of Connecticut. A questionnaire was sent to high school students in New England area. The results of their replies are summarized in the following preference table:

University	Number of Ballots Cast (169)											
	21	10	9	12	11	22	17	13	12	16	14	12
UMass	1	1	1	4	3	2	3	2	4	3	2	4
URI	2	3	4	1	1	1	4	3	2	2	4	3
UNH	3	4	3	2	4	3	1	1	1	4	3	2
UConn	4	2	2	3	2	4	2	4	3	1	1	1

49. A student council has decided to change their school mascot from a frog to the following options: Elephant, Horse, Rabbit, Knight, and Falcon due to new renovations that have been made to their school. The student council has 21 members who participated in voting. The preference schedule is presented below.

Mascot	Number of Votes (21)								
	3	2	1	3	2	2	4	1	3
Elephant	1	1	4	2	5	3	4	2	5
Horse	2	3	1	1	4	5	3	4	2
Rabbit	3	2	5	4	1	1	2	5	3
Knight	4	5	3	5	3	2	1	1	4
Falcon	5	4	2	3	2	4	5	3	1

50. A group of 23 students are voting to decide where to travel the coming summer. The top five choices are Robben Island (South Africa), New Zealand, Galapagos Islands (Ecuador), Manaus (Brazil), and Trang Islands (Thailand). Their preference schedule is presented below.

Destination	Number of Ballots Cast (23)							
	4	1	3	4	2	2	4	3
Galapagos	1	1	4	5	3	2	5	3
Manaus	2	4	3	1	2	5	4	5
New Zealand	3	5	2	4	5	4	1	2
Robben	4	3	5	2	4	1	3	1
Trang	5	2	1	3	1	3	2	4

In Exercises 51 – 60, the preference schedule of an election has been presented. Determine whether the Borda count, the Borda method with first place promotion, the Nauru score, Nanson's method, or Baldwin's method supports the Condorcet winner or the Copeland winner. If the Copeland method is inconclusive, breaking ties using pairwise comparison, followed by the Condorcet score, if needed. Do the Borda count and Nauru score confirm the Condorcet or Copeland loser?

51. Thirteen members of a search committee are in charge of hiring the new president; they have narrowed down their list to four finalists. The preference schedule is given below.

Nominee	Number of Voters (13)						
	3	3	2	2	1	1	1
Ashley	1	2	3	2	1	4	3
Beatrice	4	1	2	4	3	2	4
David	3	4	1	3	4	1	2
Paul	2	3	4	1	2	3	1

52. Students are voting to choose the best math instructor at URI. There were four nominees: Dr. A., Dr. B., Dr. C., and Dr. D. The preference rankings of 25 people are shown below.

Ranking	Percentage						
	24%	13%	15%	17%	9%	10%	12%
First choice	A	B	C	D	B	C	D
Second choice	B	C	D	A	D	B	B
Third choice	C	D	A	B	C	A	A
Fourth choice	D	A	B	C	A	D	C

53. A group of 35 college students were polled to see what sporting event they enjoyed attending the most. The choices were football, baseball, basketball, and auto racing. After all the votes were cast, the results were summarized in the following preference table.

Sport Event	Number of Voters (35)						
	3	4	4	5	6	3	4	6
Football	1	1	3	2	2	4	3	2
Baseball	4	4	1	1	3	2	2	4
Basketball	3	2	4	3	1	1	4	3
Auto Racing	2	3	2	4	4	3	1	1

54. A group of soccer fans decided to vote for their favorite player of the year. They narrowed down the list to four nominees: Cristiano Ronaldo (Real, Madrid); Lionel Messi (Barcelona, Spain); Ricardo Kaka (Real, Madrid); and Steven Gerrard (Liverpool, England). Their preference rankings are summarized below.

Sport Event	Number of Voters (25)							
	1	2	3	4	5	6	2	2
Ronaldo	2	3	4	1	3	4	1	2
Messi	4	2	1	2	4	3	3	1
Kaka	3	1	3	4	1	2	4	3
Gerrard	1	4	2	3	2	1	2	4

55. The Rock and Roll Hall of Fame is deciding which best 90's rock band will be inducted this year. There are four groups of candidates who have been critiqued in preference by 19 of the most qualified officials. Their preference schedule has been organized in the following table.

Band	Number of Voters (19)						
	2	3	4	5	1	2	2
Pearl Jam, Seattle	3	4	1	3	1	2	4
Rage Against the Machine, Los Angeles	4	3	2	1	3	4	2
Red Hot Chili Peppers, Los Angeles	2	1	4	2	4	1	3
Sublime, Long Beach	1	2	3	4	2	3	1

56. The ESPY Awards is an annual awards event created and broadcast by the American cable television network ESPN to celebrate their legacy as a sports channel. Since 1993, the event confers eponymous awards, fully styled as Excellence in Sports Performance Yearly Awards, for individual and team athletic achievement and other sports-related performances during the calendar year preceding a ceremony. Starting in 2004, the awards were chosen through voting by fans, sportswriters, broadcasters, sports executives, and ESPN personalities. The annual ESPY Awards allow sports fans worldwide to join in an online voting for their favorites in such categories as Best Male Athlete, Best Female Athlete, Best Coach/Manager, and Team of the Year. The 2009 ESPY Award Nominees for the Best Male Athlete were Michael Phelps (Swimming), Kobe Bryant (NBA), LeBron James (NBA), and Jimmie Johnson (Auto Racing). The results of voting are presented below.

Nominee	Percentage						
	9%	19%	17%	13%	18%	8%	16%
Bryant	1	2	4	1	4	1	2
James	2	3	1	4	2	3	4
Johnson	3	4	3	2	3	4	1
Phelps	4	1	2	3	1	2	3

57. Ten coaches and managers have been disagreeing as to who should represent the team in the next all-star game. They narrowed down to four nominees and agreed to decide the winner by casting preference ballots. The results of their voting on Dustin Brown (B), Michal Handzus (H), Anze Kopitar (K), and Drew Doughty (D) are presented here.

Ranking	Number of Voters (10)							
	1	1	1	2	2	1	1	1
First choice	B	B	H	H	K	K	D	D
Second choice	H	D	K	B	D	B	K	H
Third choice	K	H	D	K	H	D	H	B
Fourth choice	D	K	B	D	B	H	B	K

58. Twenty students are allowed to vote for the cafeteria lunch special each week. This week Baked Ziti, Chicken Parmesan, Camembert Chicken, and Cajun Chow Mein are the options. The winning dish is decided based on the following preference table:

Option	Number of Voters (20)							
	3	3	4	1	2	2	3	2
Baked Ziti	1	1	2	3	3	4	3	4
Chicken Parmesan	3	4	1	1	2	3	4	3
Camembert Chicken	4	2	3	4	1	1	2	2
Cajun Chow Mein	2	3	4	2	4	2	1	1

59. A group of friends wants to determine which design of import cars they like the most. They decided to use preference ballots. Their voting results are presented below.

Car	Number of Voters (35)						
	3	4	5	2	6	5	7	3
Bentley	A	A	B	D	C	C	D
BMW	D	B	C	C	B	D	C
Mercedes	B	C	D	A	D	B	B
Jaguar	C	D	A	B	A	A	A

60. Twenty-seven people were asked to conduct a survey for the most popular girl scout cookies. The five choices were Dosidos, Samoas, Tagalongs, Thin Mints, and Trefoils. The preference rankings are as follows:

Cookies	Number of Voters (27)							
	2	3	6	5	2	3	4	2
Dosidos	1	1	2	4	3	3	4	5
Samoas	2	4	3	1	5	4	5	2
Tagalongs	4	3	1	3	2	5	3	4
Thin Mints	5	2	4	5	1	1	2	3
Trefoils	3	5	5	2	4	2	1	1

Show that in Problems 61 – 64 the following elections do not have either a Condorcet winner or a Copeland winner. Apply the Condorcet score to break the ties and determine the winner.

61.

Alternative	Number of Voters (13)			
	5	4	3	1
A	1	2	3	2
B	2	3	1	1
C	3	1	2	3

62. In March 2009, the World Baseball Classic was held prior to the start of the Major League Baseball season. The first round of games is played in a round robin style with four teams in each group. The results of the games (two matchups were modified to make this competition suitable for our needs) in group C are presented in the following table.

	Canada	United States	Italy	Venezuela
Canada		5:6	2:6	7:3
U.S.	6:5		2:8	15:6
Italy	6:2	8:2		1:10
Venezuela	3:7	6:15	10:1	

63.

Ranking	Number of Voters (15)						
	3	4	3	2	1	1	1
First choice	A	A	B	D	C	C	D
Second choice	D	B	C	C	B	D	C
Third choice	B	C	D	A	D	B	B
Fourth choice	C	D	A	B	A	A	A

64. The managers of a catering service at Logan airport (Boston) meet to decide on their employee-of-the-month. Four employees are suggested: Ashley, Beatrice, David, and Timothy. The managers' preference rankings appear below.

Nominee	Number of Voters (13)										
	1	1	1	1	1	1	1	1	1	2	2
Ashley	1	1	1	2	3	4	3	2	2	3	2
Beatrice	2	3	4	1	1	1	4	3	4	2	4
David	3	4	2	3	4	2	1	1	1	4	3
Timothy	4	2	3	4	2	3	2	4	3	1	1

Which employee is the winner?

65. Consider an election taking place in Rhode Island, which suppose to determine the location of a new stadium. It could be built in only one place. Since the population in Rhode Island is concentrated in five main cities: Kingston, Providence, Newport, Woonsocket, and Bristol, the committee put this cities into the ballots. The preference rankings of the voters is summarized in the following table.

City	Number of Votes (19)							
	3	2	4	2	1	4	2	1
Bristol	2	5	3	1	2	4	4	5
Kingston	1	2	4	3	1	5	2	3
Newport	3	1	5	4	3	2	1	4
Providence	5	4	1	2	4	3	5	1
Woonsocket	4	3	2	5	5	1	3	2

66. A recent poll was taken at a local college to see how students spend their leisure time. The results were narrowed down to five options and students were asked to vote for which option they like most. The five options were playing sports, watching television, playing videogames, socializing with friends, and listening to music. The results were as follows.

Activity	Number of Votes (29)							
	5	2	6	4	4	3	3	2
Sports	1	5	4	3	2	3	5	4
Television	2	4	3	4	1	5	1	5
Video games	3	1	2	5	3	1	4	1
Socializing	4	2	5	1	5	4	2	3
Music	5	3	1	2	4	2	3	2

67. Nineteen students in a class decided to rank four cereals from least favorite to favorite.

Cereal	Number of Votes (19)							
	2	3	4	2	1	4	2	1
Cheerios	1	2	3	4	1	2	3	4
Fruit Loops	3	1	4	3	4	3	2	1
Trix	2	4	2	1	3	1	4	2
Special K	4	3	1	2	2	4	1	3

1.5   The Approval Method

The main idea of any voting method is to make a collective choice from a set of alternatives available to them, rather than having each individual act independently. However, since there are many ways to aggregate individuals' reports of their preferences in order to reach a collective choice, the gain (or loss) may depend on the procedure by which the collective choice is made. Is such an agreement possible? We saw in §1.2 that plurality methods do not usually elect a broadly supported nominee. Moreover, plurality usually penalizes centrists and promotes a confrontation between two opposing extremes. For example, the outcome of the 2000 presidential election in the United States has spurred efforts for reform of the election system. While no system is perfect, we are going to present a possible alternative to the existing plurality method.

Approval voting is a lot like plurality, except that rather than voting for a single candidate, you can vote for as many as you want. No ranking is involved, so all the votes have equal weight. The alternative that receives the most votes wins. Approval voting in public elections has a long history going back to 12th century Venice and ancient Greece when people used their voice to express support for a proposed option or candidate. However, its use has been growing in recent years since the 1970's when it was introduced by several people.

Plurality Ballot
Directions: Vote for one candidate
◯ Pat Buchanan (Reform)
◯ G. W. Bush (Republican)
◯ Al Gore (Democrat)
◯ Ralph Nader (Green)

Approval Ballot
Directions: Vote for any number of candidates
☐ Pat Buchanan (Reform)
☐ G. W. Bush (Republican)
☐ Al Gore (Democrat)
☐ Ralph Nader (Green)

Approval voting combines substantial simplicity with a strong likelihood of choosing a Condorcet winner when one exists. Compared with the plurality method, approval voting also gives a much clearer picture, in its voting results, of the political preferences of the voters. That is because plurality voting strongly discourages voting for alternative party candidates (and thus hides the true level of support for those candidates), while approval voting lets voters vote freely for whichever candidates they approve—there is never any reason to not vote for your favorite candidate. Approval voting can be easily accommodated by existing voting machines as well as by manual vote counting procedures. From the example of ballots below, we can see that the change to approval voting from a plurality ballot is very little.

Since 1970, it has been adopted by various governments and organizations around the world, most notably by the United Nations to elect the Secretary-General. Motions to adopt it have also been promoted in many other situations. In 1987 and 1988, several scientific and engineering societies inaugurated the use of approval voting in finding consensus candidates. For example, the following societies continue to use it today:

* The Mathematical Association of America (MAA), with about 32,000 members;

* The Institute of Management Science (TIMS), with about 7,000 members;

* The American Statistical Association (ASA), with about 15,000 members;

* The Institute of Electrical and Electronics Engineers (IEEE), with about 377,000 members.

Example 1.5.1. Approval Voting

A group of friends is trying to decide upon a movie to watch. From four possible choices, each person is asked to mark a $\sqrt{}$ for which movies they are willing to watch. The results of voting are as follows:

Movies	Names									
	Ann	Alice	Betty	Eve	Mary	Bob	Dave	Jim	Omar	Tom
The Matrix	$\sqrt{}$					$\sqrt{}$	$\sqrt{}$			
The Godfather	$\sqrt{}$		$\sqrt{}$		$\sqrt{}$	$\sqrt{}$		$\sqrt{}$		
Rocky			$\sqrt{}$	$\sqrt{}$		$\sqrt{}$	$\sqrt{}$	$\sqrt{}$		
Titanic					$\sqrt{}$	$\sqrt{}$	$\sqrt{}$	$\sqrt{}$	$\sqrt{}$	$\sqrt{}$

Totaling the results, we find that

The Matrix	received 3 approvals,	The Godfather	received 5 approvals,
Rocky	received 5 approvals,	Titanic	received 6 approvals.

Therefore, according to the vote, Titanic would be the winner (and The Matrix would be the loser) in this election. Moreover, Titanic is approved by the majority of voters—6 out of 10, so we could consider it as an optimal choice in this election. □

Approval opponents argue that the method is unlikely to work as well in practice as it is supposed to work in theory. Let us address some common concerns about approval voting.

- Does approval voting violate the principle of **one person, one vote**? In other words, does approval voting give more power to voters who approve more than one candidate for a given office?

  Not really, because each voter gets the same ballot, and each voter gets exactly one copy of the ballot. Each voter has equal opportunity to vote for as many or few of the candidates as he or she wishes. A ballot which approves all but one candidate for an office is canceled by a ballot which approves only that one remaining candidate. Thus, these two ballots have different strengths, but it is apparent that all voters have the same voting power.

- Approval voting forces voters to cast **equally weighted votes** for candidates they approve of. Voters cannot indicate a strong preference for one candidate and a weak preference for another. Voters in fact will almost always have different degrees of support for different candidates. In conclusion this can give strange results and is very dependent on voter tactics rather than voter preferences.

- **Approval voting does not solve the spoiler problem**. Voting for your second choice candidate can in some cases lead to the defeat of your favorite candidate. (This problem is less severe than in plurality voting, but the Hare method does a better job of addressing the spoiler problem.) Campaigns would urge their supporters to *bullet* vote for their candidate only, and approval voting would thus tend to revert back to plurality voting.

- The favorite candidate of a **majority can fail** to be elected, but only when another candidate has wider-ranging support. Try to answer the question: Which candidate deserves to win: one who is the favorite of 51% of the voters or one who is acceptable to 75%?

- The approval system might allow a candidate to win who may not have won any support in a plurality election.

Approval voting is by far the simplest, cheapest, and easiest to implement practical alternative to plurality voting with predictable correspondence between public opinion and the outcome. The winner is the candidate who is accepted by most people. Approval voting advocates predict that approval voting should increase voter participation, prevent minor-party candidates from being spoilers, and reduce negative campaigning. This method has many supporters, and it is particularly suitable when it is needed to select a wider approved candidate. Both approval voting and the Borda count method seek a nomination that has a broad support of the electorate, but approval voting is less subject to manipulation.

Example 1.5.2. The 1980 Presidential Election

In the 1980 U.S. presidential election, Ronald Reagan and Jimmy Carter faced independent candidate John Anderson. At one point in the campaign, Anderson's support was nearly 15%. However, the results of the election show different numbers.

Presidential Candidate	Political Party	Popular Vote		Electoral Vote	
Ronald Reagan	Republican	43,903,230	50.75%	489	90.9%
James Carter	Democratic	35,480,115	41.01%	49	9.1%
John Anderson	Independent	5,719,850	6.61%	0	0.0%
Edward Clark	Libertarian	921,128	1.06%	0	0.0%
Barry Commoner	Citizens	233,052	0.27%	0	0.0%
Other (+)	—	252,303	0.29%	0	0.0%

Therefore, had the election been decided by popular vote, Ronald Reagan would have won with the majority in a plurality election. An independent survey shortly before the election found that the top three candidates had the following approval rates:

Reagan   61%,   Carter   57%,   Anderson   49% .

So we see that the approval results give a more clear picture of the distribution of votes than plurality.

Example 1.5.3.   Approval Voting Chooses Plurality Loser
Consider the following election between three candidates. The preference schedule is presented here, and we assume that the first two top candidates are approved.

	Number of Ballots (6)		
Candidate	1	2	3
Alex	1 ✓	2 ✓	2 ✓
Boris	2 ✓	1 ✓	3
Chris	3	3	1 ✓

Obviously, Chris is the plurality winner, with 3 first-place votes, whereas Alex gets 1 first-place vote and Boris gets 2 first-place votes. Counting last-place votes, we see that candidates Chris and Boris are the plurality losers. In contrast, Alex is the winner under the approval system because Alex gets 6 approvals, whereas Boris and Chris get 3 approvals, each.

Example 1.5.4.   The Approval Method Suffers from Tactical Voting
Consider the following election:

	Number of Voters (5)		
Ranking	2	1	2
First choice	A	A	B
Second choice	B	C	A
Third choice	C	B	C

We assume that an alternative is approved if it did not receive a last-place ranking. To simplify our calculations, we assign 1 point for every approved alternative. So, the scores are A = 5, B = 4, C = 1; therefore, A is the winner under the approval method. Now, suppose two voters dishonestly approved only their top-ranked alternative in the last column.

The scores become A = 3, B = 4, and C = 1. Alternative B wins, and the dishonest voters have achieved a preferred outcome.

Problems.

1. During a meeting, 11 participants voted on what fast food place to go. The options were McDonald's, Wendy's, Burger King, and Taco Bell. Below are the summarized responses. Which place was approved?

Fast Food	Number of votes (11)								
	1	1	1	2	1	1	1	2	1
McDonald's	√			√	√	√		√	√
Wendy's	√				√	√		√	
Burger King		√	√	√	√			√	
Taco Bell	√				√		√		√

2. A city council needs to select a month in which to hold its annual fair. The council is going to use approval voting among four possible choices: June, July, August, and September. The results of voting are given here. Which month would be chosen?

Month	Percentage								
	3%	9%	11%	15%	17%	16%	14%	8%	7%
June		√		√	√		√	√	√
July	√			√	√				
August	√	√	√			√		√	
September		√		√			√		√

3. Eleven board members vote by approval voting on five candidates for a new position on their board as indicated in the following table. A √ indicates an approval vote. For example, Voter 1, in the first column, approves of candidates Alex, Boris, and Ellis, but disapproves Chris and Dennis. Note that the fourth voter disapproved all candidates whereas the fifth one approved all of them. Find the winner of the election.

Candidate	Voters										
	1	2	3	4	5	6	7	8	9	10	11
Alex	√	√	√		√		√	√			
Boris	√	√	√		√			√			
Chris					√	√					√
Dennis		√			√	√		√		√	
Ellis	√		√		√	√	√		√		√

4. Fifty soccer fans of Liverpool, English Premier League, vote on five nominees for the award of "most valuable player" as indicated in the following table. A √ indicates an approval vote. Which nominee gets announced as runner-up for the award?

Player	Number of votes (50)							
	9	8	7	9	7	3	4	3
Benayoun, Yossi		√	√	√	√			
Gerrard, Steven	√	√			√		√	
Kuyt, Dirk		√	√			√		√
Ngog, David		√			√		√	
Torres, Fernando	√		√	√	√	√	√	

5. A committee should decide on the minimum voting age. Which age will be approved based on the following voting?

Age	Number of votes (7)						
	1	1	1	1	1	1	1
Seventeen		√	√	√	√		
Eighteen	√	√			√		√
Nineteen		√	√			√	
Twenty			√		√		
Twenty-one	√		√	√	√		√

6. A company wants to give a promotion to one of its managers. Based on the following approval table, who will be promoted?

Nominee	Number of votes (11)						
	1	2	3	1	1	2	1
Ashley		√		√	√		√
Donna	√		√	√			
Steven		√	√			√	√
Timothy	√	√		√		√	√
Wayne	√		√		√	√	

7. A family of six is planning to buy a cat. They have narrowed the choices of breed down to six choices. The results of approval voting are presented here. Which breed would be approved?

Breed	Family Members					
	1	1	1	1	1	1
Abyssinian		√	√	√	√	
Birman	√	√			√	
Chartreux		√	√			√
Korat		√		√		
Persian	√		√	√	√	
Siberian		√	√	√	√	√

8. A farm company decides to let the public determine a new flavor for organic yogurt. The breakdown of the votes in an approval election are shown, which flavor is selected?

Flavor	Percentage of Voters									
	2%	3%	7%	11%	14%	17%	19%	13%	8%	6%
Apple		√	√	√	√			√		
Banana	√	√			√		√		√	
Cherry	√		√	√		√		√	√	√
Chocolate		√			√	√	√		√	
Coffee	√		√			√		√		√

9. Four people are running for President of their local community center. The results of the approval election is presented here.

Nominee	Number of votes (11)						
	12	7	13	6	4	3	9
Ryan	√				√	√	√
Mat		√	√		√		
Kevin		√		√			√
Tony	√	√	√		√		√

1.6 Arrow's Impossibility Theorem

In the previous sections, we discussed many voting methods. The choice of a voting system may yield many different kinds of consequences. It may bias social choices toward or away from the status quo or the interests of particular groups. We also know that each voting method has its strengths and weaknesses. Since most of the discussed methods are implemented in practical life, we observe their advantages and drawbacks. Others have only theoretical importance and wait for further scrutiny. Since elections play a persuasive and important part in our lives, it is natural to ask: what is the "right" voting method? How do we determine which of the rules should be used in a particular election from an abundance of voting methods?

To answer these questions, we need first to identify the objectives: what features of a voting method would you like to see? One of the possible criteria could be *simplicity*, resistance to *manipulability*, and *fairness*. There are many other practical voting problems that we did not address. Let us mention among them such problems as breaking ties, dealing with spoiled ballots, determine likely behavior of a particular electoral system, the balance of power between centrists and extremists, and so on.

With the growing influence of informational technolo and worldwide education, the simplicity criteria might not be so important. At least, the simplicity of the plurality and approval methods cannot be considered as dominant among other voting methods.

It is known from Gibbard (1973) and Sutterthwaite (1975) that all voting rules (except dictatorship) are vulnerable to strategic manipulations. However, some voting methods are more vulnerable than others. For instance, the Hare method is more robust and least susceptible to manipulations than any of the voting methods discussed in §§1.2 − 1.4 when the number of voters is large enough. More precisely, in an election between three or more candidates, the **Gibbard-Satterthwaite theorem** claims that one of the following outcomes may occur:

1. the rule is dictatorial (i.e., there is a single individual who can choose the winner); or

2. there is a candidate who cannot win, under the rule, in any circumstances; or

3. the voting method is susceptible to tactical voting, in the sense that there are conditions where a voter with full knowledge of how the other voters are to vote and of the rule being used would have an incentive to vote in a manner that does not reflect her preferences.

In a two-candidate race, there is only one fair voting method—the majority rule. This remarkable result was proved by Kenneth May in 1952.

Theorem 1.1 (May) *Among all two-candidate voting systems that never result in a draw, majority rule is the only anonymous one that treats all voters equally, treats both candidates equally, and is nonmanipulable.*

This theorem is a cornerstone of any voting method because its fairness can be checked in a two-candidate election. All voting methods that were discussed in this chapter give the same outcome in a two-candidate competition.

Many fairness criteria are known; two of them were presented at the beginning of this chapter, page 2. However, we discuss the following four fairness criteria that are considered by many as the most important.

- **The majority criterion:** If a majority[3] of the voters rank a candidate as their first choice, then that candidate should win the election.

- **The Condorcet criterion:** A candidate that defeats each of the other candidates in head-to-head comparisons is the winner of the election. Such a candidate is called the Condorcet winner.

- **The monotonicity criterion:** Suppose that candidate X wins the election, but for some reason, there is another election. If in a new reelection, all ballots are the same except for only changes in favor of candidate X (and only X), then X should win the new election.

- **Chernoff's criterion** or the **independence-of-irrelevant-alternatives (IIA) criterion:** Suppose that candidate X wins an election, but for some reason, there is a new election. If the only changes are that one of the nonwinning candidates is removed or disqualified, then X should win the reelection.

The monotonicity criterion for fair elections is only relevant when an election is repeated. A repeated election refers to the same number of voters choosing among the same candidates. One example of a repeated election is a straw vote followed by another vote. Frequently, preceding an election, voters discuss the candidates' strengths and weaknesses. Prior to holding an official election, voters may take a straw vote to gain a preliminary measure of the candidates' strengths. After discussion, some voters may decide to change their preferences. If a candidate gains votes at the expense of the other candidates, this candidate's chance of winning should increase. If that candidate was already leading, this candidate should still win with additional votes in his or her favor. The repeated election leads us the monotonicity criterion.

The irrelevant alternatives criterion addresses the result of removing a candidate from an election who has no chance of winning. Independence of irrelevant alternatives (IIA) is an axiom often adopted by social scientists as a basic condition of rationality.

The IIA condition is equivalent to **Chernoff's criterion:**
If alternative A is preferred to alternative B out of the choice set { A, B }, then introducing a third, irrelevant, alternative X (thus expanding the choice set to { A, B, X }) should not make B preferred to A. In other words, whether A or B is better should not be changed by the availability of X.   ■

---

[3]Majority means more than half.

Surprisingly, none of the voting methods discussed in this chapter meets all of these four criteria of fairness. So the next question should be: what voting method satisfies these four criteria of fairness? The answer to this question gives the famous theorem due to Kenneth J. Arrow (1951), called **Arrow's impossibility theorem**, which states that such a preferential voting method that can always fairly decide the outcome of an election that involves more than two candidates or alternatives does not exist. This includes not only the voting methods we know right now, but any voting method anybody might think of in the future as well. His discovery was a major factor in Arrow winning the Nobel Prize in Economics in 1972.

Note that we neither formulate this theorem explicitly nor present its proof because it involves many technical and mathematical issues that require a solid theoretical background. Actually, Dr. Arrow used some additional criteria of fairness in his theorem; later some other proofs involving modifications of the fairness criteria were found [16]. Moreover, Arrow's definition of voting systems ignores some theoretical (such as approval[4] or range voting) and practical (for instance, judging figure-skating events) methods that could be considered as possible alternatives to fair voting. Since the objective of this section is to acquaint the reader with fairness criteria that stimulate a critical point of view on the voting methods, we present some other criteria and show by examples how they can be used in a particular election.

- **Pareto[5] criterion:** If *every* voter prefers alternative A over alternative B, then B should not be declared the winner.

- **Avoidance of Condorcet losers:** Certainly, it would seem that any voting method that picked an alternative as the winner was a poor voting method. Recall that a Condorcet loser is one who is defeated by all other alternatives in paired contests.

- **Independence of clones:** Addition of a candidate identical to one already present in an election will not cause the winner of the election to change.

- **Participation criterion:** An addition to a ballot, where candidate A is strictly preferred to candidate B, to an existing tally of votes should not change the winner from candidate A to candidate B.

In voting systems theory, the independence of clones criterion is a criterion that measures an election method's robustness to strategic nomination. The criterion, first formulated by Nicolaus Tideman [34], states that the addition of a candidate identical to one already present in an election will not cause the winner of the election to change.

Yet, it is easy to show that plurality can choose a Condorcet loser. In contrast, neither the Hare nor the Coombs rule can choose a Condorcet loser. The proof is easy. In both IRV and

---

[4]Approval voting is not a voting system by the definitions of Arrow's theorem because it does not use preference ballots.

[5]The Pareto condition is named after Vilfredo Federico Damaso Pareto (1848 – 1923), an Italian industrialist, sociologist, economist, and philosopher.

Coombs, the winner gets a majority of the votes cast at some stage of the balloting and thus must be able to defeat each of the candidates still viable at that stage in a head-to-head matchup. Hence, the winner under these rules must be able to defeat at least one other candidate in a paired contest and cannot be a Condorcet loser.

In 1978, Chamberlin and Cohen [10] analyzed several voting methods by calculating the number of times they would pick a Condorcet winner when one existed. For example, given every possible set of voter preference ballots for 4 alternatives and 21 voters, these voting methods elect the Condorcet winner with the following frequency:

Plurality   53%,   Hare   75%,   Coombs   98%,   Borda   83%

As the number of voters increases, the Coombs and Borda methods elect the Condorcet winner with increasing frequency, but the plurality and Hare methods do so with decreasing frequency.

The Nanson method and the Baldwin method satisfy the Condorcet criterion: since Borda always gives any existing Condorcet winner more than the average Borda points, the Condorcet winner will never be eliminated. They do not satisfy the independence of irrelevant alternatives criterion, the monotonicity criterion, the participation criterion, and the independence of clones criterion, while they do satisfy the majority criterion, and the Condorcet loser criterion.

Plurality voting, approval voting, and the Borda count all satisfy the participation criterion. All pairwise comparison methods and the IRV fail.

Election methods that fail independence of clones can either be clone negative (the addition of an identical candidate will decrease a candidate's chance of winning) or clone positive (the addition of an identical candidate will increase a candidate's chance of winning). The Borda count is an example of a clone positive method. Plurality is an example of a clone negative method because of vote-splitting.

Instant runoff voting and approval voting meet the independence of clones criterion. The Borda count, two-round systems, and plurality fail the independence of clones criterion.

The Borda count satisfies the monotonicity criterion, the participation criterion, and the Condorcet loser criterion. It does not satisfy the Condorcet criterion, the independence of irrelevant alternatives criterion, or the independence of clones criterion.

The Borda count and approval voting do not satisfy the majority criterion, i.e., if a majority of voters rank one candidate in first place, that candidate is not guaranteed to win. This could be considered a disadvantage for the Borda count in political elections, but it also could be considered an advantage if the favorite of a slight majority is strongly disliked by most voters outside the majority, in which case the Borda or approval winner could have a higher overall utility than the majority winner.

Example 1.6.1. Plurality Chooses the Condorcet Loser

In an election between three options—A, B, and C, seven voters cast their ballots. The results are summarized in the following preference schedule:

	Number of Voters (7)		
	3	2	2
A	1	3	3
B	2	1	2
C	3	2	1

Since option A is defeated by both alternatives: A:B = 3:4 (meaning that three of the voters prefer A to B, and four of the voters prefer B to A), A:C = 3:4, it is the Condorcet loser. However, option A is the plurality winner because it has more first-place votes. A similar outcome was obtained in Example 1.4.4 on page 73.

Example 1.6.2. Plurality Violates the Independence–of–Irrelevant–Alternatives (IIA) Criterion

The county board of supervisors is voting on methods to attract business, which will result in higher local employment. The options are lowering the tax on business phone calls (P), simplifying tax and other business paperwork (S), and providing credits with lower interest rates (C). The following preference schedule shows the results of the vote.

	Number of Voters (13)				
	4	3	2	3	1
C	1	3	1	2	3
P	2	1	3	3	2
S	3	2	2	1	1

Using the plurality method, the option of providing better credits (C) is more preferable. If the least wanted alternative, P (lowering the tax on phone calls), is removed from the competition, we get

	Number of Voters (13)				
	4	3	2	3	1
C	1	2	1	2	2
S	2	1	2	1	1

We now see that option S (simplifying paperwork) wins against alternative C with a score of 7 to 6. Therefore, the plurality method does not satisfy the independence-of-irrelevant-alternatives (IIA) criterion.

Example 1.6.3. The Hare Method Violates the IIA Criterion

Consider the preference ballots cast for the president of the International Student's Organization (ISO), in which Anna Kurnikova, Boris Beckham, and Chris Pechacek are candidates. The preference schedule is given here.

	Number of Voters (37)				
	4	3	2	2	5
Anna	1	1	2	3	3
Boris	2	3	1	1	2
Chris	3	2	3	2	1

Round 1: Anna gets 7 first-place votes, Boris gets 4 first-place votes, and Chris gets 5 first-place votes. Therefore, Boris is eliminated; the 4 votes are now split evenly between Anna and Chris.

Round 2: Anna gets 9 first-place votes and Chris gets 7 first-place votes, so Anna wins the election.

Now let us imagine a similar election when one of the losers—Chris—drops out. The preference table will be as follows

	Number of Voters (16)				
	4	3	2	2	5
Anna	1	2	2	3	3
Boris	2	3	1	1	2
Chris					

or

	Number of Voters (16)				
	4	3	2	2	5
Anna	1	1	2	2	2
Boris	2	2	1	1	1

We see that Boris defeats Anna by 9 votes to 7. Because the removal of one original losing candidate changes the outcome of the election, the IRV method does not support Chernoff's criterion.

Example 1.6.4. The Hare Method Violates the Monotonicity Criterion
Consider the following election:

	Number of Voters (21)			
	8	6	2	5
A	1	2	2	3
B	2	3	1	1
C	3	1	3	2

Round 1: alternative A gets 8 first-place votes, B gets 7 first-place votes, and C gets 6 first-place votes. Therefore, alternative C is eliminated, the 6 first-place votes are now considered votes for alternative A.

Round 2: A gets 14 first-place votes and B gets 7 first-place votes, so alternative A wins the election.

Now let us imagine a similar election. All that will be changed is that some people will rank alternative B lower on their ballots and alternative A higher. No one will change their ballots in any other way.

	Number of Voters (21)			
	8	6	2	5
A	1	2	1	3
B	2	3	2	1
C	3	1	3	2

The third column shows that B has been lowered by two voters. Now we proceed with the Hare method.

Round 1: A gets 10 first-place votes, B gets 5 first-place votes, and C gets 6 first-place votes. Hence B is eliminated and five votes in the fourth column go to C.

Round 2: A gets 10 first-place votes and C gets 11 first-place votes. So alternative C becomes the winner.

We see that C is the Hare winner. This is despite the fact that the only difference between these two elections is that some people who preferred B to A in the first round, prefer A to B in the second. This is not reasonable evidence that A should win. This strange property is referred to as the non-monotonicity.

Example 1.6.5.   The Coombs Method Violates the Condorcet Criterion

Consider a competition with the following preference schedule:

Alternatives	Number of Voters (21)					
	5	4	2	4	2	4
A	1	1	2	3	2	3
B	2	3	1	1	3	2
C	3	2	3	2	1	1

Round 1: A gets 9 first-place votes, B gets 6 first-place votes, and C gets 6 first-place votes. Hence, we have no majority winner, and we apply the Coombs method to eliminate the least wanted alternative. We calculate the number of last-place vote: A has 8 third-place votes, B has 6 third-place votes, and C has 7 last-place votes. So alternative A is eliminated.

Round 2: B gets 11 first-place votes and C gets 10 first-place votes. Therefore alternative B becomes the winner.

On the other hand, alternative A is the Condorcet winner because A:B = 11:10 (meaning that eleven of the voters prefer A to B, and ten of the voters prefer B to A); A:C = 11:10. So we conclude that the Coombs method violates the Condorcet criterion.

Example 1.6.6.   The Coombs Method Violates the Participation Criterion

A committee has to decide between four nominees who should be awarded the *most gentlemanly defender* trophy. Preference ballots are summarized in the following table, and the Coombs method is then applied.

	Number of Voters (22)				
Defender	4	7	3	5	3
Agger	1	4	3	1	2
Distin	4	1	2	2	4
Sagna	3	2	1	4	3
Terry	2	3	4	3	1

Since no one gets the majority of votes, we calculate the number of last-place votes: Agger and Distin have 7 fourth-place votes, while Sagna gets 5 last-place votes, and Terry has only 3. Therefore, Agger and Distin are eliminated from this election and pairwise comparison between Sagna and Terry brings the latter one the victory.

Just before announcing Terry as a trophy holder, the committee got a phone call from their colleague who was released from the hospital and who expressed his willingness to vote. Of course, he put Terry in first place and his ballot had the following rankings: Terry $\succ$ Distin $\succ$ Sagna $\succ$ Agger. Everyone was really surprised that this ballot changed the results of the election completely: the winner becomes Distin. The Coombs method eliminates Agger first as he has more last-place votes—eight. The second round drops Terry because he will have 10 last-place votes (7 are transferred from Agger). Since Distin beats Sagna 13:10, he becomes the Coombs winner. Hence, this voting method does not satisfy the participation criterion.   □

Now we summarize our observations about fairness of plurality methods in the following table:

Criterion	Plurality	Runoff	Hare	Coombs	Approval
Majority	Yes	Yes	Yes	Yes	No
Condorcet	No	No	No	No	No
Monotonicity	Yes	No	No	No	Yes
Chernoff/IIA	No	No	No	No	Yes
Pareto	Yes	Yes	Yes	Yes	Yes
Participation	Yes	No	No	No	
Condorcet loser	No	Yes	Yes	Yes	No
Clone Independence	No	No	No	No	—

An answer "Yes" in the table means that a particular system always satisfies the indicated criterion. Whereas answer "No" indicates that the criterion may be violated. For example, the plurality method always supports the majority criterion, but may fail the Condorcet criterion. Since the independence of clone criterion is applied only for ranked voting methods, it is ambiguous for the approval system.

Example 1.6.7. Borda count violates the Majority Criterion

A *Magazine* has narrowed down nominees for the car of the year; let us label them by letters A, B, C, and D. A three-person editorial panel cast their ballots that result in the following preference table:

First choice	A	A	B
Second choice	B	B	C
Third choice	C	D	D
Fourth choice	D	C	A

To determine the winner using the Borda count method, we summarize the election results. Since first place is worth 3 points, second place is rewarded with 2 points, third place is given 1 point, and last place is worth nothing, we get the following Borda scores for each alternative:

Alternatives	Number of Points				
	1st-Place Votes$\times 3$ Points	2nd-Place Votes$\times 2$ Points	3rd-Place Votes$\times 1$ Points	4th-Place Votes$\times 0$ Points	Total Number of Points
A	$2 \times 3 = 6$	$0 \times 2 = 0$	$0 \times 1 = 0$	$1 \times 0 = 0$	6
B	$1 \times 3 = 3$	$2 \times 2 = 4$	$0 \times 1 = 0$	$0 \times 0 = 0$	7
C	$0 \times 3 = 0$	$1 \times 2 = 2$	$1 \times 1 = 1$	$1 \times 0 = 0$	3
D	$0 \times 3 = 0$	$0 \times 2 = 0$	$2 \times 1 = 2$	$1 \times 0 = 0$	2

So we see that although the Borda count winner is alternative B, the majority of first-place votes was cast for alternative A.

Example 1.6.8. Borda Count Violates the Condorcet Criterion
A toothpaste company decides to let the public determine a new flavor for children's toothpaste. The breakdown of the preference rankings are as follows:

Flavors	Number of Votes (47)					
	3	5	12	6	17	4
Grape	1	1	3	2	4	2
Cherry	2	3	1	3	2	3
Mango	3	4	2	1	1	4
Orange	4	2	4	4	3	1

In this competition, Cherry is the Condorcet winner because it beats all other alternatives: Cherry : Grape = 29 : 18,   Cherry : Mango = 24 : 23,   Cherry : Orange = 38 : 9. On the other hand, Mango is the Borda winner according to the following calculations:

Alternatives	Number of Points				
	1st-Place Votes×3 Points	2nd-Place Votes×2 Points	3rd-Place Votes×1 Points	4th-Place Votes×0 Points	Total Number of Points
Grape	$8 \times 3 = 24$	$10 \times 2 = 20$	$12 \times 1 = 12$	$17 \times 0 = 0$	56
Cherry	$12 \times 3 = 36$	$20 \times 2 = 40$	$15 \times 1 = 15$	$0 \times 0 = 0$	91
Mango	$23 \times 3 = 69$	$12 \times 2 = 24$	$3 \times 1 = 3$	$9 \times 0 = 0$	96
Orange	$4 \times 3 = 12$	$5 \times 2 = 10$	$17 \times 1 = 17$	$21 \times 0 = 0$	36

Example 1.6.9.   Borda Count Violates the IIA Criterion

An engineering team needed to select a team leader among the candidates Gilbert, Hatcher, Pirnot, and Tannenbaum. The preference rankings of the members are as follows.

Candidates	Number of Votes (18)					
	1	4	4	2	2	5
Gilbert	1	1	3	1	4	3
Hatcher	2	3	1	3	2	4
Pirnot	3	4	2	2	1	2
Tannenbaum	4	2	4	4	3	1

The Borda points are

$$
\begin{array}{llllll}
\text{Gilbert:} & 7 \times 3 & + & 0 \times 2 & + & 9 \times 1 & = 30 \\
\text{Hatcher:} & 4 \times 3 & + & 3 \times 2 & + & 6 \times 1 & = 24 \\
\text{Pirnot:} & 2 \times 3 & + & 11 \times 2 & + & 1 \times 1 & = 29 \\
\text{Tannenbaum:} & 5 \times 3 & + & 4 \times 2 & + & 2 \times 1 & = 25
\end{array}
$$

Hence Gilbert wins the election under the Borda count method. Now suppose that Tannenbaum drops the election. This yields the following preference schedule:

Candidates	Number of Votes (18)					
	1	4	4	2	2	5
Gilbert	1	1	3	1	3	2
Hatcher	2	2	1	3	2	3
Pirnot	3	3	2	2	1	1

or

Candidates	Number of Votes				
	5	2	4	2	5
Gilbert	1	1	3	3	2
Hatcher	2	3	1	2	3
Pirnot	3	2	2	1	1

New calculations of Borda points gives

$$
\begin{array}{lllll}
\text{Gilbert:} & 7 \times 2 & + & 5 \times 1 & = 19 \\
\text{Hatcher:} & 4 \times 2 & + & 7 \times 1 & = 15 \\
\text{Pirnot:} & 7 \times 2 & + & 6 \times 1 & = 20
\end{array}
$$

Now the winner becomes Pirnot. This demonstrates that the Borda count method violates Chernoff's criterion.

Example 1.6.10.   Borda Count Violates the Independence of Clones Criterion

In an election between two candidates, A and B, the following votes are cast: five votes prefer A to B and four votes prefer B to A. So A obtains 5 Borda points ($5 \times 1 + 4 \times 0 = 5$)

and B obtains 4 Borda points ($5 \times 0 + 4 \times 1 = 4$). Therefore candidate A is the Borda winner.

Suppose that a clone candidate, call it C, almost identical to candidate B is added to the candidate set. This means that candidate A cannot be placed between B and C in a preference list. Suppose the new election produces the following preference schedule:

Ranking	Number of Votes (9)	
	5	4
First choice	A	B
Second choice	B	C
Third choice	C	A

Candidate A has 10 Borda points ($5 \times 2 + 4 \times 0 = 10$), candidate B has 13 Borda points ($5 \times 1 + 4 \times 2 = 13$), and clone candidate C has 4 Borda points ($5 \times 0 + 4 \times 1 = 4$). Therefore, candidate B wins the election.

As a result of adding a clone candidate to the candidate set, the winner of the election has changed from A to B. $\square$

The plurality method and the Borda count method satisfy the monotonicity criterion because a candidate who wins an election and gains more support will win any reelection.

Example 1.6.11. The Copeland Method Violates the IIA Criterion
The main disadvantage of head-to-head comparison methods consists of failure in determination of a winner: the probability of not identifying the winner increases with the increasing number of competitors. If an election has a Condorcet winner, then elimination of some alternatives from competition cannot change the winner.

If a pairwise comparison does not determine the winner, we use the Copeland method. As an example, let us consider the following preference rankings:

Ranking	Number of Voters (18)				
	5	2	4	5	2
First choice	A	B	D	C	D
Second choice	B	C	A	B	A
Third choice	C	A	C	D	B
Fourth choice	D	D	B	A	C

This election does not have a Condorcet winner because every alternative has been defeated: $\underline{A} : B = 11 : 7$, $\underline{A} : C = 11 : 7$, $A : \underline{D} = 7 : 11$, $B : C = 9 : 9$, $\underline{B} : D = 12 : 6$, $\underline{C} : D = 12 : 6$. An expression as $\underline{A} : B = 11 : 7$ means that 11 voters prefer A to B and 7 voters prefer B against A, so A becomes a winner in this pairwise matchup. Therefore, we apply the Copeland method by calculating the scores (1.4.2):

$$C(A) = 2, \qquad C(B) = C(C) = \tfrac{3}{2}, \qquad C(D) = 1.$$

Here $C(i) = \sum_{k \neq i} s_{i,k}$ is the Copeland score of alternative i (see page 76). So alternative A is the Copeland winner. If we remove options B and C, the preference table will be as follows:

Ranking	Number of Voters (18)				
	5	2	4	5	2
First choice	A	A	D	D	D
Second choice	D	D	A	A	A

We see that alternative D beats A by a vote of 11 to 7. Due to the removal of some original losing alternatives changes the outcome of the election, this method does not satisfy the IIA criterion. □

Fairness summary is presented here.

Criterion	Borda	Nanson	Baldwin	Sequential Pairwise	Copeland
Majority	No	Yes	Yes	No	Yes
Condorcet	No	Yes	Yes	Yes	Yes
Monotonicity	Yes	No	No	Yes	Yes
Chernoff/IIA	No	No	No	No	No
Pareto	Yes	Yes	Yes	No	Yes

Problems.

1. Use the following ballots to show that plurality does not satisfy the Condorcet criterion.

	Number of Voters (9)		
	4	3	2
A	1	3	3
B	2	1	2
C	3	2	1

2. Use the following ballots to show that plurality does not satisfy the Condorcet criterion.

Ranking	Number of Voters (17)					
	4	3	5	1	3	1
First choice	A	A	B	B	C	C
Second choice	B	C	C	A	B	A
Third choice	C	B	A	C	A	B

3. Use the preference schedule from Exercise 1 to show that the plurality method nominates the Condorcet loser.

4. Use the preference schedule from Exercise 2 to show that the plurality method nominates the Condorcet loser.

5. Use the following ballots to show that plurality does not satisfy the IIA criterion.

	Number of Votes (23)					
	4	5	1	5	3	5
A	1	1	2	3	2	3
B	2	3	1	1	3	2
C	3	2	3	2	1	1

6. Use the following ballots to show that plurality does not satisfy the IIA criterion.

Ranking	Number of Voters (13)					
	3	2	3	1	1	3
First choice	A	A	B	B	C	C
Second choice	C	B	C	A	A	B
Third choice	B	C	A	C	B	A

7. Use the following ballots to show that the Hare method does not satisfy the Condorcet criterion.

	Number of Voters (41)					
	9	6	9	5	8	4
A	1	1	3	2	3	2
B	3	2	1	1	2	3
C	2	3	2	3	1	1

8. Use the following ballots to show that the IRV does not satisfy the Condorcet criterion.

Ranking	Number of Voters (29)					
	6	5	7	3	6	2
First choice	A	A	B	B	C	C
Second choice	B	C	C	A	A	B
Third choice	C	B	A	C	B	A

9. Use the following ballots to show that the Hare method does not satisfy the monotonicity criterion.

	Number of Votes (17)				
	6	2	3	4	2
A	1	2	4	2	4
B	3	3	2	1	3
C	4	1	1	3	2
D	2	4	3	4	1

10. Use the following ballots to show that the Hare method does not satisfy the monotonicity criterion.

Ranking	Number of Voters (14)			
	5	3	4	2
First choice	A	B	C	D
Second choice	B	A	B	C
Third choice	C	C	D	B
Fourth choice	D	D	A	A

11. Use the following ballots to show that the Hare method does not satisfy the IIA criterion.

	Number of Votes (15)					
	4	2	1	3	2	3
A	1	1	2	3	2	3
B	2	3	1	1	3	2
C	3	2	3	2	1	1

12. Seventeen people were asked about their favorite type of flower. The results are summarized in the following preference table.

Ranking	Number of Voters (17)				
	6	2	3	4	2
Roses	1	2	4	2	4
Daffodils	3	3	2	1	3
Tulips	4	1	1	3	2
Lilies	2	4	3	4	1

Which kind of flower wins using the IRV method? Show that the IIA criterion is violated.

13. Five people are running for office. The preference schedule is as follows.

Candidate	Number of Votes (15)					
	5	3	5	3	2	3
A	1	3	2	4	4	3
B	3	1	4	2	5	5
C	2	4	1	3	3	2
D	4	5	3	1	2	4
E	5	2	5	5	1	1

(a) Find the winner of the election under the Hare method.

(b) Suppose that before the votes are counted, candidate B withdraws from the race. Find the preference schedule for a new election held without candidate B, and then find the winner under the IRV method.

(c) The results of parts (a) and (b) show that plurality with sequential elimination violates one of the fairness criteria discussed in this section. Which one?

14. Given the following preference schedule of the election.

Ranking	Number of Votes (17)					
	5	4	1	2	3	2
First choice	B	A	C	D	B	C
Second choice	C	D	A	A	C	D
Third choice	A	C	B	C	A	A
Fourth choice	D	B	D	B	D	B

(a) Find the winner of the election under the Hare method.

(b) Explain why the winner in part (a) can be determined in the first round.

(c) Based on your observation in part (b), explain why plurality with sequential elimination satisfies the *majority* criterion.

15. A professional association executive committee is having an election to choose the location of their annual meeting. The preference rankings for the election between four alternatives (A,

B, C, and D) are given in the following table.

Ranking	Number of Votes (33)					
	7	9	5	4	3	5
First choice	C	D	B	C	C	D
Second choice	B	A	A	A	D	C
Third choice	A	B	D	B	B	A
Fourth choice	D	C	C	D	A	B

(a) Find the winner of the election under the Coombs method.

(b) Suppose that in a new election, all voters stick with their previous ballots except three voters (in the second to the last column—it is the fifth one); they switch alternatives A and B, so B becomes the last choice while A is the third one. Which alternative wins this new election?

(c) The results of parts (a) and (b) show that the Coombs method violates one of the fairness criteria. Which one?

16. The 21 members of a small precinct are preparing for the caucuses that will be held in two months. The candidates running for president are Alves, Backman, Chung, and Flavin. The results of a straw poll taken prior to the official vote are given in the following preference schedule.

Candidate	Number of Votes (21)					
	3	3	6	4	3	2
Alves	1	1	2	3	4	3
Backman	4	3	1	2	3	2
Chung	2	4	4	1	1	4
Flavin	3	2	3	4	2	1

(a) Find the winner of the election under the Hare method.

(b) In the official vote, everyone votes the same as the straw poll except for the two votes in the first column of the table—they switch their votes and move Flavin ahead of Alves in their ballots. Find the winner of the official vote under the IRV method.

(c) The results of parts (a) and (b) show that plurality with sequential elimination violates one of the fairness criteria. Which one?

17. Due to financial problems, the school board considers different options to reduce expenses. They discuss the following options: reduce sports programs, increase class size, reduce expenditures on art and music programs, and defer maintenance on buildings. Use the preference table to determine the choice that the school board recommends using the Hare method. Show that Condorcet's criterion is violated.

Option	Number of Votes (19)					
	3	2	1	5	2	6
Reduce sports programs	1	1	3	2	4	2
Increase class size	3	2	1	3	2	3
Reduce art and music programs	4	3	2	1	1	4
Defer maintenance on buildings	2	4	4	4	3	1

18. A *Magazine* is going to choose the best music program. Use the preference table to determine the best program under the Hare method. Show that Condorcet's criterion is violated.

Ranking	Number of Votes (23)					
	2	3	2	3	4	9
First choice	A	B	B	C	C	D
Second choice	C	C	A	D	A	C
Third choice	D	A	D	A	B	A
Fourth choice	B	D	C	B	D	B

19. Use the following preference schedule to show that the IRV violates the independence of clones criterion.

Candidate	Number of Votes (195)			
	65	60	40	30
A	1	4	4	4
B	2	1	2	3
C	3	2	1	2
D	4	3	3	1

20. Use the following preference schedule to show that the Coombs method does not satisfy the independence of clones criterion.

Candidate	Number of Votes (20)			
	7	6	4	3
A	1	4	3	3
B	3	1	2	4
C	4	2	1	2
D	2	3	4	1

21. Use the following preference schedule to show that the Coombs method does not satisfy the participation criterion.

Candidate	Number of Votes (11)				
	1	2	3	4	5
A	3	2	1	4	2
B	4	1	4	1	3
C	2	4	3	2	1
D	1	3	2	3	4

22. A city board is having an election to choose the best restaurant among four nominees—Alforno (A), Blue Grotto (B), Cassarino (C), and Ri Ra (R). The preference schedule for the election is given in the following table.

Ranking	Number of Votes (33)							
	4	2	9	8	3	2	1	4
First choice	C	B	R	R	A	B	A	C
Second choice	A	C	A	C	B	C	B	A
Third choice	B	A	B	A	C	R	C	B
Fourth choice	R	R	C	B	R	A	R	R

(a) Find the winner of the election under the Borda count method.

(b) Explain why this election shows that the Borda method violates the *majority* criterion.

(c) Explain why this election shows that the Borda method violates the *Condorcet* criterion.

23. In the preference table, alternative D has the majority of first-place votes. Who wins the election if we use the Borda count?

Ranking	Number of Votes (13)			
	1	7	3	5
First choice	C	D	C	A
Second choice	A	A	A	D
Third choice	B	B	D	B
Fourth choice	D	C	B	C

24. A state commission is voting on changing the legal drinking age. Use the preference schedule to determine the winner using the Borda count method. Show that Condorcet's criterion is not satisfied.

Option	Number of Votes (23)				
	1	3	8	5	6
Lower the age to 18	1	2	4	2	3
Lower the age to 19	3	3	2	3	1
Lower the age to 20	4	1	1	4	2
Keep the age at 21	2	4	3	1	4

25. A club is voting for president. Use the preference schedule to determine the winner using the Borda count method. Show that Condorcet's criterion is not satisfied.

Ranking	Number of Votes (26)				
	2	13	4	1	6
First choice	C	D	C	A	A
Second choice	A	A	B	D	B
Third choice	B	B	D	B	C
Fourth choice	D	C	A	C	D

26. A company is considering Atlanta, Boston, Chicago, and Detroit for a new facility to be built. A group of senior managers voted to determine where the facility will be located. Use the preference table to determine the city that was chosen using the Borda count method. Show that the independence–of–irrelevant–alternatives criterion is not satisfied.

City	Number of Votes (23)					
	1	3	4	7	6	2
Atlanta	1	3	2	3	2	3
Boston	3	1	1	2	4	2
Chicago	2	4	3	1	1	4
Detroit	4	2	4	4	3	1

27. A company is going to build a new factory in either Arkansas, Colorado, Indiana, or Kansas. The results of voting of the board of directors is shown in the preference schedule. Determine the state chosen using the Borda count method. Show that Chernoff's criterion is violated.

State	Number of Votes (26)					
	1	5	4	7	4	2
Arkansas	1	3	2	3	2	3
Colorado	3	1	1	2	4	2
Indiana	2	4	3	1	1	4
Kansas	4	2	4	4	3	1

28. An election is held to choose the department chair at a local college. The candidates are professors named Bisshopp, Dafermos, Falb, Gidas, and McClure. The preference rankings for the election are as follows:

Professor	Number of Votes (19)					
	5	3	1	5	2	3
Bisshopp	1	3	5	4	3	4
Dafermos	3	2	1	5	5	1
Falb	2	5	3	1	4	5
Gidas	4	1	4	3	2	3
McClure	5	4	2	2	1	2

(a) Find the winner of the election under the Nanson method.

(b) Find the winner of the election under the Nauru score system.

(c) Suppose that before the votes are counted, Professor Dafermos withdraws from the race. Find the preference schedule for a new election held without Professor Dafermos, and then find the winner under the Nanson and Nauru methods.

(d) Explain why this election shows that the Nanson and Nauru methods violate the *independence–of–irrelevant alternatives* criterion.

29. The town of Kingston is having an election for "Miss Kingston." The candidates are Allison, Ivana, Kristie, Nadia, and Susie. The election resulted in the following preference schedule:

Nominee	Percentage								
	21%	7%	13%	16%	3%	18%	3%	11%	8%
Allison	1	5	3	4	2	2	5	4	3
Ivana	3	1	1	2	5	4	3	5	2
Kristie	2	4	5	1	1	3	4	3	5
Nadia	5	2	4	3	4	1	1	2	4
Susie	4	3	2	5	3	5	2	1	1

(a) Find the winner of the election under the Nauru and Baldwin methods.

(b) Suppose that before the votes are counted, Allison is found to be ineligible because of her grades. Find the preference schedule for a new election held when Allison's name is removed, and then find the winner under the Nauru and Baldwin methods.

(c) Explain why this election shows that the Nauru and Baldwin methods violate the *independence–of–irrelevant alternatives* criterion.

30. In the world speed skating competition, the standing between five countries was as follows:

Country	Number of Votes (14)			
	2	3	4	5
Canada	1	2	4	5
Korea	2	1	5	4
U.S.	3	4	1	3
China	4	3	3	2
The Netherlands	5	5	2	1

Which country wins the competition using the Baldwin method? However, the last skater from The Netherlands illegally changed lanes during his run and was disqualified. His disqualification led to the overall elimination of The Netherlands team. Does it change the results of the competition?

31. Use the Copeland method to determine the winner of the election summarized in the following table. Is Chernoff's criterion satisfied?

Alternative	Number of Votes (29)				
	5	2	3	3	1
A	1	4	4	2	3
B	2	1	2	4	4
C	4	3	1	1	2
D	3	2	3	3	1

32. The 28 members of a soccer team are holding an election to choose a senior captain. The five candidates are Berbatov, Fletcher, Giggs, Owen, and Rooney. The results of the election are shown in the following preference schedule.

Candidate	Number of Votes (28)					
	6	4	8	2	5	3
Berbatov	1	3	3	3	2	5
Fletcher	4	1	2	5	4	4
Giggs	5	2	1	2	5	2
Owen	3	5	4	1	1	3
Rooney	2	4	5	4	3	1

(a) Find the winner of the election using the Copeland method.

(b) Just after the election, it is discovered that Berbatov failed MTH 106; he is therefore ineligible to be captain of the soccer team. Find the preference rankings for a recount without Berbatov, and then find the winner of this recount under the Copeland method.

(c) The results of (a) and (b) show that the Copeland method violates the fairness criteria discussed in §1.6. Which one?

33. Fourteen students in a Film Studies class have a choice of four movies to watch from the year 1994. The students rank the films from favorite to least favorite. The teacher decides to use the Borda count method in order to choose what movie to watch: Shawshank Redemption, Pulp Fiction, Forrest Gump, and Hoop Dreams. The preference ranking of the votes are listed here. Determine the winner and the loser using the Borda method in two similar elections. What criterion is violated?

Movie	Number of Votes (14)			
	5	4	3	2
Shawshank	3	2	1	3
Pulp Fiction	1	4	2	2
Forrest Gump	2	1	4	4
Hoop Dreams	4	3	3	1

Movie	Number of Votes (14)			
	5	4	3	2
Shawshank	3	2	3	4
Pulp Fiction	1	4	1	1
Forrest Gump	2	1	2	2
Hoop Dreams	4	3	4	3

34. LaSalle Academy is holding a music competition in their auditorium. Each candidate plays a different instrument and are judged based on their talent. At night, Mike's supporters wine and dine the four judges who ranked the ballots shown in the last two columns. They convince them to rearrange their rankings, placing Mike first. Who wins the competition?

Movie	Number of Votes (29)				
	7	8	10	2	2
Andy	1	3	3	1	1
Boris	2	1	4	4	3
Mike	4	2	1	2	4
Tommy	3	4	2	3	2

1.7   Two Ranking Methods

So far, we considered voting methods that intended to determine a single winner (if any). There are known cases when we not only need to know who wins the election but also need to determine who comes in second, third, and so on. In sport activities, it is a custom to reward the winner with a gold medal, the second-place finisher with a silver medal, and the third-place holder with a bronze medal. Here are some examples of situations in which it would be good to know how to construct societal rankings from individual rankings.

- Suppose a condominium association needs to elect a president, a vice president, and a treasurer. Of course, it is convenient for everybody to vote only once instead of conducting three separate elections. So the optimal approach is to use one of the voting methods that results in ranking candidates. In other words, the winner of the election gets to be the president, the second-place candidate gets to be the vice president, and the third-place candidate gets to be the treasurer.

- Imagine that we need to elect a board of trustees consisting of seven people among eighteen candidates. A good way of proceeding would be to rank all eighteen candidates, then pick the first seven as required members.

- Management wants to have a team of new employees ready to work on a special task. So far, it is not known exactly how many people are needed—it depends on various circumstances, including available funds. However, management wants to have a list of two to five candidates for the team.

- When a new position is available, usually, the hiring committee receives quite a large number of applications. To narrow down the set of candidates, the committee selects approximately 5 to 12 of them for a second round.

- A TV sports program wants to choose three hockey stars of the week. A natural way is to rank all candidates and then pick up the three best players.

Therefore, we consider some modifications of previously discussed voting methods that allow us to rank alternatives in a competition. There are actually two natural approaches—**extended** and **recursive**—that can be used to determine several finishers in an election or rank alternatives. Plurality and point distribution methods can be easily adopted for these purposes when the preference schedule of the election is known. However, two-stage methods (runoff or Nanson) are not applicable in the extended approach. We show how we can achieve ranking of candidates by examples.

Example 1.7.1.   Extended Plurality
A vacation club is trying to determine in what order the following destinations should be visited this year. The choices are Acapulco, Cancun, Hawaii, or Orlando. The club decided to use the plurality method, but due to time constraints, it was applied only once. The preference rankings of the members are as follows:

	Number of Votes (20)					
Destination	3	4	4	3	4	2
Acapulco	1	2	3	2	4	3
Cancun	2	1	4	3	3	4
Hawaii	3	4	1	4	2	1
Orlando	4	3	2	1	1	2

Counting first-place votes gives

Acapulco:	3 first-place votes,
Cancun:	4 first-place votes,
Hawaii:	6 first-place votes,
Orlando:	7 first-place votes.

We know that Orlando is the plurality winner. Clearly Hawaii is second with 6 first-place votes. Likewise, Cancun is third (4 first-place votes) and Acapulco is last. The ranking of all four destinations under the *extended* plurality method is as follows:

$$\text{Orlando} \succ \text{Hawaii} \succ \text{Cancun} \succ \text{Acapulco}$$

Example 1.7.2. Extended Hare Method

A company wants to involve its employees in sport activities. For this purpose, the management decides to hire instructors who will give 50-minute classes twice per week. In order to schedule classes, the management needs to rank activities. For this purpose, the company asked its employees to cast their ballots, and applied the extended Hare method to the following preference lists presented here.

	Number of Votes (31)							
	4	5	3	4	7	1	3	4
Aerobics	1	1	2	2	3	3	4	4
Martial Arts	2	4	3	4	2	1	1	3
Swimming	3	3	4	1	1	4	3	2
Tennis	4	2	1	3	4	2	2	1

The number of first-place votes are

Aerobics:	9 first-place votes,
Martial Arts:	4 first-place votes,
Swimming:	11 first-place votes,
Tennis:	7 first-place votes.

Since Martial Arts received the least number of first-place votes, this option is eliminated. The new count yields

Aerobics: 9 first-place votes,
Swimming: 11 first-place votes,
Tennis: 11 first-place votes.

In the second round, Aerobics is dropped leaving two options in the final pool. Since Tennis defeats Swimming with the score 16:15, we obtain the final ranking to be

$$\text{Tennis} \ \succ \ \text{Swimming} \ \succ \ \text{Aerobics} \ \succ \ \text{Martial Arts} \ .$$

This example shows that the IRV method may lead to different rankings obtained by the plurality.

Example 1.7.3. Extended Coombs Method

A soccer team wants to rank colors for their jerseys to wear during matchups. The preference schedule of the team members is listed in the following table:

	Number of Votes (20)					
	2	1	4	1	5	7
Green	1	1	2	2	3	4
Red	2	2	1	4	2	3
Yellow	3	4	4	1	4	1
White	4	3	3	3	1	2

First we check whether there is a majority winner. Calculating the first-place votes, we see that

Green has 3 first-place votes,   Red has 4 first-place votes,
Yellow has 8 first-place votes,   White has 5 first-place votes.

Therefore, there is no majority winner. To apply the Coombs method, we need to know the number of last-place votes:

Green has 7 last-place votes,   Red has 1 last-place vote,
Yellow has 10 last-place votes,   White has 2 fourth-place votes.

We eliminate alternative Yellow from this competition because this option has more last-place votes than all the others. Hence, Yellow is the loser and we assign the fourth place to it. The ten fourth-place votes go to White and Red and we get the new distribution:

	Number of Votes (20)					
	2	1	4	1	5	7
Green	1	1	2	1	3	3
Red	2	2	1	3	2	2
White	3	3	3	2	1	1

or

	Number of Votes			
	3	4	1	12
Green	1	2	1	3
Red	2	1	3	2
White	3	3	2	1

Since White is the majority winner (with 12 first-place votes out of 20) and Red is the loser (with 12 last-place votes out of 20), we get the following rankings

$$\text{White} \quad \succ \quad \text{Green} \quad \succ \quad \text{Red} \quad \succ \quad \text{Yellow} .$$

Note that Yellow is the plurality winner and the plurality loser at the same time.

Example 1.7.4. Extended Borda Count

A company board voted on a distribution of \$50,000 donated by a retired employee. The board suggested to spend money on books, computers, presents, and gym equipment. Their preference rankings are as follows:

Option	Number of Votes (31)								
	4	3	2	4	5	3	2	4	4
Books	1	1	2	2	3	3	3	4	4
Computers	3	2	1	3	4	1	2	3	2
Gym equipment	4	4	3	4	1	2	1	2	1
Presents	2	3	4	1	2	4	4	1	3

Ranking options using the extended Borda count method is quite simple: we need to count the points that each option earns. Recall that in our case of four alternatives first place is worth 3 points, second place is rewarded with 2 points and third place gets 1 point. The tallies become

Books: $\quad 7 \times 3 + 6 \times 2 + 10 = 43$ points,
Computers: $\quad 6 \times 3 + 9 \times 2 + 12 = 48$ points,
Gym equipment: $\quad 10 \times 3 + 7 \times 2 + 2 = 46$ points,
Presents: $\quad 8 \times 3 + 9 \times 2 + 7 = 49$ points.

Now we can rank these options:

$$\text{Presents} \quad \succ \quad \text{Computers} \quad \succ \quad \text{Gym equipment} \quad \succ \quad \text{Books} .$$

It is a pity that people do not read books but prefer presents.

Example 1.7.5. Extended Borda System with the First-place Promotion

An engineering department of twenty one members are selecting a new chair, undergraduate adviser, schedule negotiator, and treasurer supervisor. Five members of the department are eligible and willing to run for these positions. The preference schedule of the department members is given here.

Candidate	Number of Votes (21)									
	1	1	1	2	1	3	2	3	3	4
Besio, Walter	1	1	2	2	3	3	4	4	5	5
Fischer, Godi	2	3	5	4	1	4	5	3	1	2
Huang, Helen	3	2	1	5	2	5	1	5	2	3
Jackson, Leland	4	5	4	3	5	2	3	2	4	1
Kay, Steven	5	4	3	1	4	1	2	1	3	4

Recall that the Borda system with the first-place promotion assigns the following number of points for each place: the first place worth 5 points, the second place is rewarded with 3 points, the third place gets 2 points, and the fourth place earns 1 point. Counting the tally for each candidate, we get

Candidate	Number of Points					
	1st-Place Votes×5 Points	2nd-Place Votes×3 Points	3rd-Place Votes×2 Points	4th-Place Votes×1 Points	5th-Place Votes×0 Points	Total Number of Points
Besio	$2 \times 5 = 10$	$3 \times 3 = 9$	$4 \times 2 = 8$	$5 \times 1 = 1$	$5 \times 0 = 0$	32
Fischer	$4 \times 5 = 20$	$5 \times 3 = 15$	$4 \times 2 = 8$	$5 \times 1 = 4$	$3 \times 0 = 0$	48
Huang	$3 \times 5 = 15$	$5 \times 3 = 15$	$5 \times 2 = 10$	$0 \times 1 = 0$	$8 \times 0 = 0$	40
Jackson	$4 \times 5 = 29$	$6 \times 3 = 18$	$4 \times 2 = 8$	$5 \times 1 = 5$	$2 \times 0 = 0$	51
Kay	$8 \times 5 = 40$	$2 \times 3 = 6$	$4 \times 2 = 8$	$6 \times 1 = 2$	$1 \times 0 = 0$	60

The end ranking (in linked form), based on the *extended* Borda count method becomes:

$$\text{Kay} \quad \succ \quad \text{Jackson} \quad \succ \quad \text{Fischer} \quad \succ \quad \text{Huang} \quad \succ \quad \text{Besio} .$$

Example 1.7.6. Extended Nauru System

Four candidates—Anna, Boris, Carlos, and Dimitry—are running for president, vice president, and treasurer of a service department. The preference schedule is as follows.

Candidate	Number of Votes (22)						
	7	5	1	4	1	3	1
Anna	1	2	2	3	3	4	4
Boris	2	3	1	4	2	3	1
Carlos	3	1	3	2	1	2	3
Dimitry	4	4	4	1	4	1	2

Calculating the Nauru scores, we get

Candidate	Number of Points				
	1st-Place Votes×1 Points	2nd-Place Votes×$\frac{1}{2}$ Points	3rd-Place Votes×$\frac{1}{3}$ Points	4th-Place Votes×$\frac{1}{4}$ Points	Total Number of Points
Anna	$7 \times 1 = 7$	$6 \times \frac{1}{2} = 3$	$5 \times \frac{1}{3} = \frac{5}{3}$	$4 \times \frac{1}{4} = 1$	$12\frac{2}{3}$
Boris	$2 \times 1 = 2$	$8 \times \frac{1}{2} = 4$	$8 \times \frac{1}{3} = \frac{8}{3}$	$4 \times \frac{1}{4} = 1$	$9\frac{2}{3}$
Carlos	$6 \times 1 = 6$	$7 \times \frac{1}{2} = \frac{7}{2}$	$2 \times \frac{1}{3} = \frac{2}{3}$	$7 \times \frac{1}{4} = \frac{7}{4}$	$11\frac{11}{12}$
Dimitry	$7 \times 1 = 7$	$1 \times \frac{1}{2} = \frac{1}{2}$	$7 \times \frac{1}{3} = \frac{7}{3}$	$7 \times \frac{1}{4} = \frac{7}{4}$	$11\frac{7}{12}$

This allows us to rank candidates (in linked form), based on the *extended* Nauru system:

$$\text{Anna} \quad \succ \quad \text{Carlos} \quad \succ \quad \text{Dimitry} \quad \succ \quad \text{Boris} .$$

Example 1.7.7. Extended Copeland Method

A committee is trying to award scholarships to four students—Anton, Brian, Crista, and Donna—according to their rankings. The votes' summary is shown below:

Candidate	Number of Votes (22)						
	7	5	1	4	1	3	1
Anton	1	1	2	3	3	4	4
Brian	2	3	4	1	4	1	3
Crista	3	4	1	2	2	3	2
Donna	4	2	3	4	1	2	1

Obviously, Anton meets the majority criterion—he got 12 first-place votes out of 22. However, we apply a pairwise comparison method. Compare Anton with the three other candidates, we get the following results:

Anton : Brian = 12:10;   Anton : Crista = 4:18;   Anton : Donna = 13:9.

Since the Copeland method assigns 1 point for a win $\frac{1}{2}$points for a draw, and 0 points for a loss, Anton gains 2 points (two wins and 1 loss). Comparing Brian with two other candidates, we see that he lost all competitions: Brian : Crista = 10:12 and Brian : Donna = 9:13. Therefore, Brian is the Condorcet loser (with 0 points). Crista defeats all candidates, which brings her 3 points. Donna lost two comparisons with Crista and Anton, which gives her 1 point. Now we can rank the candidates:

$$\text{Crista (3 points)} \quad \succ \quad \text{Anton (2 points)} \quad \succ \quad \text{Donna (1 point)} \quad \succ \quad \text{Brian (0 points)} .$$

Example 1.7.8. Extended Approval Voting

The preference table of a class of college students for their favorite "ethnic" food is presented here.

Candidate	|	Number of Votes (30)											
	|	1	2	3	1	1	2	2	1	2	1	1	2	1
Chinese	|	√	√	√		√		√	√					
French	|	√	√	√		√								√
Indian	|							√	√		√			√
Italian	|		√		√	√		√	√	√	√	√	√	√
Mexican	|	√			√	√	√	√	√	√	√		√	
Russian	|	√			√	√	√	√	√	√	√	√		

To rank favorite food, the approval vote count results in the following point distribution: Italian – 14 (approval winner),   Mexican – 13,   Russian – 12,   Chinese – 10, French – 8,   Indian – 5.   □

The ranking of options may be achieved by another strategy, called the **recursive approach**. The idea behind a recursive process consists of repeated applications of a particular voting method followed by eliminating the leader. At every stage, the winner's name is removed from the preference schedule, followed by recounting votes, and application of the voting method to the new reduced set of options.

Example 1.7.9. Recursive Plurality

Let us consider a competition between five people who are running for four positions—mayor, tax collector, police chief, and treasurer. The preference schedule of the election is presented here.

Candidate	|	Number of Votes (25)									
	|	1	2	3	1	3	2	4	2	4	3
Baudet, Gerard	|	1	1	2	2	3	3	4	4	2	5
DiPippo, Lisa	|	5	3	1	4	5	4	5	5	5	1
Kowalski, James	|	3	2	4	5	1	5	2	1	3	4
Lamagna, Edmund	|	4	5	3	1	2	1	3	2	4	3
Peckham, Joan	|	2	4	5	3	4	2	1	3	1	2

Calculating first-place votes for every candidate, we see that Peckham is the plurality winner with 8 first-place votes (Baudet has 3, DiPippo gets 6, Kowalski has 5, Lamagna has 3). By eliminating Peckham from competition, we obtain a new table:

Candidate	Number of Votes (25)									
	1	2	3	1	3	2	4	2	4	3
Baudet, Gerard	1	1	2	2	3	2	3	3	1	4
DiPippo, Lisa	4	3	1	3	2	3	2	4	4	1
Kowalski, James	2	2	4	4	1	4	1	1	2	3
Lamagna, Edmund	3	4	3	1	4	1	4	2	3	2

Since Kowalski gets 9 first-place votes (Baudet has 7, DiPippo has 6, and Lamagna has 3), this candidate is eliminated, and we have a new preference table:

Candidate	Number of Votes (25)									
	1	2	3	1	3	2	4	2	4	3
Baudet, Gerard	1	1	2	2	2	2	2	2	1	3
DiPippo, Lisa	3	2	1	3	3	3	3	3	3	1
Lamagna, Edmund	2	3	3	1	1	1	1	1	2	2

Eliminating the leader, Lamagna, with 12 first-place votes, we see that Baudet defeats DiPippo in head-to-head comparison with the score 19:6. Therefore, we can rank candidates:

$$\text{Peckham} \succ \text{Kowalski} \succ \text{Lamagna} \succ \text{Baudet} \succ \text{DiPippo} . \qquad \square$$

A two-round voting system such as runoff cannot be used in ordering candidates as the basis for an extended approach because all competitors but two are eliminated after the first round. However, this method can be used for a recursive approach as the following example shows.

Example 1.7.10. Recursive Runoff
A school board was facing a dilemma on what kind of fundraiser they should hold. The five options were narrowed down to an Auction, Bingo, a Car Wash, selling Discount Cards, and Lollipops selling. The board decided to rank all options using the runoff method as a basis for a recursive approach. After all preference ballots were cast, the results of voting are summarized in the following table.

Option	Number of Votes (15)					
	3	2	2	2	2	4
Auction	1	4	5	3	4	3
Bingo	2	1	1	2	3	5
Car Wash	3	5	3	1	2	4
Discount Cards	4	2	2	5	1	2
Lollipops	5	3	4	4	5	1

Comparing the two best options—Lollipops and Bingo both have four first-place votes—we see that the latter defeats Lollipops with the score 11:4 (which means that 11 voters prefer Bingo while 4 prefer Lollipop). So we put Bingo in first place and eliminate this option from further competition.

In the second round, four first-place votes of Bingo go to Discount Cards, and the second round between Discount Cards and Lollipops results in the victory of Discount Cards with the score 9:6. Putting Discount Cards in the second position and eliminating it from competition, we arrive at the following preference schedule:

Option	Number of Votes (15)					
	3	2	2	2	2	4
Auction	1	2	3	2	2	2
Car Wash	2	3	1	1	1	3
Lollipops	3	1	2	3	3	1

Now the Auction has 3 first-place votes, Car Wash has 6 first-place votes, and Lollipops also has 6 first-place votes. Comparison of these two options yields the superior Car Wash with the score 9:6, which we place in the third position. The last round of elections between Auction and Lollipops results in the following final ranking:

$$\text{Bingo} \succ \text{Discount Cards} \succ \text{Car Wash} \succ \text{Lollipops} \succ \text{Auction}.$$

Example 1.7.11. Recursive Borda Count

The 15 members of a basketball team were asked to rank the most valuable players among four candidates. The preference schedule of the nominees are given in the table.

Player	Number of Votes (15)					
	3	2	2	2	2	4
Allen, Ray	1	4	4	4	3	3
Davis, Glen	2	1	2	3	4	2
Pierce, Paul	3	3	1	1	2	4
Rondo, Rajon	4	2	3	2	1	1

Calculating points (based on 3-2-1-0 point distribution), the management got

Allen:	$3 \times 3 + 0 \times 2 + 6 \times 1 = 15,$
Davis:	$2 \times 3 + 9 \times 2 + 2 \times 1 = 26,$
Pierce:	$4 \times 3 + 2 \times 2 + 5 \times 1 = 21,$
Rondo:	$6 \times 3 + 4 \times 2 + 2 \times 1 = 28.$

So we set apart the Borda winner, Rondo, and consider a new preference table:

Player	Number of Votes (15)					
	3	2	2	2	2	4
Allen, Ray	1	3	3	3	2	2
Davis, Glen	2	1	2	2	3	1
Pierce, Paul	3	2	1	1	1	3

The Borda points are

Allen:	$3 \times 2 + 6 \times 1 = 12,$
Davis:	$6 \times 2 + 7 \times 1 = 19,$

Pierce: $6 \times 2 + 2 \times 1 = 14.$

Again, we remove the winner, Davis, and leave the pool with two candidates—Allen and Pierce. Since Pierce defeats Allen with score 8:7, we get the required ranking:

$$\text{Rondo} \succ \text{Davis} \succ \text{Pierce} \succ \text{Allen} .$$

$\square$

It is very common that an election may result in a draw. Therefore, **breaking ties** is an essential procedure that should be established before any election. In many cases ties can be avoided all together by a special requirement, for instance, by considering only an odd number of votes in pairwise contests. For example, some sport activities (boxing, water diving) as well as art competitions (Van Cliburn international piano competition) use an odd number of judges. Others have developed their own tie-breaking procedures.

In some cases, ties are impossible to break. If two runners finish simultaneously so the electronic equipment fails to determine who finished first, then both competitors are declared to be winners and awarded with gold medals. One of the ways to break ties in an election is to use different voting methods that may determine the winner uniquely.

Breaking a tie can be achieved by either making an arbitrary choice, such as flipping a coin, or by bringing in an additional voter. For example, under Robert's Rules of Order, the president of any group votes only when there is a tie. In 1876, General Henry M. Robert (1837 – 1923) set up rules for the publication of the *Pocket Manual of Rules of Order*. It sold half a million copies before its revision was published in 1915, which became the most widely used manual of parliamentary procedure and remains today the most common parliamentary authority in the United States. There are other ways of breaking a tie that are less arbitrary than flipping a coin. If a tie results from using the Borda count method, it could be broken by choosing the candidate with the most first-place votes. If a tie results from the pairwise comparison method, it could be broken by choosing the winner of the one-to-one comparisons between the candidates involved in the tie followed, if needed, by the Condorcet score. Different tie-breaking methods could produce different winners. Therefore, to remain fair, the tie-breaking method should be decided upon in advance.

Example 1.7.12.   UEFA Champions League 2008/2009

The results of matchups in group H during the 2008/2009 season in the UEFA Champions League are summarized in the following table:

Team	Juventus	Real	Zenit	BATE	**Points**
Juventus Turin	–	2:1	1:0	0:0	12
Real Madrid	0:2	–	3:0	2:0	12
Zenit St Petersburg	0:0	1:2	–	1:1	5
BATE Borisov	2:2	0:1	0:2	–	3

In the UEFA competition, a team is given 3 points for a win, 1 point for a draw, and 0 point for a loss. This scoring system fails to determine a winner in group H because two teams had the same number of points. However, Juventus was given first place because of their pairwise matchups, where Juventus beat Real twice (in a home game and away).

Problems. In all problems, for the given preference schedule, rank the candidates/options using

- **(a)** the extended plurality method;
- **(b)** the extended Hare method;
- **(c)** the extended Coombs method;
- **(d)** the extended Borda count method;
- **(e)** the extended Nauru score system;
- **(f)** the extended Copeland method (when there is a tie between two competitors, use their pairwise matchup to determine the winner followed by the Condorcet score, if needed);
- **(g)** extended approval voting (when the first two choices are considered to be approved).

After that, rank candidates using the recursive seven methods.

1. Rank candidates in the election.

Ranking	Number of Votes (22)				
	5	4	6	2	5
First choice	A	B	C	A	D
Second choice	C	D	B	B	C
Third choice	D	C	A	C	B
Fourth choice	B	A	D	D	A

2. Rank candidates in the election.

Ranking	Number of Votes (18)				
	2	3	1	5	7
First choice	A	B	B	C	D
Second choice	D	C	D	A	B
Third choice	B	A	C	D	C
Fourth choice	C	D	A	B	A

3. Rank candidates in the election.

Preference	Number of Votes (19)				
	7	5	4	2	1
First choice	A	B	C	D	A
Second choice	D	A	D	B	B
Third choice	C	C	B	A	C
Fourth choice	B	D	A	C	D

4. Rank alternatives in the election.

Ranking	Number of Votes (33)				
	9	8	7	6	3
First choice	C	D	B	A	A
Second choice	A	B	C	D	B
Third choice	B	C	D	B	C
Fourth choice	D	A	A	C	D

5. Rank options in the election.

Option	Number of Votes (17)					
	1	4	3	2	3	4
A	1	1	2	3	2	4
B	2	4	1	4	1	3
C	3	3	4	1	4	2
D	4	2	3	2	3	1

6. Rank candidates in the election.

Ranking	Number of Votes (33)							
	4	3	9	5	6	3	2	1
First choice	A	A	B	B	C	C	D	D
Second choice	B	D	C	D	A	A	C	A
Third choice	C	C	D	A	B	D	A	C
Fourth choice	D	B	A	C	D	B	B	B

7. Rank candidates in the election.

Preference	Percentage				
	27%	15%	12%	26%	20%
First choice	A	B	B	C	D
Second choice	D	D	A	B	A
Third choice	B	A	C	D	C
Fourth choice	C	C	D	A	B

8. Rank candidates in the election.

Ranking	Percentage				
	9%	24%	22%	26%	19%
First choice	A	A	B	C	D
Second choice	B	D	A	D	C
Third choice	D	C	C	A	B
Fourth choice	C	B	D	B	A

9. Rank candidates in the election.

Candidate	Percentage					
	23%	22%	13%	14%	24%	4%
A	1	4	4	2	3	2
B	3	1	2	4	2	1
C	2	3	1	1	4	3
D	4	2	3	3	1	4

10. Rank candidates in the election.

Candidate	Percentage				
	27%	22%	26%	9%	16%
A	1	2	3	4	3
B	2	1	2	3	4
C	3	4	1	2	2
D	4	3	4	1	1

Chapter 1 Review

Important Terms, Symbols, and Concepts

1.1 Preference Ranking

A **ballot** is a record of how someone voted. A **preference ballot** is a type of ballot structure used in several electoral systems in which voters rank a list or group of candidates/alternatives in order of preference. A **preference table** or schedule is a record that summarizes the results of all the individual preference ballots in an election.

Majority criterion: if a majority (at least 50% plus one vote) of voters regard a candidate to be the best, then that candidate wins the election.

1.2 The Plurality Methods

Plurality method: each person chooses his or her favorite candidate and the candidate with the most first-place votes wins.
Runoff method: a two-round voting system based on the plurality method. If no candidate is the majority winner, the two best finishers proceed into the second round and all other alternatives are eliminated. The winner is determined by the majority of votes in the second round between the best two alternatives.
Single transferable vote: the system where voters rank the candidates in order of preference. The winner is determined in a sequence of rounds. If at any step, a candidate has the majority of votes, then this candidate is the winner and iteration stops. Otherwise, the worst candidate is eliminated and iteration proceeds until the winner is found.
The **Hare method** or **plurality with sequential eliminations** or **instant run-off voting** (IRV) is a multiple-round single transferable voting method. The worst option to be eliminated is the alternative having the least number of first-place votes.
Coomb method: another variation of the single transferable voting method. The worst option is the alternative having the largest number of last-place votes.

1.3 Point Distribution Systems

The Borda count: each candidate (or alternative) gets 0 points for each last place vote received, 1 point for each next-to-last point vote, etc., all the way up to $n-1$ points for each first place vote (where n is the number of candidates/alternatives). The candidate with the largest point total wins the election.
The Borda count with first-place promotion: each first-place vote is rewarded with n points; all other points are distributed as in the Borda count.
The Nauru count: a system of preferential voting in which voters vote for the candidates in their constituency in order of their preference. A first preference vote counts as a whole vote, a second vote counts as half a vote, a third preference vote as one-third of a vote, and so on. The candidate with the largest point total wins the election.
The modified Borda count: similar to the Borda system; however, the number of points given for a voter's first and subsequent preferences is determined by the total number of

candidates the voter has actually ranked, rather than the total number running. The candidate with the highest average preference score is the winner.

Nanson's method: is a multiple-round voting system. At each stage, it eliminates those choices from a Borda count tally that are at or below the average Borda count score; then the ballots are re-tallied as if the remaining candidates were exclusively on the ballot. This process is repeated if necessary until a single winner remains.

Baldwin's method: is a multiple-round voting system based on Borda count. In each round, the candidate with the fewest point tally is eliminated, and the points are re-tallied as if that candidate were not on the ballot.

Average Rating: each alternative is given a score (for example, from 0 to 100 or 0 to 10) by each voter. The alternative with the highest total score wins.

1.4 Head-to-Head Comparisons

The **Condorcet** or **pairwise comparison method** is any single-winner election method that meets the Condorcet criterion, that is, which always selects the Condorcet winner, the candidate who would beat each of the other candidates in a head-to-head comparison, if such a candidate exists. A **weak Condorcet** winner (if any) is an alternative that is not defeated by any other competitor. As usual, alternative A beats alternative B if more voters rank A over B.

Copeland's method is a head-to-head round robin competition in which the winner is determined by finding the candidate with the most pairwise victories. The Copeland winner can be determined by counting scores: for every victory in pairwise comparison, the winner is rewarded with 1 point, for a draw—$1/2$ points, and for a loss—nothing. Then the Copeland winner is the alternative with the most overall points (if any).

Agenda is an ordering of the alternatives to be considered in pairwise comparison. **Sequential pairwise voting** compares pairs of candidates according to some agenda to determine which one defeats another. After such comparisons, a loser is eliminated, and the winner moves on to confront the third candidate in the list. This process continues throughout the entire agenda, and the one remaining at the end is declared to be the winner.

1.5 Approval Method.
A voter can vote for as many alternatives as he or she wishes. The alternative that receives the most votes wins. In other words, voters can cast either 0 or 1 vote for each alternative. The alternative with the largest point total is declared the winner.

1.6 Arrow's Impossibility Theorem

Arrow's impossibility theorem states that a preferential voting method subject to some fairness conditions that can always fairly decide the outcome of an election that involves more than two candidates or alternatives does not exist.

Majority criterion: if a majority of the voters rank a candidate as their first choice, then that candidate should win the election.

A voting system satisfies the **Condorcet winner criterion** if it nominates the Condorcet winner if it exists.

Monotonicity criterion: raising a winner's placement on a ballot should not prevent it from winning an election.

Independence of irrelevant alternatives (IIA) or **Chernoff's condition:** Determination of whether alternative A or alternative B is better should not be changed by the availability of an additional candidate X, which is irrelevant to the choice between A and B.

May theorem: among all two-candidate voting systems that never result in a draw, majority rule is the only anonymous method that treats all voters equally, treats both candidates equally, and is nonmanipulable.

1.7 Two Ranking Methods

Extended voting method is obtained from plurality or point distribution methods by assigning points to every alternative. This allows us to rank candidates in any election.

Recursive voting sequentially determines the winner at each round and then eliminates this alternative from the election and proceeds again with recursive voting.

Review Exercises.

1. Suppose there are 140 votes cast in an election among five candidates—Euler, Gauss, Kolmogorov, Hilbert, and Leibniz—to be elected by plurality. After the first 100 votes are counted, the tallies are as follows:

   Euler – 12,   Gauss – 23,   Kolmogorov – 19,   Hilbert – 29,   Leibniz – 17

   What is the minimal number of the remaining votes each of the candidates can receive and be assured of a win?

2. Suppose there are 150 votes cast in an election among 5 jazz pianists—Geri Allen, Patricia Barber, Roberto Carnevale, Carsten Dahl, and Eliane Elias—to be elected by plurality. After the first 127 votes are counted, the tallies are as follows:

   Allen – 21,   Barber – 24,   Carnevale – 26,   Dahl – 29,   Elias – 27

   What is the minimal number of the remaining votes each of the candidates can receive and be assured of a win?

3. If 303 votes are cast, what is the smallest number of votes a winning candidate can have in a 4-candidate race that is to be decided by plurality?

4. If 304 votes are cast, what is the smallest number of votes a winning candidate can have in a 5-candidate race that is to be decided by plurality?

5. Four candidates running for town office receive first-place votes as follows: Arrighi, 1,541; Bertram, 1,230; Cahill, 981; and D'Angelo, 539. Is there a candidate who earns a majority? Who wins the election using the plurality method?

6. Four candidates running for town council receive first-place votes as follows: Huggins, 3,234; Lum, 2,193; Mack, 1,479; and Patria, 642. Is there a candidate who earns a majority? Who wins the election using the plurality method?

7. A scientific society conducted a survey regarding issues they want to address. The choices were Research funding, Computer updates, Equality in the workplace, and Attracting more

women to science. Their preferences are summarized in the table. What option is their first choice using the Hare method?

Options	Number of Votes (1000)					
	291	103	143	67	297	99
Research funding	3	3	1	4	4	3
Computer updates	1	2	2	1	2	4
Equality in the workplace	4	4	3	2	3	1
Attracting more women to science	2	1	4	3	1	2

8. Members of the chamber of commerce have been asked to vote on their preference for a topic for the keynote address. The choices are Government regulation, Social justice, Education, and Professional training. Their preferences are summarized in the table. What opinion is their first choice using the Hare method?

	Percentage of Votes					
	30.14%	20.29%	16.8%	8.6%	16.74%	7.43%
Social justice	4	4	4	2	1	1
Education	2	1	3	3	3	4
Government regulation	1	3	2	4	4	2
Professional training	3	2	1	1	2	3

In Problems 9 – 15, determine the winner and the loser using

(a) the plurality method;

(b) the plurality method followed by a runoff between the top two finishers;

(c) the Hare method;

(d) the Coombs method.

9. Determine the winner and the loser in the following election:

Number of Votes (9)					
2	1	2	1	2	1
A	A	B	B	C	D
D	C	D	C	D	A
B	D	C	A	B	C
C	B	A	D	A	B

10. Determine the winner and the loser in the following election:

	Number of Votes (18)						
	2	1	3	1	3	2	6
A	1	1	4	3	4	2	2
B	2	4	1	1	2	3	4
C	3	2	2	4	1	1	3
D	4	3	3	2	3	4	1

11. Determine the winner and the loser in the following election:

Number of Votes (9)					
3	2	1	3	2	2
A	B	A	C	D	E
D	C	E	E	B	D
E	A	B	D	E	B
C	D	C	B	A	C
B	E	D	A	C	A

12. Determine the winner and the loser based on the following set of preference lists

	Number of Votes (11)					
Rank	2	2	1	1	2	3
First	A	E	C	B	B	D
Second	B	B	E	A	D	A
Third	C	D	D	C	A	E
Fourth	D	C	A	E	E	C
Fifth	E	A	B	D	C	B

13. Suppose that a city must decide whether to build a soccer field, a tennis court, a baseball field, a basketball court, or a football field. The residents of the city are polled and their preference rankings are as follows:

	Percentage of Votes					
Option	15%	20%	9%	21%	16%	19%
Soccer field	1	1	3	4	5	3
Tennis court	2	4	1	5	3	4
Baseball field	3	2	5	1	4	5
Basketball court	4	5	2	3	1	2
Football field	5	3	4	2	2	1

Find the winner and the loser.

14. College band directors across the Northeast are trying to set up a showcase to promote marching bands in college to high school students. Four schools are willing to host it next years. They are the University of Massachusetts at Amherst, the University of Rhode Island, the University of New Hampshire, and the University of Connecticut. A questionnaire was sent to high schools in all of the New England area to see where they wanted to go. The 160 high schools replied.

	Number of Votes (160)							
University	23	19	30	22	13	9	27	17
UMass	1	1	2	2	4	3	4	3
URI	2	3	3	4	1	2	3	1
UNH	3	2	4	3	2	1	2	4
UConn	4	4	1	1	3	4	1	2

15. A toothpaste company made a survey where it asked customers to rank new toothpaste flavors. Find the winner and the loser based on the results presented here.

Flavor	Percentage of Votes							
	14%	12%	19%	13%	9%	5%	17%	11%
Lemon Ice	1	2	3	4	1	2	3	4
Sweet Berry Punch	2	3	4	1	2	3	4	1
Tropicana Exotica	3	4	1	2	4	4	1	2
Serpylli	4	1	2	3	3	1	2	3

In Problems 16 – 21, Determine the winner/s and the loser using the Borda count, the Nauru score, and Nanson's and Baldwin's methods.

16. The Rock and Roll Hall of Fame and Museum is a building located on the shores of Lake Erie in downtown Cleveland, Ohio, United States. One day, 55 of the most highly qualified musical critics gathered together to choose the best rock and roll band of all the time. Their preference rankings are summarized in the following table:

Nominee	Number of Votes (55)					
	15	11	14	7	7	1
Led Zeppelin	1	2	3	4	2	3
The Beatles	2	1	4	3	4	2
The Rolling Stones	3	4	1	2	3	4
The Who	4	3	2	1	1	1

17. A class was surveyed for their favorite activities. The options for these activities were cycling, dancing, skiing, and video gaming.

Activity	Number of Voters (41)									
	10	4	1	9	2	5	6	2	1	1
Cycling	1	2	4	2	3	3	4	2	4	1
Dancing	3	4	2	1	4	2	3	4	1	4
Skiing	2	3	1	3	1	4	2	1	3	2
Video gaming	4	1	3	4	2	1	1	3	2	3

18. Use the given preference table to determine the winner of the election.

Preference	Number of Votes (29)						
	4	3	5	2	6	1	8
First	A	A	B	B	C	C	D
Second	B	C	D	A	B	D	A
Third	C	D	A	C	D	B	C
Fourth	D	B	C	D	A	A	B

19. A store is holding an employee of the month survey. Four employees have been nominated: Tessa, Jess, Kim, and Annie. The results of preference rankings are summarized in the following table.

Nominee	Number of Votes (25)											
	2	2	1	4	1	2	3	2	2	1	2	3
Annie	1	1	2	2	3	3	4	4	1	2	2	4
Jess	4	3	3	1	1	2	3	3	4	4	3	1
Kim	2	4	1	4	4	1	2	1	2	3	4	2
Tessa	3	2	4	3	2	4	1	2	3	1	1	3

Who wins the election?

20. Who will win the election based on the following set of preference lists?

	Number of Votes (13)					
Preference	2	2	3	3	1	2
First	A	B	C	D	E	E
Second	B	C	D	E	B	A
Third	C	D	E	B	A	C
Fourth	D	E	A	A	D	B
Fifth	E	A	B	C	C	D

21. Members of a basketball college team ranked their favorite fruits according to the following schedule:

	Number of Votes (21)											
Fruit	2	2	1	1	1	2	3	2	1	1	2	3
Apple	1	1	2	2	3	3	4	4	5	2	1	5
Banana	5	4	3	1	5	2	3	1	4	5	3	1
Mango	3	5	1	4	4	1	5	2	1	3	4	3
Orange	4	2	5	3	2	4	1	5	3	1	5	2
Plum	2	3	4	5	1	5	2	3	2	4	2	4

Which fruit would win the election?

In Problems 22 – 25, find a winner using the modified Borda count based on a 3-2-1 distribution of points. Also determine the number of points each of three best finishers get in this election.

22. In 2009, the MVP football player was determined using the modified Borda method based on 3-2-1 distribution of points. Find the winner and the number of points if the voting breakdowns are as follows.

	First	Second	Third	Total
Tim Tebow, Florida	462	229	113	
Darren McFadden, Arkansas	291	483	120	
Colt Brennan, Hawaii	54	114	242	
Chase Daniel, Missouri	25	84	182	
Dennis Dixon, Oregon	17	31	65	

23. Who is the winner in the following election?

	Number of Voters (27)									
	4	1	3	2	3	3	2	3	4	2
A	1	1	2	3	4	5	5	4	3	2
B	2	5	1	1	3	4	3	2	5	4
C	3	4	5	2	1	1	2	5	4	3
D	4	3	4	5	2	3	1	1	2	5
E	5	2	3	4	5	2	4	3	1	1

24. In November 2010, twenty sport journalists cast their ballots in order to determine a prospective Heisman Trophy winner. The results are as follows.

	First	Second	Third	Forth	Fifth	Total
Toby Gerhart	9	2	3	1	5	
Mark Ingram	6	8	4	0	2	
Ndamukong Suh	7	5	6	2	0	
Colt McCoy	5	4	9	1	1	
Others	4	6	7	2	1	

25. Who is the winner in the following election?

	3	4	2	1	1	3	2	3	2	4
A	1	2	3	4	5	1	2	3	4	5
B	2	4	5	3	1	5	4	5	2	1
C	3	1	2	5	4	2	5	1	3	4
D	4	5	1	2	3	4	3	2	1	3
E	5	3	4	1	2	3	1	4	5	2

Number of Voters (25)

26. An election is held among six candidates (A, B, C, D, E, and F). Using the Copeland method, A gets 5 points, B gets 3 points, C gets 2.5 points, D gets 1.5 points, and E gets 1 point. How many points does F get?

In problems 27 – 32, use the Copeland method to determine the winner. Break ties, if needed, by applying the pairwise comparison, followed by the Condorcet score.

27. Using the given preference table, who wins the election using the Copeland method?

Preference	4	2	7	6	3	4
First	A	B	C	D	A	B
Second	B	C	D	B	D	C
Third	C	D	A	A	C	D
Fourth	D	A	B	C	B	A

Number of Votes (26)

28. Using the given preference table, who wins the election using the Copeland method?

Ranking	8	7	4	3	6	2
First	A	B	C	C	D	D
Second	B	D	D	A	C	B
Third	C	A	B	B	A	A
Fourth	D	C	A	D	B	C

Number of Votes (30)

29. Use the given preference table to determine who wins the election using the Copeland method.

Option	3	26	22	1	14	4
A	1	3	2	1	2	1
B	3	1	4	2	3	4
C	2	4	1	3	4	2
D	4	2	3	4	1	3

Number of Votes (70)

30. Use the given preference table to determine who wins the election using the Copeland method.

Alternative	Number of Votes (20)					
	1	7	3	4	3	2
A	1	2	3	4	1	2
B	3	4	1	2	3	1
C	2	3	4	1	2	4
D	4	1	2	3	4	3

31. By the end of the season, a selected panel comes together to look over the players in the NBA to decide which of them deserves the title Most Valuable Player (MVP). Together they agreed on the following five players: Kobe Bryan, Dwayne Wayne, Lebrun James, Rajah Rondo, and Steve Nash. Which one of these people will be nominated based on the following preference schedule?

Option	Number of Voters (35)							
	3	4	5	2	6	5	7	3
Bryan	1	2	3	4	5	1	4	3
James	2	5	4	1	4	5	1	2
Nash	3	4	5	2	1	2	5	1
Rondo	4	1	2	3	2	3	2	5
Wayne	5	3	1	5	3	4	3	4

32. A group of fifteen people are debating on what genre of music is the best. After narrowing the choice to cardas, tango, fado, jazz, punk, and swing, they decided to vote on this issue using preference ballots. The results are summarized in the following table. Which genre is the best using the Copeland method?

Genre	Number of Voters (15)						
	1	4	3	2	1	3	1
Cardas	1	2	6	4	6	6	1
Tango	2	1	2	3	4	3	2
Fado	3	6	5	6	1	4	5
Jazz	4	3	4	5	2	1	6
Punk	5	5	3	1	5	5	3
Swing	6	4	1	2	3	2	4

33. Ten board members vote by approval voting on eight candidates for new positions on their board as indicated in the following table. A $\sqrt{}$ indicates an approval vote.

Candidates	Voters									
	1	2	3	4	5	6	7	8	9	10
A	√		√		√	√	√	√		√
B		√	√	√		√	√	√	√	
C			√					√		
D	√	√		√	√		√	√	√	√
E	√		√		√		√		√	
F	√		√	√	√	√	√			√
G	√	√	√	√	√			√		
H		√		√		√		√		√

(a) Which candidate is chosen for the board if just one of them is to be elected?

(b) Which candidates are chosen if the top four are selected?

34. Twelve members of a board of directors must select a new head forthe local orchestra. Who will be nominated based on the following approval table?

	Number of Voters						
	1	2	4	1	2	1	1
Cortese	√		√		√		√
Davis		√		√		√	√
Malina	√	√	√		√	√	√
Zinman	√		√		√		

35. A professional society wants to hold its next annual convention in one of the following cities: New Orleans, Las Vegas, San Francisco, or New York. The executive council, consisting of 12 members, decides to make the decision by approval ballot, and the results are listed in the table. Which city wins the vote?

City	Number of Voters (16)						
	5	1	4	2	2	1	1
New Orleans	√			√	√		√
Las Vegas		√		√			√
San Francisco	√				√	√	√
New York	√		√		√		

36. A family of seven is planning to buy a dog. They have narrowed the choices of breed down to six choices. The results of approval voting are presented here. Which breed would be approved?

Breed	Family Members						
	1	1	1	1	1	1	1
Affenpinscher		√	√	√	√		
Bergamasco	√	√			√	√	
Cavachon		√	√		√		
Doxiepoo		√			√		√
Enga-Apso	√		√	√	√	√	
Foodle		√	√	√	√		

37. A club decided to use an approval vote in the election for president of the club. The five candidates for the position were Agu, Brown, Chernova, Finkel, and Gennert. Who won the election if their votes were summarized in the following table?

	Number of Voters (15)									
	1	2	1	1	2	1	1	3	1	2
Agu		√			√	√		√	√	
Brown		√	√		√			√		
Chernova	√				√	√	√	√		√
Finkel			√				√		√	
Gennert	√				√	√		√		

38. The members of a marching band vote to choose what time they would like to gather for an evening rehearsal. The choices were 4 p.m., 5 p.m., 6 p.m., or 7 p.m. The results of their votes are summarized in the following table. Which option would be chosen?

	Number of Voters (21)								
	1	2	3	4	2	1	2	3	1	2
4 p.m.	√	√	√		√	√		√		
5 p.m.	√		√					√		
6 p.m.		√				√	√	√	√	√
7 p.m.						√		√		

39. A student's committee has narrowed the location of their college reunion down to: Denver, New York, and Providence. Use the Hare method to determine the winner of the election using the preference schedule. Is the independence–of–irrelevant–alternatives criterion satisfied?

City	Number of Votes (25)			
	13	10	5	7
Denver	3	1	3	2
New York	2	3	1	1
Providence	1	2	2	3

40. Consider the following five preference ballots:

Preference					
First	D	D	D	A	A
Second	A	C	B	B	C
Third	C	A	A	C	B
Fourth	B	B	C	D	D

Who is the winner of this election using the Borda count method? Does this election satisfy the majority criterion?

41. In a competition for the best comedy movie, only four 20-minute fragments from the movies labeled by A, E, N, and S, were shown to the committee's members in order to avoid any bias in judging. Use the preference schedule to determine the winner using the Borda count method. Is Condorcet's criterion satisfied in this election?

Preference	Number of Votes (19)				
	3	4	7	1	4
First	E	S	E	A	N
Second	N	A	N	E	S
Third	S	N	A	N	A
Fourth	A	E	S	S	E

42. Use the preference table to determine the winner using the Borda count method and the Nauru system. Is the independence–of–irrelevant–alternatives criterion satisfied?

Option	Number of Votes (44)					
	14	4	7	10	5	4
A	1	3	4	2	3	2
B	3	2	2	4	1	1
C	2	1	3	1	4	4
D	4	4	1	3	2	3

43. The girl's soccer team, coaches, and a select number of fans from a university were asked to vote on their favorite player for the season to be named the most valuable player (MVP) for the year. The four candidates were Allie Johnson, Samantha Smith, Carrie Hauser, and Liz Gorman. The preference rankings are presented here.

	Number of Votes (60)					
Player	11	6	16	13	10	4
Johnson	1	3	3	4	3	3
Smith	4	1	2	1	2	4
Hauser	3	4	1	3	4	1
Gorman	2	2	4	2	1	2

(a) Find the Coombs winner in this election.

(b) Before the committee announces the winner, they find out that Carrie Hauser had been using steroids during the soccer season and therefore should be disqualified from the competition. Find the Coombs winner after taking her out of the election.

(c) The results of parts (a) and (b) show that the Coombs method violates one of the fairness criteria discussed in §1.6. Which one?

44. Use the preference schedule to determine the winner using the Borda count method. Is the Condorcet criterion satisfied in this election?

	Percentage					
Preference	18.47%	12.75%	17.28%	21.34%	10.18%	19.98%
First	A	B	C	D	A	B
Second	D	A	D	C	B	C
Third	C	C	A	B	D	D
Fourth	B	D	B	A	C	A

45. Use the Hare method to determine the winner of the election. Is the independence–of–irrelevant–alternatives criterion satisfied?

	Number of Votes (127)				
Preference	25	61	26	11	4
First	A	B	C	D	D
Second	D	A	D	A	C
Third	B	D	A	C	B
Fourth	C	C	B	B	A

46. Use the following ballots to show that the Hare method violates the monotonicity criterion.

	Number of Votes (23)						
Preference	3	2	8	4	2	2	2
First	A	A	B	C	D	D	C
Second	D	B	D	A	A	C	A
Third	C	D	C	D	B	A	D
Fourth	B	C	A	B	C	B	B

47. Use the following ballots to show that the IRV does not satisfy the monotonicity criterion.

	Number of Voters (36)			
Ranking	5	12	8	11
First choice	A	B	A	D
Second choice	B	C	D	C
Third choice	C	D	C	B
Fourth choice	D	A	B	A

48. Use the following ballots to show that the IRV does not satisfy the monotonicity criterion.

Ranking	Number of Voters (15)				
	4	5	3	2	1
First choice	A	B	C	D	E
Second choice	B	C	D	E	C
Third choice	C	A	A	C	D
Fourth choice	D	E	B	A	B
Fifth choice	E	D	E	B	A

49. Use the preference table to determine the winner using the Nauru count. Is the independence–of–irrelevant–alternatives criterion satisfied?

Preference	Number of Votes (54)					
	8	16	12	6	5	7
First	A	B	C	A	B	D
Second	D	C	D	D	A	C
Third	C	D	A	B	C	B
Fourth	B	A	B	C	D	A

50. Use the following preference table to show that Nanson's method violates the IIA criterion.

Option	Number of Votes (31)							
	5	6	2	4	7	3	3	1
First	A	D	A	C	B	C	B	A
Second	C	C	D	D	D	B	C	D
Third	B	A	B	A	A	A	D	C
Fourth	D	B	C	B	C	D	A	B

51. Use the following preference table to show that Nanson's method violates the IIA criterion.

Option	Percentage					
	4%	21%	15%	18%	12%	30%
A	1	2	3	5	4	1
B	2	1	5	2	3	4
C	3	5	4	1	2	3
D	4	4	2	3	1	5
E	5	3	1	4	5	2

52. As a publicity stunt for its soon-to-be-published cookbook, the Culinary Club of Smallville decided to have a *Best Pie* contest. The entries were narrowed down to four pies for the final round of the contest. Here is a summary of preference rankings:

Option	Percentage					
	16%	17%	19%	18%	16%	14%
Apple Pie	1	4	3	3	2	1
Chocolate Pie	2	1	4	4	1	2
Carrot Pie	3	2	1	2	3	4
Peach Pie	4	3	2	1	4	3

Which pie wins the election under Baldwin's method? But wait! The chairman of the election overlooks how the committee destroyed the ballots before the results had been certified by the administration. You guessed it. The survey had to be repeated. In the new survey, everyone voted exactly as the original survey except for 6 voters in the fifth column who decided to jump on the bandwagon and vote for Apple Pie instead of Chocolate Pie. Does the new survey support the original winner?

53. There was a dispute between 19 friends about the most popular foreign car on the market. They decided to determine the most popular car by voting and determine the winner using either the Borda count method or the Nauru score system. After narrowing down the cars' makers in the contest to be Ferrari, Mercedes, Lamborghini, and Bentley, they voted. The results are summarized in the following preference table.

Option	Number of Votes (19)					
	4	3	2	4	5	1
Ferrari	4	3	1	2	1	4
Mercedes	1	1	2	4	4	2
Lamborghini	2	2	3	3	2	3
Bentley	3	4	4	1	3	1

Which car is the most popular? If some ballots containing Mercedes were spoiled, and a new voting was conducted without Mercedes, which car wins the new election?

54. Use the Borda count method and the following preference table to show that the majority winner is not the Borda winner.

Alternative	Number of Votes (29)				
	8	4	10	5	2
A	2	3	2	4	1
B	1	2	3	2	2
C	3	4	4	1	4
D	4	1	1	3	3

55. In the preference table, alternative A has the majority of first-place votes. Who wins the election if we use the Borda count?

Ranking	Number of Votes (25)			
	6	9	3	7
First choice	A	B	C	A
Second choice	B	C	B	D
Third choice	C	A	D	B
Fourth choice	D	D	A	C

56. During the regular 2009/2010 season, the following four soccer teams in the English Premier League were nominated for the *most popular team:* Liverpool, Chelsea, Arsenal, and Everton. The results of votes are summarized in the following table.

Team	Percentage					
	6%	9%	15%	18%	25%	27%
Arsenal	1	3	1	2	4	2
Liverpool	2	4	3	4	2	1
Everton	3	2	4	3	1	3
Chelsea	4	1	2	1	3	4

Just before announcing the winner, it was found that some Chelsea players were involved in inappropriate behavior and this team was dropped from consideration. Who wins the new election under the IRV method? Which fairness criterion is violated?

57. Consider the election given by the following preference schedule.

	Number of Votes (13)			
Option	5	3	2	3
A	1	3	3	3
B	2	1	1	4
C	3	4	2	2
D	4	2	4	1

(a) Find the Condorcet winner in this election.

(b) Find the winner of this election under the Borda count method.

(c) Suppose that D drops out of the race. Find the winner under the Borda count method when D is removed from the preference schedule.

(d) The results of (a), (b), and (c) show that the Borda count method violates several of the fairness criteria discussed in §1.6. Which ones?

58. Consider the election given by the following preference schedule.

	Number of Votes (22)				
Preference	9	5	4	3	1
First	A	B	B	C	D
Second	C	D	C	A	C
Third	B	C	A	D	B
Fourth	D	A	D	B	A

(a) Find the Condorcet winner in this election.

(b) Find the winner of this election under the Hare method.

(c) Suppose that D drops out of the race. Find the winner under the Hare method when D is removed from the preference schedule.

(d) The results of (a), (b), and (c) show that the Hare method violates several of the fairness criteria discussed in §1.6. Which ones?

59. The results in the central division of NHL (National Hockey League) are presented below

Team	Chicago	Nashville	Detroit	St Louis	Columbus	**Total**
Chicago Blackhawks	–	1	$\frac{1}{2}$	0	1	$2\frac{1}{2}$
Nashville Predators	0	–	1	1	0	2
Detroit Red Wings	$\frac{1}{2}$	0	–	$\frac{1}{2}$	1	2
St Louis Blues	1	0	$\frac{1}{2}$	–	$\frac{1}{2}$	2
Columbus Blue Jackets	0	1	0	$\frac{1}{2}$	–	$1\frac{1}{2}$

Use this table to show that the Copeland method violates the IIA criterion.

60. Some results in the National Basketball Association between four teams are presented in the following table.

Team	Cleveland	Denver	LA Lakers	Orlando	**Total wins**
Cleveland	–	94:103	114:116	99:89	2
Denver	90:103	–	100:103	102:112	2
LA Lakers	100:75	86:99	–	104:108	4
Orlando	101:96	120:101	75:100	–	4

Since every team played two games with every other team—one game at home and another one away—the table contains the results of 12 matchups. Use this table to show that the Copeland method violates the IIA criterion.

61. Use the following preference schedule to show that Nanson's method violates the IIA criterion.

Ranking	Number of Votes (31)							
	5	6	2	4	7	3	3	1
First	B	D	A	C	B	C	A	A
Second	A	C	D	B	D	B	C	D
Third	C	A	B	A	A	D	B	C
Fourth	D	B	C	D	C	A	D	B

In Problems 62 – 64, rank the candidates/options using

(a) the extended plurality method;

(b) the extended Hare method;

(c) the extended Coombs method;

(d) the extended Borda count method;

(e) the extended Nauru score system;

(f) the extended Copeland method (when there is a tie between two competitors, use their pairwise matchup to determine the winner followed by the Condorcet score, if needed);

(g) extended approval voting (when the first two choices are considered to be approved).

After that, rank candidates using the recursive seven methods.

62. Rank options in the election.

Option	Number of Votes (22)					
	6	5	3	3	4	1
A	1	4	2	3	4	2
B	2	1	3	4	3	4
C	4	3	1	1	2	3
D	3	2	4	2	1	1

63. Rank candidates in the election.

Ranking	Number of Votes (17)							
	2	3	1	4	3	1	2	1
First choice	A	A	B	B	C	C	D	D
Second choice	B	D	C	D	A	B	A	B
Third choice	D	C	A	C	D	A	B	C
Fourth choice	C	B	D	A	B	D	C	A

64. Rank candidates in the election.

Candidate	Percentage					
	13%	18%	21%	12%	26%	10%
A	1	4	2	1	3	4
B	3	2	1	4	2	3
C	4	1	3	2	4	1
D	2	3	4	3	1	2

References and Further Reading

[1] Abramowitz, Alan I., *Voice of the People: Elections and Voting in the United States*, McGraw-Hill Humanities/Social Sciences/Languages, New York, 2003.

[2] Arrow, Kenneth J., *Social Choice and Individual Values*, Yale University Press; 2nd edition, New Haven, 1970.

[3] Borda, J. C. de, *Mémoire sur les élections au Scrutin*, Histoire de l'Académie Royale des Sciences, Paris, 1781.

[4] Börgers, Christoph, *Mathematics of Social Choice*, SIAM, Philadelphia, 2010.

[5] Brams, Steven J., *Mathematics and Democracy: Designing Better Voting and Fair-Division Procedures*, Princeton University Press, Princeton, 2007.

[6] Brams, Steven J., *Approval Voting*, Springer; 2nd edition, New York, 2007.

[7] Brams, Steven J., *The Presidential Election Game*, A K Peters, Ltd.; 2nd edition, Wellesley, MA, 2007.

[8] Brams, Steven J., Gehrlein, William V., Roberts, Fred S., *The Mathematics of Preference, Choice and Order: Essays in Honor of Peter C. Fishburn*, Springer, New York, 2008.

[9] Caplan, Bryan, *The Myth of the Rational Voter: Why Democracies Choose Bad Policies*, Princeton University Press, Princeton, 2008.

[10] Chamberlin, John R. and Cohen, Michael D., Toward Applicable Social Choice Theory: A Comparison of Social Choice Functions Under Spatial Model Assumptions, *American Political Science Review*, **72**, No. 4, 1341 – 1356, 1978.

[11] Condorcet, Marquis de, *Essai sur l'application à la possibilitè des dècisions rendues à la pluralitè de voix*, Imprimerie Royale, Paris, 1785.

[12] Cool, Thomas, *Voting Theory for Democracy*, Thomas Cool Consultancy & Econometrics, 2007.

[13] Copeland, A.H., *A 'reasonable' social welfare function*, Notes from a seminar on applications of mathematics to the social sciences. University of Michigan, 1951.

[14] Coughlin, Peter J., *Probabilistic Voting Theory*, Cambridge University Press, Cambridge, 1992.

[15] Fishburn, Peter, Paradoxes of Voting, *American Political Science Review*, **68**, 537 – 546, 1974.

[16] Geanakoplos, John, Three Brief Proofs of Arrow's Impossibility Theorem, *Economic Theory*, **26**, 211 – 215, 2005.

[17] Hodge, Jonathan K. and Kilma, Richard E., *The Mathematics of Voting and Elections: A Hands-On Approach*, American Mathematical Society, Providence, 2005.

[18] Lipschutz, Seymour, Schiller, John J., Srinivasan, R. Alu, *Schaum's Outline of Beginning Finite Mathematics*, McGraw-Hill, New York, 2004.

[19] May, Kenneth, A Set of Independent Necessary and Sufficient Conditions for Simple Majority Decision, *Econometrica*, **20**, 680 – 684, 1952.

[20] Merlin, V.R. and Saari, D.G. Copeland Method. II. Manipulation, Monotonicity, and Paradoxes, *Journal of Economic Theory*, **72**, No. 1; January, 148 – 172, 1997.

[21] Mueller, Dennis C., *Public Choice III*, Cambridge University Press; 3 edition, Cambridge, 2003.

[22] Nurmi, Hannu, *Voting Procedures under Uncertainty*, Springer, New York, 2002.

[23] Poundstone, William, *Gaming the Vote: Why Elections Aren't Fair (and What We Can Do About It)*, Hill and Wang, New York, 2009.

[24] Saari, Donald G., *Chaotic Elections! A Mathematician Looks at Voting*, American Mathematical Society, Providence, 2001.

[25] Saari, Donald G., *Decisions and Elections: Explaining the Unexpected*, Cambridge University Press, Cambridge, 2001.

[26] Saari, Donald G., *Basic Geometry of Voting*, Springer, New York, 2003.

[27] Saari, Donald G., *Disposing Dictators, Demystifying Voting Paradoxes: Social Choice Analysis*, Cambridge University Press, Cambridge, 2008.

[28] Saari, Donald G. and Merlin, V.R. The Copeland Method. I. Relationships and the Dictionary, *Economic Theory*, **8**, No. l; June, 51 – 76, 1996.

[29] Schofield, Norman, General Instability of Majority Rule, *Review of Economic Studies*, **50**, 695 – 705, 1983.

[30] Simeone, Bruno and Pukelsheim, Friederich (editors), *Mathematics and Democracy*, Springer, New York, 2006.

[31] Straffin, Jr. Philip D. *Topics in the Theory of Voting*, Birkhuser, Boston, 1980.

[32] Taylor, Alan, *Social Choice and the Mathematics of Manipulation*, Cambridge University Press, Cambridge, 2005.

[33] Taylor, Alan and Pacelli, Allison M., *Mathematics and Politics: Strategy, Voting, Power, and Proof*, Springer; 2nd edition, New York, 2008.

[34] Tideman, T. Nicolaus, Independence of Clones as a Criterion for Voting Rules, *Social Choice and Welfare* 4, No. 3, 185 – 206, 1987.

[35] Young, H. P. and Levenglick, A., A Consistent Extension of Condorcet's Election Principle, *SIAM Journal on Applied Mathematics*, **35**, No. 2, 285 – 300, 1978.

[36] Young, H. P., Condorcet's Theory of Voting, *American Political Science Review*, **82**, no. 2, 1231 – 1244, 1988.

The Internet Sites

http://aceproject.org/
*ACE Electoral Knowledge Network Expert site providing encyclopedia on electoral systems and management, country by country data, a library of electoral materials, latest election news, the oportunity to submit questions to a network of electoral experts, and a forum to discuss all of the above
http://www.idea.int/publications/esd/index.cfm
*A handbook of electoral system design from international IDEA
http://wiki.electorama.com/wiki/Election-methods_mailing_list
*A mailing list for technical discussions about election methods.
http://en.allexperts.com/e/v/vo/voting_system.htm
*Encyclopedia of voting systems.
http://theorem.ca/ mvcorks/code/voting_methods.html
*Evaluating voting methods by Matt Corks
http://www.dmoz.org/Society/Politics/Campaigns_and_Elections/Voting_Systems/
*Open directory project category on voting systems
http://www.openstv.org/
*OpenSTV – Software for computing a variety of voting systems including IRV, STV, and Condorcet.
http://www.maa.org/editorial/knot/LiberalArts.html
*Student's social choice by Alex Bogomolny. Illustrates various concepts of choice using Java applets.
http://xaravve.trentu.ca/pivato/Teaching/voting.pdf
*Voting, arbitration, and fair division (pdf) by Marcus Pivato.
http://fc.antioch.edu/ james_green-armytage/voting.htm
*Voting methods: tutorial and essays by James Green-Armytage
http://pj.freefaculty.org/Ukraine/PJ3_VotingSystemsEssay.pdf
*Voting systems (pdf) by Paul E. Johnson. A textbook-style overview of voting methods and their mathematical properties.

Chapter 2

Apportionment Problems

The roots of a fair division are hard to trace—we have observed division problems from ancient times. One of the best known examples gives a biblical story about King Solomon who needed to decide how to divide a baby between two women claiming to be the mother of this same child. Serious analysis of these problems began in the 18th century when the founding fathers of the United States faced the problem of the allocation of seats between states in the House of Representatives. This history is fascinating and involves many colorful and powerful figures. The United States is a very unusual country in having a legislative branch with two independent parts, the Senate and the House of Representatives. Its origin was a compromise between small states and large states: the Senate has two senators from each state, and the House comprises representatives from states in accordance with their population. Who would have thought that the U.S. Constitution would be the source of so much work for mathematicians!

There are two kinds of division problems. One of them involves the allocation of items of equal value between members of a group. It could be distribution of seats in the House between states or allocation of fair trucks between different districts. Such problems are usually called apportionment problems. The term apportionment stems from the Latin word *portio* (share) that means distribution or allotment in proper shares. Apportionment problems involve distribution and assignment of things between members of a group on a proportional basis.

Another kind of division problem arises when we need to divide items of unequal value. A typical problem of this type is an inheritance problem or a divorce problem. For example, how to divide a house, a boat, and a car between three children inheriting these items from their parents? This kind of problem is usually referred to as a fair division problem.

These two kinds of division problems involve different mathematical tools and approaches. This chapter deals with the mathematical aspects of apportionment problems, originating from the allocation of seats in the House of Representatives. Fair division problems are considered in other books (see, for instance, [9]).

2.1   Introduction

The objective of this section is to give motivational examples of apportionment problems, outline the terminology that will be used in the chapter, and discuss problems of division, including paradoxes.

We will utilize terminology stemming from the U.S. government, such as the House of Representatives, because apportionment problems had originated as a result of allocating a fair number of seats between U.S. states. Consequently, we will label the items to distribute as **seats** and the total number of items as the **house size**. Therefore, the number of representatives to be allocated is called the house size. The various members or parties to whom we will assign the items are referred to as **states**. A state's **population** is the measurement of its size on which allocation of seats will be made. The sum of the states' populations is called the **total population**.

In allocating a group of representatives among several states, the **natural divisor** (also called **standard divisor**) is defined as

$$\text{Natural (Standard) Divisor} = \frac{\text{total population}}{\text{house size}}, \tag{2.1.1}$$

and a state's **natural quota** (also called **standard quota**) as

$$\text{Natural Quota} = \frac{\text{state's population}}{\text{natural divisor}} = \frac{\text{state's population}}{\text{total population}} \times \text{house size}. \tag{2.1.2}$$

The natural quota for a state is the exact fair share that would be allowed (if a whole number was not required). In practical applications of the above formula, the natural divisor is usually not an integer, but a rational number that has infinite decimal representation. Since we cannot use such exact values and operate with strings of infinitely many digits, such numbers are truncated (all digits in their decimal representations are eliminated except the first few) resulting in a rounding error. Dividing a state's population by the truncated natural divisor will also provide an approximation to the exact value of the natural quota. Computers and calculators use very sophisticated algorithms to perform arithmetic operations with great accuracy. Let us allow them to do their job without human interference. Therefore, we strongly recommend using whole numbers (state's population, house size, and total population) for numerical calculations in Eq. (2.1.2) instead of using a truncated value for the natural divisor.

A state should receive its fair share but, most of the time, the fair share is not an integer. Using whole numbers instead of fractions so that their sum keeps the same value is called an **apportionment problem** that has concerned many mathematicians and noted statesmen because of both its complexity and its profound political consequences. The rule(s) for rounding summands so that their sum is maintained at its original value is called an **apportionment method**. Apportionment methods are used to round the population proportions of electoral districts (or the vote proportions of parties) to integer numbers of seats in a representative body. The effect of the rounding process is a central issue in the theory of apportionment.

Example 2.1.1. In the Boston Bruins 2008/2009 regular hockey season, the team won the Northeast Division with 53 wins, 19 regulation losses, and 10 overtime losses. The Bruins' coach wanted to present these results to the audience in percentages using only whole numbers. Initially he made the following table:

Games	Numbers	Percentage
Games won	53	$\frac{53}{82} \times 100\% \approx 64.63\%$
Games lost	19	$\frac{19}{82} \times 100\% \approx 23.17\%$
Overtime game's lost	10	$\frac{10}{82} \times 100\% \approx 12.19\%$
Total games played	82	100%

In this table, the expression $\frac{53}{82} \times 100\% \approx 64.63\%$ means that the standard quota of the "state" *Games won* is about 64.63%, the population of this "state" is 53, with the total population of 82, and the house size 100.

Since the coach's objective is to round numbers in the last column to whole numbers, he came to the problem of allocation. Hence, there are three states with populations of 53, 19, and 10, respectively, and the house size is 100. Rounding numbers in a conventional way (if the fractional part is 0.5 or above round up, if the fractional part is below 0.5 round down), the manager reported that his team won 65% of the matchups, lost 23%, and lost 12% of overtime games.

Does it work? Let us look at the Bruins' performance next season in 2009/2010. The team won 39 games, lost 30, and lost in overtime 13 matchups. Making a similar table, the coach would get

Games won: $\frac{39}{82} \times 100\% \approx 47.56\%$,

Games lost: $\frac{30}{82} \times 100\% \approx 36.58\%$,

Overtime game lost: $\frac{13}{82} \times 100\% \approx 15.85\%$.

After rounding these numbers to the nearest integer, we claim that Boston Bruins won 48% of the games, lost 37% of the matchups, and lost in overtime 16%, making the total of 101%. This is not right! Therefore, just rounding numbers in a conventional way can lead to a wrong allocation. □

The number of seats of a state should (in any fair distribution) be proportional to its ratio in the population. As we see from the previous example, we need a rounding procedure to obtain the required allocation. In discrete mathematics, there are two kinds of rounding: rounding up to the nearest integer, called the **ceiling**, and rounding down, called the **floor**. They are reflected by two special notations (which we apply, for example, to $\pi = 3.1415926\ldots$):

$$\lfloor \pi \rfloor = \lfloor 3.1415926\ldots \rfloor = 3 \qquad \text{(floor)},$$
$$\lceil \pi \rceil = \lceil 3.1415926\ldots \rceil = 4 \qquad \text{(ceiling)}.$$

Recall that the irrational number $\pi = 3.1415926\ldots$ has 3 as the integer part and $0.1415926\ldots$ as the fractional part. Note that floor and ceiling of a whole number give the same value. For instance, $\lfloor 4 \rfloor = \lceil 4 \rceil = 4$. When these notations are applied to apportionment, we use the term quota instead of floor or ceiling:

The floor, $\lfloor q \rfloor$, of a real number q is the largest integer smaller than or equal to q. If the natural quota is rounded down to the nearest integer, we call this number the **lower quota**. The ceiling, $\lceil q \rceil$, of a real number q is the next integer larger than or equal to q. If the natural quota is rounded up to the nearest integer, this number is called the **upper quota**.

When comparing states' representations in the House, we need a clear measure of how fairly these states are apportioned. There are known two such measures, and both of them are presented below. The **constituency size** (also called average constituency or district population) of a state is the quotient:

$$\textbf{Average constituency} = \frac{\text{population of the state}}{\text{number of representatives from the state}}, \qquad (2.1.3)$$

if the denominator is not zero. When the number of a state's representatives is zero, we assign ∞ to be its constituency size. Its reciprocal is called the **representative share**:

$$\textbf{Representative share} = \frac{\text{number of representatives from the state}}{\text{population of the state}}. \qquad (2.1.4)$$

So it represents the share of a congressional seat given to each citizen of the state. On the other hand, the constituency size of a state is defined to be the number of persons per representative in this state. Comparing the representations of two states A and B, we say that state A is **more poorly represented** than state B if the constituency size of A is greater than B's constituency size. In other words, state A is more poorly represented than state B if its representative share is less than the representative share of state B. For example, suppose state A has a population of 1,000,000 people and six representatives and state B has 600,000 citizens and four representatives. Each of state A representatives have the average of $\frac{1000000}{6} \approx 166,666.6\overline{6}$ constituents, whereas each representative from state B averages $\frac{600000}{4} = 150,000$ constituents. (As usual, we write a line over repeated integers.) Their reciprocals provide the values of representative shares: state A has 6×10^{-6} seats or 6 microseats (a microseat is one-millionth of a seat), whereas state B has $6.\overline{6}$ microseats. Because a representative from state A serves more constituents than a representative from state B, or because its representative share is less than one of state B, it is fair to say that state A is more poorly represented in the House than state B.

Example 2.1.2. Three campus organizations are to contribute members to the student government based on their relative size. There are 7 seats to be distributed among the three organizations. The number of students in each organization are **A**: 15, **B**: 30, and **C**: 45, with total of 90. Some calculations are made that can be summarized in the following table.

Campus Organization	Population	Natural Quota	Natural Quota Approximation	Lower Quota	Upper Quota
A	15	$\frac{15}{90} \times 7$	$1.16666\overline{6}$	1	2
B	30	$\frac{30}{90} \times 7$	$2.33333\overline{3}$	2	3
C	45	$\frac{45}{90} \times 7$	3.5	3	4
Total:	90			6	9

Here $\overline{6}$ is an abbreviation for repetition of the digit 6 in the decimal representation. So we see that lower quotas underestimate the required allocation, and the upper quotas overestimate it. Let us consider their constituency sizes to identify which state deserves either adding a seat or eliminating a seat. Calculations show that for lower quotas, we have the following constituency sizes that are obtained by dividing their lower quotas into their populations:

$$\mathbf{A} \rightsquigarrow \frac{15}{1} = 15, \qquad \mathbf{B} \rightsquigarrow \frac{30}{2} = 15, \qquad \mathbf{C} \rightsquigarrow \frac{45}{3} = 15 \qquad \text{(lower quotas)}.$$

Hence, every campus organization is equally represented, but lower quotas do not provide the required allocation of seven members. What should we do? It seems that we have to take a step back from the fair distribution and promote one of the organizations with a seat. This means that we cannot fairly allocate seats between organizations and preserve the average constituencies—at least one campus organization would be more poorly represented.

Let us look at upper quotas that lead to the following constituency sizes (that are obtained by dividing their upper quotas into their populations):

$$\mathbf{A} \rightsquigarrow \frac{15}{2} = 7.5, \qquad \mathbf{B} \rightsquigarrow \frac{30}{3} = 10, \qquad \mathbf{C} \rightsquigarrow \frac{45}{4} = 11.25 \qquad \text{(upper quotas)}.$$

Since organization **C** is more poorly represented than any of other groups, it make sense to give a seat to organization **C**; but this would lead to even larger number of seats. Thus, we are forced to eliminate a seat from organization **A**, which gives us the following average constituencies:

$$\mathbf{A} \rightsquigarrow \frac{15}{1} = 15, \qquad \mathbf{B} \rightsquigarrow \frac{30}{3} = 10, \qquad \mathbf{C} \rightsquigarrow \frac{45}{4} = 11.25.$$

However, such seat distribution, $1 + 3 + 4 = 8$, has one extra seat. Dropping one seat from campus organization **B** (because **B** has the lowest constituency size), we get

$$\mathbf{A} \rightsquigarrow \frac{15}{1} = 15, \qquad \mathbf{B} \rightsquigarrow \frac{30}{2} = 15, \qquad \mathbf{C} \rightsquigarrow \frac{45}{4} = 11.25.$$

Is such a distribution of seats optimal? The answer is positive because it leads to the lowest discrepancies among the average constituencies. However, do we have a method to use?

Most likely not because we have to check all possible allocations of seats between states. A similar analysis could be done using representative shares.

Now consider another example with a different distribution of students:

Campus Organization	Population	Natural Quota	Lower Quota	Upper Quota
A	15	$1.16666\overline{6}$	1	2
B	16	$1.24444\overline{4}$	1	2
C	59	$4.58888\overline{8}$	4	5
Total:	90		6	9

The constituency sizes for lower quotas would be as follows

$$\mathbf{A} \rightsquigarrow \frac{15}{1} = 15, \qquad \mathbf{B} \rightsquigarrow \frac{16}{1} = 16, \qquad \mathbf{C} \rightsquigarrow \frac{59}{4} = 14.75 \qquad \text{(lower quota)}.$$

The total of lower quotas is less than the required number of the student government body by 1. Suppose we want to give one seat to one of the groups, preserving the total of 7. By giving one seat to campus organization **B**, which has the highest constituency size, we come to the following constituency sizes:

$$\mathbf{A} \rightsquigarrow \frac{15}{1} = 15, \qquad \mathbf{B} \rightsquigarrow \frac{16}{2} = 8, \qquad \mathbf{C} \rightsquigarrow \frac{59}{4} = 14.75 \qquad \text{(lower quota)}.$$

For upper quotas, we have

$$\mathbf{A} \rightsquigarrow \frac{15}{2} = 7.5, \qquad \mathbf{B} \rightsquigarrow \frac{16}{2} = 8, \qquad \mathbf{C} \rightsquigarrow \frac{59}{5} = 11.8 \qquad \text{(upper quotas)}.$$

Since upper quotas provide two extra seats, we need to eliminate them from our allocation. So, we drop one seat from campus groups **A** and **B** to obtain the following average constituencies:

$$\mathbf{A} \rightsquigarrow \frac{15}{1} = 15, \qquad \mathbf{B} \rightsquigarrow \frac{16}{1} = 16, \qquad \mathbf{C} \rightsquigarrow \frac{59}{5} = 11.8 \qquad \text{(upper quotas)}.$$

Is this apportionment better than the one we get from lower quotas? It is hard to judge. We need rules to identify the "best" allocation. This is our goal for the following sections.

□

The problem can be abstracted as follows. Let n be the number of states. The populations of these states will be denoted by $p_1, p_2, p_3, \ldots, p_n$. Thus, the total population is $p = p_1 + p_2 + p_3 + \ldots + p_n$. Let

- h be the size of the house. This is the number of seats to be distributed among the states.

- $q_i = p_i * h/p$ be the natural (standard) quota for state i, with population p_i, which is the exact fair share that would be allowed.

- **ND** be the natural (standard) divisor, the size of an average population that should correspond to one seat: $\mathbf{ND} = \dfrac{\text{total population}}{\text{house size}} = \dfrac{p}{h}$. It is a unit of measurement ($\mathbf{ND}$ people correspond to 1 seat) required for allocation of seats.

The apportionment problem consists of rounding natural quotas q_i, $i = 1, 2, \ldots, n$, to whole numbers in such a way that their rounded values, $\tilde{q}_i$, sum to the house size:

$$q_1 + q_2 + \cdots + q_n = \tilde{q}_1 + \tilde{q}_2 + \cdots + \tilde{q}_n = \text{house size}.$$

The population of each state is proportional to the natural quota:

$$p_i = q_i\,(\mathbf{ND}) = q_i\,\frac{p}{h}. \tag{2.1.5}$$

Example 2.1.3. Which State is More Poorly Represented

According to U.S. Bureau of the Census estimates, in 2009, the state of Rhode Island had a population of 1,053,209 with 2 representatives allocated in the House, and Massachusetts had a population of 6,593,587 and was allocated 10 representatives. The constituency size of Rhode Island was

$$\frac{\text{population of Rhode Island}}{\text{number of representatives allocated to Rhode Island}} = \frac{1,053,209}{2} = 526604.5,$$

whereas the constituency size of Massachusetts was

$$\frac{\text{population of Massachusetts}}{\text{number of representatives allocated to Massachusetts}} = \frac{6,593,587}{10} = 659358.7.$$

The representative share of Rhode Island was $1.898958326 \times 10^{-6}$ or 1.8989 microseats (a microseat is one-millionth of a seat), while the representative share of Massachusetts was 1.5166 microseats. Therefore, Massachusetts is more poorly represented in the U.S. House of Representatives than Rhode Island. $\qquad\square$

It would be ideal, of course, to have an apportionment in which the constituency sizes or representative shares are the same for all states because it would be consistent with the "one person one vote" concept. However, it is tremendously difficult and not often possible to obtain such an ideal allocation. Therefore, we should instead strive for having their measures of fairness as close to each other as possible. One measure of how close we come to this goal is called **absolute unfairness**, defined as

$$|\text{constituency size of state A} - \text{constituency size of state B}|.$$

Another measure of unfairness is

$$|\text{representative share of state A} - \text{representative share of state B}|.$$

The choice of which of these two measures to use may affect the results of apportionments. It is surprising that these two measures of fairness could disagree in allocations, as Example 2.1.6 shows. Later, in §2.4, we will consider the apportionment method that minimizes discrepancies in representative shares.

Example 2.1.4. Finding Absolute Unfairness

Suppose that club A, with 534 members, has six delegates, and club B, with 425 members, has five delegates. The constituency size of state A is $\frac{534}{6} = 89$ and the constituency size of state B is $\frac{425}{5} = 85$. So state A is more poorly represented than state B; the absolute unfairness of this apportionment is $89 - 85 = 4$.   □

Although absolute unfairness measures the imbalance of an apportionment between two states, it is inadequate to compare the unfairness of two apportionments. More appropriate would be the the ratio of absolute unfairness of an apportionment over the smallest average constituency, called

$$\text{\textbf{relative unfairness}} = \frac{\text{the absolute unfairness of the apportionment}}{\text{the smaller average constituency of the two states}}. \qquad (2.1.6)$$

E. V. Huntington [5], a mathematician from Harvard University, showed that if relative differences are compared instead of absolute differences, then both representative shares or average constituencies would lead to identical apportionments. His method is presented in §2.5.

Example 2.1.5. Finding Relative Unfairness

We reconsider Example 2.1.3; the absolute unfairness of this apportionment is $659358.7 - 526604.5 = 132754.2$. The relative unfairness of this apportionment then becomes

$$\text{relative unfairness} = \frac{132754.2}{526604.5} \approx 0.252.$$

Example 2.1.6. Rounding Real Numbers

Consider an apportionment problem of rounding the following numbers in the sum to whole numbers, preserving the total of 60.

$$4.47 + 11.54 + 15.76 + 22.81 + 5.42 = 60.$$

In this sum, every entry can be considered as a state quota, with 5 being the total number of states: $q_1 = 4.47$, $q_2 = 11.55$, $q_3 = 15.76$, $q_4 = 22.81$, and $q_5 = 5.41$. The number of seats to be allocated is $h = 60$ (house size). In this problem, we do not know the total population, p, and the populations of the five states. However, the population of each state is proportional to its natural quota, see Eq. (2.1.5). Therefore, all constituency sizes as well as representative shares will be proportional to the natural divisor $(= p/h)$. For simplicity, we choose the value of the natural divisor to be 1. Calculating the lower quotas, we obtain
$$\lfloor q_1 \rfloor = 4, \quad \lfloor q_2 \rfloor = 11, \quad \lfloor q_3 \rfloor = 15, \quad \lfloor q_4 \rfloor = 22, \quad \lfloor q_5 \rfloor = 5, \text{ with total of 57.}$$

For upper quotas, we have
$$\lceil q_1 \rceil = 5, \quad \lceil q_2 \rceil = 12, \quad \lceil q_3 \rceil = 16, \quad \lceil q_4 \rceil = 23, \quad \lceil q_5 \rceil = 6, \text{ with total of 62.}$$
So none of these allocations gives the correct one. Calculating the representative shares for lower $(\lfloor q \rfloor / q)$ and upper $(\lceil q \rceil / q)$ quotas, we get

Numbers	4.47	11.54	15.76	22.81	5.42
Representative share for $\lfloor q \rfloor$	0.89485	0.9532	0.951776	0.964489	0.9225
Representative share for $\lceil q \rceil$	1.1185	1.03986	1.01522	1.00832	1.1070

For instance, the "state" **4.47** has the representative share $\frac{4}{4.47} \approx 0.894854586$ if it is allotted with 4 "seats" and $\frac{5}{4.47} \approx 1.118568233$ if it is allotted with 5 "seats." Which five of the combined lower or upper quotas can we choose to preserve the total of 60? There are $2^5 = 32$ possible choices and it is not easy to make a decision. The largest representative share for ceilings is 1.1185 for $\lceil 4.47 \rceil$ and the lowest representative share for floors is 0.89485, again for $\lfloor 4.47 \rfloor$. To minimize the range of discrepancies in the representative shares, we could eliminate either the largest representative share or the lowest representative share, but we cannot drop both of them because it will eliminate "state" **4.47**. Let us start with one of them, say we eliminate the largest one. Then we have to drop the second largest representative share (since the lowest representative share is untouchable), which corresponds to $\lceil 5.42 \rceil$ leaving 1.03986 to be the largest one.

Numbers	4.47	11.54	15.76	22.81	5.42
Representative share for $\lfloor q \rfloor$	0.89485				0.9225
Representative share for $\lceil q \rceil$		1.03986	1.01522	1.00832	

For the other three "states" we have no choice but to drop their lower quotas' representative shares in order to preserve the total of 60. This leads to apportionment

$$4 + 12 + 16 + 23 + 5 = 60, \tag{2.1.7}$$

with the range of discrepancies: $1.03986 - 0.89485 \approx 0.145$. Suppose that we start with elimination of the smallest representative share, which corresponds to "state" **4.47**. Since we cannot drop the largest one, we eliminate the two next smallest shares to obtain

Numbers	4.47	11.54	15.76	22.81	5.42
Representative share for $\lfloor q \rfloor$		0.9532		0.964489	
Representative share for $\lceil q \rceil$	1.1185		1.01522		1.1070

This gives us the allocation $5 + 11 + 16 + 22 + 6 = 60$ having the range of discrepancies:

$$1.1185 - 0.9532 \approx 0.16536,$$

which is larger than the previous one. Therefore we stick with the allocation (2.1.7) that provides the minimum range of discrepancies. This apportionment can be obtained by rounding the real numbers in the sum to the nearest whole numbers. The method that minimizes the discrepancies in representative shares is presented in §2.4.

Now we turn our attention to constituency sizes that are organized in the following table:

Numbers	4.47	11.54	15.76	22.81	5.42
Average constituency for $\lfloor q \rfloor$	1.1175	$1.04\overline{90}$	$1.050\overline{6}$	$1.036\overline{81}$	1.084
Average constituency for $\lceil q \rceil$	0.894	$0.9591\overline{6}$	0.985	0.99173	$0.90\overline{3}$

Eliminating the three largest average constituencies from the table, we get

Numbers	4.47	11.54	15.76	22.81	5.42
Average constituency for $\lfloor q \rfloor$		$1.04\overline{90}$		$1.036\overline{81}$	
Average constituency for $\lceil q \rceil$	0.894		0.985		$0.90\overline{3}$

This leads to the apportionment

$$5 + 11 + 16 + 22 + 6 = 60 \tag{2.1.8}$$

with the range of discrepancies

$$1.04\overline{90} - 0.894 = 0.155\overline{09}.$$

If we drop the two smallest average constituencies, this yields

Numbers	4.47	11.54	15.76	22.81	5.42
Constituency size for $\lfloor q \rfloor$	1.1175				1.084
Constituency size for $\lceil q \rceil$		$0.9591\overline{6}$	0.985	0.99173	

So we get the allocation

$$4 + 12 + 16 + 23 + 5 = 60,$$

with the range of discrepancies

$$1.1175 - 0.9591\overline{6} = 0.1558\overline{3}.$$

However, we reject this apportionment because allocation (2.1.8) has a lower range of discrepancies. Therefore, the allocations of (2.1.7) and (2.1.8) that minimize the range of discrepancies based on average constituencies and on representative shares are different. □

From its beginning in 1789, the U.S. Congress was faced with deciding how to apportion the House of Representatives. The controversy continued until 1941, with the enactment of the Hill-Huntington method (see §2.5). During congressional debates on apportionment in the intervening 150 years, Congress approved and implemented four different apportionment methods:

1790 − 1830	The Jefferson method (see §2.3).
1840	The Webster method (see §2.4).
1850 − 1900	The Hamilton method, known also as the Vinton method (see §2.2).
1910 − 1930	The method of major fractions assigned seats similarly to the Webster.
1940 − now	The Hill-Huntington method (see §2.5).

Note that the table above does not identify the method precisely: some methods were modified (as in 1860), some were used in conjunction with another method (1850, 1880, 1890, and 1930), and in 1920 there were no new apportionments. As a result of these debates, it was agreed that a perfect method (if such exists) would of course satisfy the following five properties.

1. The first property, which is called the **quota rule** (or quota property), requires that an apportionment method should always operate either with lower quotas or with upper quotas. In other words, the final allocation for each state is equal to its natural quota, rounded either up or down to a nearest whole number.

2. The number of representatives for a state should not decrease if the house size increases.

3. All states should abide by the same formula for allocation.

4. Methods should not artificially favor large states at the expense of the smaller ones and vice versa.

5. Every state should have at least one representative, which is required by U.S. Constitution.

Since 1790, all apportionment methods that have been used in the USA after the adoption of the Constitution, including rejected apportionment methods, violate at least one of the five properties above. Moreover, some of the methods are susceptible to the apportionment paradoxes discussed below.

In 1881, the U.S. Congress was surprised that, when using Hamilton's method (see §2.2), Alabama was entitled to 8 representatives in a House having 299 members but would receive only 7 representatives in a 300-member house. Under Hamilton's method, Alabama would receive fewer representatives in a larger House, although no state had a change in population. This strange situation occurred again, following the 1890 census, when Arkansas lost a representative as the House increased from 359 to 360 members. Since that time, the following property has been referred to as the **Alabama paradox**:

> An increase in the total number of seats to be apportioned causes a state to lose a seat. Such a property is referred to as non-monotonic.

In the early 1900s, it was discovered that Hamilton's method had experienced another serious flaw, called the **population paradox**:

> When state A's population is growing faster than state's B's population, yet state A loses a representative to state B. It is assumed that the total number of representatives remains the same.

When Oklahoma joined the union in 1907, the House of Representatives had to be reapportioned in order to accommodate the new member. Congress decided to increase the size of the House by five and give Oklahoma its fair share of seats and leave the apportionment of the other states unchanged. However, when the House was reapportioned, New York was required to give one of its seats to Maine. This phenomenon is called the **new state paradox**:

When a new state is added and its share of seats is added to the legislature, it causes a change in allocation of seats previously given to another state.

In 1980, Michael Balinski and H. Peyton Young proved a surprising theorem, called the **Balinski and Young's impossibility theorem**, which states:

There is no apportionment method that avoids all paradoxes and at the same time satisfies the quota rule.

In particular, they looked in detail at fairness issues growing out of apportionment problems. Specifically, they stated various axioms or rules that an apportionment method should obey. Actually, Balinski and Young devised a method, called the quota method, that does not allow the Alabama paradox to happen while also satisfy the quota property. However, their method (which is a refinement of the Hill-Huntington method) still favors large states since Jefferson's method is used to compare the states.

The U.S. Constitution does not specify exactly the number of seats in the House of Representatives and the method of allocation to be used. The first apportionment of 65 seats was made in 1789, and it was not based on a census. Starting in 1790, a new apportionment of representatives was made every 10 years with each census. For the first 130 years, the House of Representatives grew every decennial census (with one exception in 1840). In 1910, the House grew to its present size of 435 members after the census. The Reapportionment Act of 1929 permanently froze the size of the House of Representatives at 435 members; later in 1941, a permanent method for apportionment was established.

Not all countries are consistent with their apportionment methods. For example, in Greece, since 1926, very rarely are there two consecutive elections with exactly the same electoral system. In the last several decades, there has been a "tradition" of changing the electoral system. Typically the changes take place a few months before the elections, so the government's chances for reelection are maximized.

Problems.

1. A bus company operates 6 bus routes (A, B, C, D, E, and F) using 77 buses. The buses are apportioned among routes based on the average number of daily passengers per route, given in the following table.

Route	A	B	C	D	E	F
Passengers	2770	3510	7630	1870	4920	5634

   (a) Describe the "states" and the "seats" in this apportionment problem.

   (b) Find the natural divisor.

   (c) Find each state's natural quota, upper quota, and lower quota.

   (d) Find constituency sizes for lower and upper quotas.

   (e) Find representative shares for lower and upper quotas.

2. A country has five states (A, B, C, D, and E). There are 78 seats in the house of representatives. The population of each state (in millions) is given in the following table. Find

State	A	B	C	D	E
Population	2.8	3.1	6.3	8.7	4.9

   (a) the natural divisor;

   (b) each state's natural quota;

   (c) each state's lower and upper quota.

   (d) Find constituency sizes for lower and upper quotas.

   (e) Find representative shares for lower and upper quotas.

3. A country of 13.7 million has 5 states (A, B, C, D, and E). According to the country's constitution, the seats in the legislature are apportioned to states based on their populations. The natural quota of each state is given in the following table.

State	A	B	C	D	E
Natural Quota	12.38	7.45	9.72	6.93	13.52

   (a) Find the number of seats in the country's legislature.

   (b) Find the natural divisor.

   (c) Find the population of each state, upper quota, and lower quota.

   (d) Find constituency sizes for lower and upper quotas.

   (e) Find representative shares for lower and upper quotas.

4. A country of 25 million has 5 states (A, B, C, D, and E). According to the country's constitution, the seats in the legislature are apportioned to states based on their populations. The natural quota of each state is given in the following table.

State	A	B	C	D	E
Natural Quota	3.72	4.66	5.13	2.84	3.65

   (a) Find the number of seats in the country's legislature.

   (b) Find the natural divisor.

   (c) Find the population of each state, upper quota, and lower quota.

   (d) Find constituency sizes for lower and upper quotas.

   (e) Find representative shares for lower and upper quotas.

5. Consider an apportionment problem in rounding the following numbers in the sum to a whole number, preserving the total of 10.

$$0.45 + 1.78 + 3.76 + 2.41 + 1.62 = 10.$$

   (a) Find the number of seats in the given apportionment problem.

   (b) Determine the upper and lower quotas for each number.

   (c) By choosing the natural divisor to be 1, find the pair of upper quotas that has the largest relative unfairness.

6. Consider an apportionment problem in rounding the following numbers in the sum to a whole number, preserving the total of 23.

$$1.78 + 8.35 + 7.27 + 2.46 + 3.14 = 23.$$

    (a) Find the number of seats in the given apportionment problem.

    (b) Determine the upper and lower quotas for each number.

    (c) By choosing the natural divisor to be 1, find the pair of upper quotas that has the largest relative unfairness.

7. If the American Dental Association has 378,056 members and 8 representatives on the National Health Board, what is the association's constituency size? What is the association's representative share?

8. If a state plumber union has 14,565 members and 15 representatives on the federal council, what is its average constituency? What is the union's representative share?

9. Which state is more poorly represented in 2009: state of Vermont with a population of 621,270 and 1 representative, or state of Indiana with a population of 6,376,792 and 9 representatives? What is the absolute unfairness of this apportionment in terms of representative share and constituency size? What is the relative unfairness of this apportionment?

10. Which state is more poorly represented in 2009: state of Alaska with a population of 686,293 and 1 representative, or Tennessee with a population of 6,214,888 and 9 representatives? What is the absolute unfairness of this apportionment in terms of representative share and constituency size? What is the relative unfairness of this apportionment?

11. Which state is more poorly represented in 2009: state of Wisconsin with a population of 5,627,967 and 8 representatives, or Nebraska with a population of 1,783432 and 3 representatives? What is the absolute unfairness of this apportionment in terms of representative share and constituency size? What is the relative unfairness of this apportionment?

12. Two companies formed a consortium. On a 12-member board, company A with 1,756 stockholders, received 5 members, and company B, with 2,324 stockholders, received 7 members. Calculate the absolute and relative unfairness of this apportionment.

13. Two companies formed a consortium. On a 9-member board, company A with 845 stockholders, received 5 members, and company B, with 396 stockholders, received 4 members. Calculate the absolute and relative unfairness of this apportionment.

14. According to U.S. Bureau of the Census estimates, in 2010, Utah had a population of 2,784,572 people while Wyoming had 544,270 people in July 2009. Utah was allotted three seats in the House, and Wyoming was allotted one. Calculate the absolute and relative unfairness of this apportionment.

15. In the sum
$$1.5 + 2.12 + 3.37 + 4.51 + 5.5 = 17,$$
round numbers either up (ceiling) or down (floor) to whole numbers preserving the total of 17. Find the apportionment that minimizes average constituencies, and compare it with the allocation that minimizes representative shares.

16. In the sum
$$1.18 + 1.26 + 1.37 + 1.55 + 1.54 = 7,$$
round numbers either up (ceiling) or down (floor) to whole numbers preserving the total of 7. Find the apportionment that minimizes average constituencies, and compare it with the allocation that minimizes representative shares.

2.2 Hamilton's Method

The first census of the United States was taken in 1790, less than three years after the ratification of the Constitution. Once the numbers were in, Congress had to decide how to use the data to apportion the representatives. They also had to decide how many representatives the House should have. In the spring of 1792, Congress passed a bill to apportion the House, using a method proposed by Alexander Hamilton (1757 – 1804), the Secretary of the Treasury, now known as Hamilton's method. His picture is on the $10 bill, so sometimes we abbreviate/identify his name with $10.

The congressional bill was vetoed by President George Washington, which was his first exercise of this power and one of only two that he exercised in eight years as President. Hamilton's procedure, which might be described as the largest fractional remainders method, was used by Congress from 1851 to 1901. However, due to some flaws and undesirable features, it was never strictly implemented because changes were made in the apportionments that were not consistent with the method. The Hamilton method is used to apportion (at least part of) the legislatures of Austria, Belgium, Costa Rica, El Salvador, Indonesia, Madagascar, Namibia, and Sweden. It has generally been known as the Vinton method (for Samuel Vinton (1792 – 1862), a representative of the state of Ohio, its chief advocate after the 1850 census). Assuming a fixed House size, the Hamilton-Vinton method can be described as follows:

1. Calculate the natural quota for each state according to Eq. (2.1.2), page 150.

2. For the initial allocation, assign to each state its *lower quota*—the natural quota rounded down to the nearest integer.

3. If the initial allocation is less than the correct number of seats to be apportioned (house size), give the surplus seats, one per state, to the states in descending order of the fractional parts of their natural quotas until the final allocation is achieved.

In rare cases, Hamilton's method can be inconclusive. If the initial allocation is m seats less than the house size, and the mth and the $(m+1)$th largest fractional parts are the same, the method does not work. The procedure guarantees that every state will get at least the lower quota ($\lfloor q \rfloor$, where q is the natural quota). Some states would be apportioned with their upper quotas ($\lceil q \rceil$). Therefore, the Hamilton method satisfies the quota rule (see page 158): the method hands out to each state either its lower quota or upper quota. It can be seen as a distribution of m bonus seats to the states when the sum of lower quotas is m seats short.

To start the Hamilton allocation, divide each state's population by the total population (which gives you the fair share) and multiply by the house size. This number is referred to as the natural (standard) quota. Award each state the floor, called lower quota, which is a whole number obtained by rounding down (= discard the fractional parts) the natural quota. If the number of seats assigned using the lower quotas is less than the House total (this will be true if at least one natural quota is not an integer), rank the fractional remainders of state's quotas and award seats in order from highest to lowest until the house size is reached. The Hamilton-Vinton method has simplicity in its favor, but its downfalls were suffering from the three paradoxes—Alabama, new state, and population (see §2.1).

Hamilton's method may leave one or more states without a seat. When the method is applied to apportion legislatures in countries, as, for instance, the United States, where every state should have at least one seat due to constitutional requirement, the Hamilton method requires an amendment to satisfy this condition. We do not discuss these amendments because they are far from the scope of this book.

Example 2.2.1. Hamilton's Method

Suppose that at an art show, there is room for 18 booths that could be used by artists. The guild has decided that the booths should be assigned in proportion to the types of artists: there are 75 painters, 34 sculptors, and 41 weavers. When applying Hamilton's method, it is convenient to organize calculations in the following table:

Members	Popu-lation	Natural Quota	Hamilton I. A.	Fractional Part	Bonus	Hamilton F. A.
Painters	75	9	9	0	0	9
Sculptors	34	4.08	4	0.08	0	4
Weavers	41	4.92	4	0.92	1	5
Total:	150	18	17			18

Using Hamilton's method, we give "Weavers" an extra seat (bonus) because its fractional part is the largest.

Example 2.2.2. Rounding Numbers

Here is a typical apportionment problem of rounding numbers to integers in the sum:

$$0.34 + 1.79 + 2.18 + 3.57 + 2.12 = 10.$$

Each summand can be considered as the quota: $q_1 = 0.34$, $q_2 = 1.79$, $q_3 = 2.18$, $q_4 = 3.57$, and $q_5 = 2.12$. According to Hamilton, the initial allocation is obtained by rounding the quotas down to the nearest integers, called floors or lower quotas. This initial allocation is two seats less than required; therefore, we assign the upper quotas to two states with the largest fractional parts. This leads to the following table:

Natural Quota	Hamilton Initial Allocation	Fractional Part	Rank	Bonus	Hamilton Final Allocation
0.34	$\lfloor 0.34 \rfloor = 0$	0.34	3	0	0
1.79	$\lfloor 1.79 \rfloor = 1$	0.79	1	1	$2 = \lceil 1.79 \rceil$
2.18	$\lfloor 2.18 \rfloor = 2$	0.18	4	0	2
3.57	$\lfloor 3.57 \rfloor = 3$	0.57	2	1	$4 = \lceil 3.57 \rceil$
2.12	$\lfloor 2.12 \rfloor = 2$	0.12	5	0	2
Total:	8				10

The fourth column presents an ordering of fractional parts of the natural quotas (summands), from largest to smallest. This example demonstrates that Hamilton's method may assign zero seats to a state (in our case, it is the first one with the natural quota 0.34). □

The Hamilton method, which was used from 1850 to 1890, has the undesirable property of not being monotonic, meaning that increasing the size of the House can cause a decrease in representation for some states. This phenomenon became known as the Alabama paradox when it appeared after the 1880 census, because representatives were not being assigned to Alabama in a monotonic fashion for House sizes between 298 and 302. On October 25, 1881, C. W. Seaton, the chief clerk of the Census Office, wrote the Congress: "While making these calculations I met with the so-called Alabama paradox where Alabama was allotted 8 Representatives out of a total of 299, receiving 7 when the total became 300. With 299 seats in the House, Alabama's quota was 7.646 seats, and was allocated eight seats based on this quota. When a House size of 300 was used, Alabama's quota increased to 7.671, but Illinois and Texas now had larger fractional remainders than Alabama: 18.702 and 9.672, respectively."

This phenomenon had actually been noticed ten years earlier, when the Hamilton method was applied to various House sizes between 268 and 283; the troublesome state then was Rhode Island. The final exasperating appearance of this paradox occurred after the 1900 census when Congress had to decide on increasing the House size. This time, both Colorado and Maine were in fluctuation. For instance, Maine would get 4 seats for a house size between 350 and 356, but for $h = 357$ Maine's apportionment would go down to 3 seats. The next increase in house size between 358 to 381 would cause Maine to again have only 3 seats. A Congressional bill was proposed, which would have enlarged the House to 357 (from 356 in 1890). The resulting debate was bitter and partisan, leading Congress to scrap the Hamilton method. In its place, the modification of the Webster method was applied to a House size of 386.

Later, another flaw in Hamilton's method was found—it created a systematic bias in favor of large states over smaller ones.

Example 2.2.3. Alabama Paradox
Consider four states with the populations given in the table below with 68 representatives in the House.

States	Population	Natural Quota	Fractional part	Hamilton I. A.	Hamilton Final Allocation
A	220	1.496	0.496	1	2
B	590	4.012	0.012	4	4
C	4,330	29.444	0.444	29	29
D	4,860	33.048	0.048	33	33
Total:	10,000	68		67	68

When the number of representatives increases to 69, one would get the following table:

States	Popu- lation	Natural Quota	Lower Quota	Hamilton F. A.
A	220	1.518	1	1
B	590	4.071	4	4
C	4,330	29.877	29	30
D	4,860	33.534	33	34
Total:	10,000	69	67	69

So we see that state A lost one representative when the House size increased.   □

One could argue that the Alabama paradox should not be an important consideration in apportionments, since the House size was fixed at 435, but the Hamilton-Vinton method is subject to other anomalies: the population paradox and the new states paradox. The population paradox occurs when a state that grows at a greater percentage rate than another one has to give up a seat to the slower growing state.

The new states paradox works in much the same way as an apportionment would after a new state enters the Union: any increase in House size caused by the additional seats for the new state may result in seat shifts among states that otherwise would not have happened. The paradox was discovered in 1907 when Oklahoma became a state. After giving Oklahoma 5 seats—it was the state's fair share—the size of the House of Representatives was increased from 386 to 391. The innocent intent to leave the number of seats unchanged for all other states was violated: Maine gained one seat (4 instead of 3) and New York lost a seat (38 to 37). Finding a formula that avoided the paradoxes was a goal when Congress adopted a rounding, rather than a ranking, method when the apportionment law was changed in 1911.

Example 2.2.4. Population Paradox

There were 6 hospitals and 17 available doctors in 2010. The allocation of doctors between hospitals is based on the number of certified beds each center has. Every year the manager collects the data and assigns doctors to medical centers using Hamilton's method. In the previous year there was the following allocation, presented in the last column of the table.

Hospital	Number of beds	Natural Quota	Hamilton I. A.	Hamilton F. A.
Butler	127	1.8006	1	2
Fatima	289	4.0975	4	4
Kent	388	5.5012	5	5
Landmark	19	0.2693	0	0
Miriam	247	3.5020	3	4
Newport	129	1.8290	1	2
Total:	1199	17	14	17

This year the number of beds is different in 2 hospitals because Kent is able to add 20 beds and Miriam adds 15 beds. The percent increase in the number of beds of Miriam hospital is determined as follows:

$$\text{percent increase} = \frac{15}{247} \times 100\% \approx 6.07287\%.$$

On the other hand, the percent increase in the number of beds of Kent hospital is

$$\text{percent increase} = \frac{20}{388} \times 100\% \approx 5.1546\%.$$

Therefore, the population, which is the number of beds, of Miriam hospital is growing faster by increasing its capacity by 6.07% while Kent hospital grows by 5.15%. This leads to the new allocation of doctors presented in the table below. Since Miriam hospital lost one doctor while its population grows faster than Kent hospital, we observe the population paradox.

Hospital	Number of beds	Natural Quota	Lower Quota	Hamilton F. A.
Butler	127	1.7495	1	2
Fatima	289	3.9813	3	4
Kent	408	5.6207	5	6
Landmark	19	0.2617	0	0
Miriam	262	3.6094	3	3
Newport	129	1.7771	1	2
Total:	1234	17	13	17

Example 2.2.5. New States Paradox

A graduate school at a university needs to distribute 19 graduate assistant jobs among 4 colleges: science, education, engineering, and liberal arts. The school used the Hamilton method to allocate the jobs between colleges based on their undergraduate enrollments, as shown in the table. Since science's natural quota has the largest fractional part, this college is entitled to its upper quota.

College	Enroll-ments	Natural Quota	Lower Quota	Hamilton Final Allocation
Science	317	1.3688	1	2
Education	2139	9.2365	9	9
Engineering	497	2.1461	2	2
Liberal Arts	1447	6.2484	6	6
Total:	4400	19	18	19

However, this year the university added a new college—medical school. Now we need to reapportion assistantships between five colleges. Since the standard divisor is 227.5, the medical school has $150/227.5 \approx 0.6593$ fair shares, which is rounded to 1. So the total number of assistantships is now 20 instead of 19. Comparison of the final allocations for these two years indicates that after adding a new college the science college lost one seat while the education college gains the upper quota.

College	Enroll-ments	Natural Quota	Hamilton I. A.	Hamilton Final Allocation
Science	317	1.3934	1	1
Education	2139	9.4021	9	10
Engineering	497	2.1846	2	2
Liberal Arts	1447	6.3604	6	6
Medical	150	0.6593	0	1
Total:	4550	20	18	20

Example 2.2.6. Hamilton's Method may be Inconclusive
Let us consider the rounding problem which is where the Hamilton method fails:

$$1.51 + 2.51 + 1.27 + 2.75 + 1.96 = 10.$$

The sum of lower quotas gives us $1 + 2 + 1 + 2 + 1 = 7$, which is three seats less than the house size, $h = 10$. After assigning the upper quotas to the last two states (with $q_4 = 2.75$ and $q_5 = 1.96$) with largest fractional parts, we come up with a dilemma: which of the first two states deserves a seat. Since they both have the same fractional part, 0.51, we cannot allocate integers according to the Hamilton rule.

2.2.1 Lowndes' Method

As we know from Example 2.2.2, the Hamilton method may allocate zero seats to one or more states. This property is a violation of the U.S. Constitution, and when Hamilton's method was in use for fifty years, it was modified to fix this flaw. Since the Hamilton method relies entirely on the sizes of the fractional parts of natural quotas instead of taking into account the states' populations, it favors large states. To fix this flaw, a similar method, which is a modification of the Hamilton method, was proposed by South Carolina Congressman William Lowndes (1782 – 1822). In 1822, Congress discussed Lowndes' method, but it was never adopted, most likely because it favors small states.

Lowndes' method allocates at least one seat to every state. It works exactly the same way as the Hamilton method, but it has one amendment—instead of fractional parts the method looks at relative fractional parts (R.F.P.). Recall that the relative fractional part of a real number is its fractional part divided by its integer part if it is not zero. When the integer part of a state's natural quota is zero, such a state gets zero seats initially. To assure that this state receives one seat, we assign ∞ to be its relative fractional part. For instance, the relative fractional part of the irrational number $\pi = 3.1415926\ldots$ is $\dfrac{0.141592654\ldots}{3} \approx 0.047197551$. On the other hand, the relative fractional part of 0.5 is infinity. Note that a fractional part of a real number q can be written mathematically as the difference $q - \lfloor q \rfloor$. The algorithm includes the following steps:

1. For each state, calculate its natural (standard) quota: $q_i = p_i \times h/p$, which is the

state's population over total population times the house size, see Eq. (2.1.2), page 150.

2. Make an initial allocation by rounding down all natural quotas, so each state gets $\lfloor q_i \rfloor$ seats.

3. Calculate the relative fractional part (R.F.P.) for each, q_i, the natural quota of state i:

$$\textbf{Relative Fractional Part} = \begin{cases} \frac{q_i - \lfloor q_i \rfloor}{\lfloor q_i \rfloor}, & \text{if } \lfloor q_i \rfloor \neq 0, \\ \infty, & \text{if } \lfloor q_i \rfloor = 0. \end{cases} \qquad (2.2.1)$$

4. Rank/order all relative fractional parts, from largest to smallest, and assign upper quotas to states with largest relative fractional parts until the required allocation is achieved. ∎

Example 2.2.7. Rounding Numbers
We consider the following rounding problem:

$$0.17 + 1.53 + 3.72 + 4.31 + 0.27 = 10.$$

Each of these five numbers can be considered as a natural quota for some state. Since the Hamilton method works only with quotas, actual values of states' populations do not matter. Both methods, Hamilton's and Lowndes', tentatively assign lower quotas to each state. Then these methods distribute surplus seats, one per state, until the house size is reached. The Hamilton method assigns the first and the last numbers zero seats because their fractional parts are the smallest, which means that these states would not get a seat. In contrast, Lowndes' method always awards at least one seat to every state. Let us look at the table:

Numbers	Natural Quota	$10 I. A.	Hamilton F. A.	R.F.P.	Lowndes' Final Allocation
A	0.17	0	0	∞	1
B	1.53	1	2	$0.53 = \frac{0.53}{1}$	1
C	3.72	3	4	$0.24 = \frac{0.72}{3}$	3
D	4.31	4	4	$0.0775 = \frac{0.31}{4}$	4
E	0.27	0	0	∞	1
Total:	10	8	10		10

In this rounding problem, populations of five states are not given, but only the house size, $h = 10$. However, we know from Eq. (2.1.5), page 155, that their populations are proportional to the natural quotas, q_i, $i = 1, 2, 3, 4, 5$. The population of the first state, $p_1 = 0.17 \, (p/h)$, is nine times less than the population of the second state, $p_2 = 1.53 \, (p/h)$, but both states get one seat. Is it fair?

Example 2.2.8. Deuce
Consider the following four states, call them A, B, C, and D, with the given populations.

If we try to allocate thirty one seats between them, we will get two states with the same relative fractional parts, as the following table shows.

States	Popu- lation	Natural Quota	$10 I. A.	Hamilton F. A.	R.F.P.	Lowndes' F. A.
A	3125	19.3711	19	19	0.0195329	19
B	1042	6.4591	6	7	0.0765180	?
C	521	3.2295	3	3	0.0765180	?
D	313	1.9402	1	2	0.9402119	2
Total:	5001	31	29	31		??

Two states, B and C, compete for one seat, and Lowndes' method is inconclusive.

Example 2.2.9. Lowndes' Method versus Hamilton's Method

A bank with six branches wants to set up a total of fifteen ATM machines at the different branches. The ATMs will be apportioned according to the average number of customers per day at each branch given in the table.

Branch	A	B	C	D	E	F
Average Number of of Customers per Day	317	286	231	174	148	64

To apportion these machines, we will use Hamilton's method and compare its final allocation with Lowndes' distribution. The table of values below speaks for itself.

States	Popu- lation	Natural Quota	$10 I. A.	Hamilton F. A.	R.F.P.	Lowndes' F. A.
A	317	3.834677419	3	+1 = 4	0.2782258063	4
B	309	3.737903226	3	?	0.2459677420	3
C	231	2.794354839	2	+1 = 3	0.3971774195	3
D	174	2.104838710	2	2	0.052419355	2
E	148	1.790322581	1	+1 = 2	0.790322581	2
F	61	0.737903226	0	?	∞	1
Total:	1240	15	11	—		15

The initial allocation is four machines short, so according to Hamilton's procedure, four branches with largest fractional parts should be rewarded with bonuses. Since $\dfrac{15 \times 309}{1240} - 3 = \dfrac{15 \times 61}{1240} = \dfrac{183}{248}$, the Hamilton method is inconclusive. On the other hand, the Lowndes method is applicable while also providing the required allocation.

Problems.

1. According to U.S. Bureau of the Census estimates, in 2009, Mississippi's population was 2,951,996; Colorado's population was 5,024,748; and Arizona's population was 6,595,778. Allocate 19 members of the U.S. House of Representatives to these 3 states using the Hamilton method.

2. According to U.S. Bureau of the Census estimates, in 2009, Michigan's population was 9,969,72; Nevada's was 2,643,085; and New Mexico's was 2,009,671. Allocate 23 members of the U.S. House of Representatives to these 3 states using the Hamilton method.

3. Use the Hamilton method to round each of the following numbers in the sum to a whole number preserving the total of 10.

$$0.09 + 0.31 + 1.9 + 3.78 + 3.92 = 10.$$

4. Repeat the previous exercise, but now use Lowndes' method.

5. Use the Hamilton method to round each of the following numbers in the sum to a whole number preserving the total of 20.

$$0.91 + 5.73 + 6.42 + 4.58 + 2.36 = 20.$$

6. Repeat the previous exercise, but now use Lowndes' method.

7. Use the Hamilton method to round each of the following numbers in the sum to a whole number preserving the total of 30.

$$0.12 + 1.56 + 6.27 + 8.53 + 13.52 = 30.$$

8. Repeat the previous exercise, but now use Lowndes' method.

9. Students at a university have a thirty-one-member student council, which represents the 2,400 members of the student body. There are 732 seniors, 658 juniors, 531 sophomores, and 479 freshmen. How do you think the seats on the council should be distributed to the classes? Use Hamilton's method first, and then compare the outcome with Lowndes' method.

10. The employees of a company are negotiating a new contract. There are 274 cashiers, 43 drivers, 517 maintenance workers, 126 security officers, and 50 electricians. The eleven-person negotiations committee has members in proportion to the number of employees in each of the five groups. Assign members to the negotiations committee using Hamilton's method. If one of the groups is not presented in the committee, use Lowndes' method instead.

11. A campus has 4 freshman dormitories. Building A has 27 units, building B has 9 units, building C has 28 units, and building D has 38 units. A 10-person resident council will set rules governing the complex. Membership in the council is to be proportional to the number of units in each building. Is allocation of representatives to this council the same for Hamilton's method and Lowndes' method?

12. A hospital has a nursing staff of 167 nurses working in 4 shifts: No. 1: 6:00am to 12 (noon), No. 2: 12:00 to 6pm, No. 3: 6pm to 12 (midnight), and No. 4: midnight to 6am. The number of nurses is allocated to each shift according to the average number of patients per shift, given in the following table.

Shift	No. 1: 6am – 12	No. 2: 12 – 6pm	No. 3: 6pm - 12	No. 4: 0 – 6am
Patients	587	738	252	83

(a) Describe the "states" and the "seats" in this apportionment problem.

(b) Use Hamilton's method to determine how many nurses each shift will get.

(c) Use Lowndes' method to determine how many nurses each shift will get.

13. Four companies decide to merge into one new company that will be governed by 30-member board of directors. The seats in the board of directors are allocated in proportion to the net worth of each company.

Company	Alpha	Beta	Gamma	Delta
Net Worth (in thousands)	5810	3735	1657	491

(a) Use Hamilton's method to determine how many members will come from each company.

(b) Use Lowndes' method to determine how many members will come from each company.

14. A county is divided into five districts with the following populations:

District	South	North	East	West	Center
Population	3660	1820	4270	2110	8930

There are nine seats to be apportioned. Allocate these seats between districts using
(a) Hamilton's method;   (b) Lowndes' method.

15. The Boston subway has five lines: Blue, Orange, Green, Red, and Silver. Allocate 56 cars to each line based the average number of passengers who commute on each line:

State	Blue	Orange	Green	Red	Silver
Natural Quota	16.93%	18.06%	28.36%	31.83%	4.82%

Distribute cars using
(a) Hamilton's method;   (b) Lowndes' method.

16. A labor council is being formed from the members of five unions. The carpenters union has 62 members, the electricians union has 49 members, the painters have 43 members, the plumbers have 34 members, and the general workers have 29 members. The council will have 19 representatives. Find the number of representatives from each union in the council using
(a) Hamilton's method;   (b) Lowndes' method.

17. The Republic of Nicaragua elects a head of state—the president—and a legislature. The President of Nicaragua and his or her vice-president are elected on one ballot for a five-year term. The National Assembly has 92 members: 90 deputies elected for a five-year term by proportional representation, the outgoing president, and the runner-up in the last presidential election. The summary of the 5 November 2006 Nicaragua National Assembly election results is as follows.

Party	Number of Votes
Sandinista National Liberation Front	840,851
Nicaraguan Liberal Alliance	597,709
Constitutionalist Liberal Party	592,118
Sandinista Renovation Movement	194,416
Alternative for Change	12,053

Allocate 90 seats between these five parties using
(a) Hamilton's method;   (b) Lowndes' method.

18. A country has five states, call them A, B, C, D, and E. The total population of the country is 5 million. According to the constitution, the seats in the legislature are apportioned in accordance to their populations. The natural quota of each state is given in the following table.

State	A	B	C	D	E
Natural Quota	6.77	3.06	0.43	5.78	3.96

(a) Find the number of seats in the legislature.

(b) Distribute seats using Hamilton's method.

(c) Allocate seats using Lowndes' method.

19. A university is made up of five different schools: Agriculture (A), Business (B), Engineering (E), Medical (M), and Science (S). The total number of students is 8,350. The faculty positions at the university are apportioned to the various schools based on the school's respective enrollments. The natural quota of each school is given in the following table.

School	A	B	E	M	S
Natural Quota	14.74	28.01	3.68	71.45	5.12

(a) Find the number of faculty positions at the university.

(b) Distribute positions using Hamilton's method.

(c) Allocate positions using Lowndes' method.

20. A bus company operates five bus routes (A, B, C, D, and E). The 87 available buses are apportioned among the routes on the basis of average number of daily passengers per routes, given in the following table.

Route	Daily average number of passengers
A	18,510
B	21,073
C	16,657
D	14,441
E	7,835

(a) Use Hamilton's method to determine how many members should come from each company.

(b) Use Lowndes' method to determine how many members should come from each company.

21. A city police department employs 214 patrol officers. The officers are to be apportioned among the six precincts based on the number of crimes reported in each precinct over the past year as shown in the table.

Precinct	1	2	3	4	5	6
Number of Crimes	2671	592	1374	928	1837	864

Apportion the officers using
(a) Hamilton's method;   (b) Lowndes' method.

22. A small country is comprised of five states, A, B, C, D, and E. The population of each state, in thousands, is given in the following table.

State	A	B	C	D	E
Population	127	286	354	409	533

According to the country's constitution, the congress will have 92 seats, divided among the five states according to their respective populations. Apportion these seats among five states using
(a) Hamilton's method;   (b) Lowndes' method.

23. The Austrian Parliament consists of two chambers: the National Council and the Federal Council. The National Council is made up of 183 representatives and is elected for a five-year legislative term. In contrast, the Federal Council is not elected directly. Its members are delegated by the provincial diets in accordance with the proportional distribution of political parties in each provincial diet.

    There is 4% barrier for a party to have a seat in the Council. The 2008 election to Austrian National Council resulted in the following distribution of votes:

Party	Number of Votes
Social Democratic Party of Austria (SPO)	1,430,206
Austrian People's Party (OVP)	1,269,656
Freedom Party of Austria (FPO)	857,029
Alliance (BZO)	522,933
The Greens	509,936
Others	297,549
Total:	4,887,309

    Allocate these 183 seats between the five parties using
    (a) Hamilton's method;   (b) Lowndes' method.

24. Parliamentary elections were held in Chile on December 13, 2009. The Chamber of Deputies consists of 120 seats, and they are allocated according to the number of popular votes in the elections. Apportion 120 seats between five parties using
    (a) Hamilton's method;   (b) Lowndes' method.

Party	Number of Votes
Concentration for More Democracy	2,901,503
Coalition for Change	2,841,314
New Majority for Chile	298,765
Clean Chile Vote Happy	353,325
Independents	144,663
Total:	6,539,570

25. The German federal election in 2009 was intended to elect the 323 members of the Bundestag, the federal parliament of Germany. More than 12 parties participated in the election. However, according to the law, the party is assigned a seat if it has at least 5% of population votes. It narrows down to six parties that are represented in the parliament/Bundestag. Summary of the 27 September 2009 German Bundestag election results are as follows:

Party	Number of Votes
Christian Democratic Union (CDU)	13,852,743
Christian Social Union (CSU)	3,190,950
Social Democratic Party (SDP)	12,077,437
Free Democratic Party (FDP)	4,075,115
The Left	4,790,007
The Greens	3,974,803
Others	
Total:	43,235,817

    Allocate these 323 seats between the six parties using
    (a) Hamilton's method;   (b) Lowndes' method.

26. The Metropolitan Police Department (MPD) in Washington D.C. has a very difficult task in ensuring the safety of its citizens. It is even more difficult for Metro P.D. especially because

they control much of the traffic restrictions due to the movement of the president and other important officials throughout the city. To accomplish this task, the department has to allocate extra officers just as traffic and crowd control on certain days so that they can keep everything in check. Washington D.C. is a hot spot for crime as well as diplomatic heat. While federal agencies are in charge of protecting the actions of federal employees, Metro still has the overreaching hand on crimes committed in the city that are not against the federal government. The MPD is split up into six districts, which control different portions of the city. The hot spot, however, is the First District which is in charge of the national mall including the White House and Capital Hill. On special days, the MPD has to pull 30 additional officers from other districts to help in the control of large scale events in or around Capital Hill, the White House, and the National Mall. The chief of police in Washington D.C. is charged with the difficult job of deciding how many officers should be pulled out from each district. He decides to allocate 30 officers in proportion to the number of citizens in each district. Apportion the officers using

(a) Hamilton's method;   (b) Lowndes' method.

District	#1	#2	#3	#4	#5	#6
Population	17,068	21,813	28,654	31,152	13,729	22,409

27. If you were President Washington, how would you distribute the 120 seats in the House of Representatives to the 15 states listed below? Use Hamilton's method and Lowndes' method. The table shows the results of the 1790 census.

State	1790 Population	State	1790 Population
Connecticut	236841	Delaware	55540
Georgia	70835	Kentucky	68705
Maryland	278514	Massachusetts	475327
New Hampshire	141822	New Jersey	179570
New York	331589	North Carolina	353523
Pennsylvania	432879	Rhode Island	68446
South Carolina	206236	Vermont	85533
Virginia	630560	**Total:**	3,615,920

28. Show that Hamilton's method leads to the Alabama paradox as the house size increases from 65 to 66 seats, with populations given by the following table

State	A	B	C
Population	3454	814	25

29. There are 83 local police, 14 federal agents, and 25 state police involved with drug enforcement in a city. A special 22-person task force will be formed to investigate a particular case. Officers will be assigned to this task force with membership proportional to the number in each of the three types of law enforcement officers.

(a) Allocate officers using Hamilton's method.

(b) Allocate officers using Lowndes' method.

(c) Now increase the task force's size to 23, and redo the apportionment using Hamilton's method. Do you observe the Alabama paradox? What group loses an officer when the size of the task force increases?

30. Suppose a company has four divisions: Acquisition (A), Business Relations (B), Computers (C), and Digital (D). Division A has 37 employees, Division B has 125 employees, C has 64,

and D has 29. Assume that a 21-member quality improvement council has membership on the council proportional to the number of employees in the four divisions.

   (a) Apportion this council using Hamilton's method.

   (b) Apportion this council using Lowndes' method.

   (c) Now increase the council's size to 22, under what method does an Alabama paradox occur?

31. A hospital has four emergency clinics throughout the city. Statistics have been gathered regarding the number of patients seen on the night shift for each clinic.

Clinic	South Side	North Side	West Side	East Side
Patients	235	73	341	109

   (a) If there are 14 emergency physicians available, how should they be apportioned using Hamilton's rule?

   (b) Increase the number of physicians until an Alabama paradox occurs. At what increase does the paradox occur and what clinic loses a doctor?

32. Show that Hamilton's method leads to the Alabama paradox as the house size increases from 21 to 22 seats, with populations given by the following table. Show that Lowndes' method leads to the Alabama paradox as the house size increases from 31 to 32 seats.

State	A	B	C	D	E
Population	235	387	509	1261	1261

33. Show that Lowndes' method leads to the Alabama paradox as the house size increases from 57 to 58 seats, with populations given by the following table.

State	A	B	C	D	E
Population	71	164	209	367	432

34. Thirteen representatives are apportioned among three states A, B, and C, with the populations (in thousands) shown in the following table.

State	A	B	C
Population	763	5,138	8,609

   Five years later, the population of A has increased by 38,000 and the population of B has increased by 205,000, whereas C's population remained the same. The 13 representatives are reapportioned using the Hamilton method. Does this example illustrate the population paradox? Redo the problem using Lowndes' method to allocate 10 representatives.

35. Southeastern Pennsylvania Transportation Authority (SEPTA) has three lines: the Broad Street Line, the Market-Frankford Line, and the Regional Rail. The numbers of passengers who use each line per day are given in the following table:

Line	Broad Street	Market-Frankford	Regional Rail
Passengers per Day	35,200	6,760	87,310

   Assume that 88 new cars are apportioned among the 3 lines according to the number of passengers using each line. One year later, the Authority is able to purchase another eighty eight cars and again apportions them among the three lines using either Hamilton's method or Lowndes' method. Which of these two apportionment methods illustrates the population paradox?

36. A group of three regional airports has hired a company to increase security. The contract calls for 10 security personnel to be apportioned to the airports according to the number of passengers using each airport each day. One year later, the number of passengers per day increases at the three airports, so the ten security personnel are reallocated using Lowndes' method. This data is shown in the table. Does this example illustrate the population paradox?

Airport	A	B	C
Passengers per Day (now)	763	5,138	8,609
Passengers per Day (one year later)	801	5,343	8,609

37. The University of Pennsylvania is one of the oldest universities in America. The current enrollments at the university's three campuses are shown in the table. The administration has decided to upgrade 11 computer labs according to the enrollments at the 3 campuses using the Hamilton method. One year later, the enrollments have increased at each campus and again it is decided to upgrade eleven labs. Does this example illustrate the population paradox?

Campus	Altoona	Berks	DuBois
Students (now)	12,765	2,981	36,042
Students (one year later)	13,147	3,100	36,044

38. Redo the previous problem using Lowndes' method for 17 computer labs. Do you observe the population paradox?

39. There are three emergency clinics in a city. Statistics have been gathered regarding the annual number of patients seen on the night shift for each clinic. Apportion 17 emergency physicians using Hamilton's method.

Clinic	A	B	C
Patients	245,709	11,036	47,813
Patients (one year later)	252,080	11,587	47,813

The average number of patients one year later is shown in the last row. Do you observe the population paradox? Redo the problem using Lowndes' method for 27 physicians.

40. Consider Example 2.2.3, page 165. Suppose the population of state A grows by 10, but the population of state C grows by 100, other states keep the same populations. Reapportion 68 seats between these 4 states using the Hamilton method. What kind of paradox do you observe?

41. Eleven representatives are apportioned among three states A, B, and C, which have the populations (in thousands) 6543, 4789, and 498, respectively. The next census shows that the population of state A was unchanged, but the populations of states B and C grew by 30,000 and 10,000, respectively. Reapportion 11 representatives to observe the population paradox.

42. In 2008, the Community College of Rhode Island experienced the highest student enrollment, as shown below. Currently, the college has 33 computer labs located in 3 branch campuses. The college is opening a new campus, with one computer lab in accordance with predicted enrollment. Does this example illustrate the new states paradox? Does the answer depend on the method of allocation?

Campus	Knight	Flanagan	Liston	New
Enrollment	9,702	5,398	1,795	717

2.3   Jefferson's Method

Many of the Founding Fathers of the United States came up with their own methods for apportionment. One of them, Thomas Jefferson (1743 – 1826), proposed the method now bearing his name, which was the first apportionment method used in the United States. Jefferson's method is also known as the method of greatest divisors, which is a representative of the class of apportionment methods we will discuss in the next three sections.

Thomas Jefferson was the third president of the United States (1801 – 1809), the principal author of the Declaration of Independence (1776), and one of the most influential Founding Fathers for his promotion of the ideals of republicanism in the United States. His sympathy for the French Revolution led him into conflict with Alexander Hamilton when Jefferson was Secretary of State in President Washington's Cabinet. Both friends, Jefferson and Washington, were from a large state—Virginia. It happens that in the first apportionment, in 1792, the largest state, Virginia (population of 630,558,) was awarded an additional representative, at the expense of the smallest state, Delaware (population of 55,538). Jefferson's method was followed more or less for about half of a century, until 1841. His picture is on the $2 bill. We often abbreviate/identify Jefferson's allocation with $2.

Jefferson's method is an example of a **divisor method** that replaces the natural divisor with a different number, called the **modified divisor**. Then using this modified divisor, we find the **modified quotas** by computing the ratios of states' populations over the modified divisor for every state, followed by a rounding procedure to obtain the required allocation. All divisor methods differ from each other in the rounding of modified quotas. Such a modified divisor is not unique—there are infinite possible values. Any number from a specific interval (in exceptional cases, this interval could be empty, which indicates that the divisor method is inconclusive) can be chosen as a modified divisor, and each of them will produce the correct final allocation. Therefore, the goal of a divisor method is to determine the interval from which a modified divisor should be picked, and then to verify that this choice leads to the correct apportionment.

For the initial allocation, the Jefferson method uses the same approach as the Hamilton method: it rounds natural quotas down, called floors or lower quotas. Except in rare cases when all standard quotas are integers, this initial allocation is always less than the required allocation. Therefore, to obtain larger quotas, we need to choose a modified divisor as a number smaller than the natural divisor.

Jefferson's method, which is a divisor method with rounding down, is also known as the method of d'Hondt or the Hagenbach-Bischoff method. Legislatures using this system include those of Albania, Argentina, Austria, Belgium, Brazil, Bulgaria, Chile, Colombia, Croatia, Czech Republic, Denmark, East Timor, Ecuador, Estonia, Finland, Hungary, Iceland, Israel, Japan, Republic of Macedonia, Republic of Moldova, Montenegro, the Netherlands, Northern Ireland, Paraguay, Poland, Portugal, Romania, Scotland, Serbia, Slovenia, Spain, Turkey, Venezuela, and Wales.

Let us outline the procedure (called an algorithm) needed to implement the Jefferson method.

1. Calculate natural (standard) quotas for each state according to Eq. (2.1.2), page 150.

2. Initially assign each state its lower quota—the floor of its natural quota. This allocation is called the initial or tentative allocation.

3. Check to see whether the sum of the lower quotas is equal to the correct number of seats to be apportioned. If the sum of the lower quotas is equal to the house size, then apportion to each state the number of seats equal to its quota. If at least one natural quota is not an integer, the initial allocation is less than required, calculate the number of surplus seats needed.

4. When the initial allocation is NOT equal to the correct number of seats to be apportioned, then calculate the threshold divisor for every state according to the formula

$$\textbf{Threshold Divisor} = \frac{\text{state's population}}{\text{initial allocation} + 1} = \frac{p_i}{\lfloor q_i \rfloor + 1}, \qquad (2.3.1)$$

where p_i is the population of state i and q_i is the natural quota of this state. We abbreviate the threshold divisor as **TD**.

5. Rank all threshold divisors from largest to smallest. Such an ordering breaks the axis of real numbers into subintervals having endpoints as threshold divisors.

6. If the initial allocation is m seats short (the number of surplus seats), choose a modified divisor (**MD**) as any number between the $(m + 1)$th divisor and the mth divisor from the ordered list obtained in the previous part.

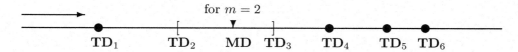

7. For each state, calculate the modified quota:

$$\textbf{Modified Quota} = \frac{\text{state's population}}{\text{modified divisor}} = \frac{p_i}{\textbf{MD}}. \qquad (2.3.2)$$

Rounding down modified quotas to the nearest integers (floors of modified quotas) are called **modified lower quotas**. Assignment of that number (= modified lower quota) of representatives to each state is called the final allocation. If the sum of seats in the final allocation is the exact number of seats to be apportioned, you are done.

8. If the sum of rounded down modified quotas exceeds the house size, this indicates that either the quota property is violated and at least one state gains more than one seat or more than m (the number of surplus seats) states get surplus seats. In this case, calculate the threshold divisors for each additional seat each state gains. Thus, if a state gains s ($s \geqslant 1$) additional seats compared to the upper quota, calculate

$$\mathbf{TD}_1 = \frac{p_i}{\lfloor q_i \rfloor + 1}, \quad \mathbf{TD}_2 = \frac{p_i}{\lfloor q_i \rfloor + 2}, \quad \ldots, \quad \mathbf{TD}_s = \frac{p_i}{\lfloor q_i \rfloor + s}. \qquad (2.3.3)$$

9. Rank all threshold divisors (from parts 4 and 8); choose a modified divisor (**MD**) as any number between the $(m+1)$th divisor and the mth divisor from the ordered list of all threshold divisors.

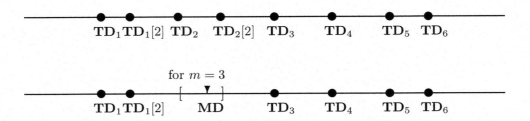

10. Repeat part 7 to obtain modified quotas. By rounding them down, you should obtain the required final allocation. Assign to every state the modified lower quota of seats.

    ■

Note that the modified divisor is not unique and could be any number in the range indicated in parts six or nine of the algorithm. It will always be smaller than the natural divisor. By assigning to each state its modified lower quota we obtain the final allocation, which may allot no seat to a state. The method is inconclusive when the mth and the $(m+1)$th threshold divisors are the same.

Jefferson's method (known also in Europe as d'Hondt[1]) is clearly generous to large states at the expense of small states, which is not necessarily a bad thing. In the European democracy context, if a country uses Jefferson's method, then parties that get relatively large votes are likely to get more than their fair share of seats. This tendency, some believe, means trading stability to some extent for equity and fairness. If it is more likely for a single party to get a majority in parliament, or to more easily form a coalition of parties to govern, it may be better for society than having an unstable coalition form. Therefore, Jefferson's method encourages a formation of coalition and the merging of parties. A coalition with many small partners may result in many changes of government, which may not be healthy in the long term. Political scientists have done a variety of empirical studies related to these issues.

---

[1]Victor d'Hondt (1841 – 1901) was a Belgian lawyer, salesman, jurist of civil law at Ghent University, and mathematician.

Example 2.3.1. Jefferson's Method

New England is a region in the northeastern corner of the United States, bordered by the Atlantic Ocean, Canada, and the state of New York. It consists of the modern U.S. states of Maine, New Hampshire, Vermont, Massachusetts, Rhode Island, and Connecticut. Suppose these six states decide to convene 25 representatives for a conference to make recommendations on further land reservations of wild nature in this region. The conference organizers decide that the number of representatives from each state should be proportional to its population. According to the U.S. Bureau of the Census estimates for 2008, Maine had 1,315,809 people; New Hampshire had 1,316,456 people; Vermont had 621,270 people; Massachusetts had 6,497,967 people; Rhode Island had 1,050,788 people; and Connecticut had 3,501,252 people.

State	Popu-lation	Nat. Quota	$\lfloor q \rfloor$ I. A.	Threshold Divisor	Rank	M. Quota (MD = 510000)	$2 F. A.
NH	1315809	2.2998	2	438603	5	2.58	2
ME	1316456	2.3009	2	438818.$\overline{6}$	4	2.58	2
VT	621270	1.0858	1	310635	6	1.2	1
MA	6497967	11.3573	11	541497.25	1	12.7	12
RI	1050788	1.8366	1	525394	2 ≺MD	2.06	2
CT	3501252	6.1195	6	500178.8571	3	6.86	6
Total:	14303542	25	23				25

Let us review the calculations given in the table. Natural quotas for each state are determined according to the formula (2.1.2), page 150. When these quotas are rounded down (the resulting integers are called lower quotas), we end up with two surplus seats. Then we calculate the threshold divisors, see Eq. (2.3.1). For example, New Hampshire's threshold divisor is $1315809/3 = 438603$. After ranking/ordering all threshold divisors, we pick a modified divisor as any real number between the threshold divisor of RI (which is the second in the ordered list) and CT (which is the third). So the modified divisor should be chosen from the interval

$$500178.8571 < \textbf{Jefferson's Modified Divisor} \leqslant 525394.$$

Any number from this interval can be used as a modified divisor, including the upper bound, 525394. For instance, we chose **MD = 510000**. When we divide it into the state's populations, we obtain modified quotas, presented in the seventh column of the table. These modified quotas are rounded down to the nearest integers (called modified lower quotas) that are written in the last column, which shows the final allocation of seats. Another modified divisor from the interval will lead to different modified quotas; however, the final allocation will be always the same.

Example 2.3.2. Rounding Numbers

Here is a typical apportionment problem of rounding numbers to integers in the sum:

$$0.17 + 1.21 + 2.57 + 3.39 + 2.66 = 10.$$

Each number in the sum can be considered as the natural quota: $q_1 = 0.17$, $q_2 = 1.21$, $q_3 = 2.57$, $q_4 = 3.39$, and $q_5 = 2.66$. The initial allocation is the same as the Hamilton one, so we round the quotas down to the nearest integer:

$$\lfloor 0.17 \rfloor + \lfloor 1.21 \rfloor + \lfloor 2.57 \rfloor + \lfloor 3.39 \rfloor + \lfloor 2.66 \rfloor = 0 + 1 + 2 + 3 + 2 = 8.$$

This allocation is two seats less than required. To apply Jefferson's method, first we must calculate the threshold divisors for every state and rank them. Since we need to add two seats, we choose a modified divisor as a number between the second and the third threshold divisors after ranking.

In this rounding problem, the state's populations are not given. However, this information is not important because we can calculate them using the formula (2.1.5), page 155. Since the state's population is proportional to the state's quota, we can pick up the natural divisor to be any number we wish, for instance, we could take it to be 1. This leads to the following table (we use abbreviation **I.A.** for the initial allocation, **MD** for modified divisor, and **F.A.** for the final allocation), which speaks for itself.

#	Natural Quota	$\lfloor q \rfloor$ I. A.	Threshold Divisor	Rank	Modified Quota (with **MD** = 0.85)	Jefferson F. A.
1	0.17	0	0.17	5	0.2	0
2	1.21	1	0.605	4	1.4235	1
3	2.57	2	0.856$\overline{6}$	2$\prec$MD	3.0235	3
4	3.39	3	0.8475	3	3.9882	3
5	2.66	2	0.886$\overline{6}$	1	3.1294	3
Total:	10	8				10

Note that we could choose a modified divisor as any number from the interval: $0.8475 <$ modified divisor (**MD**) $\leqslant 0.856\overline{6}$. The final allocation will be the same, independently of another choice for **MD** within this interval. The lower bound of this interval (0.8475) cannot be used because it would lead to a wrong apportionment of 11. However, the upper bound is a legitimate value for the modified divisor.

Now assume that we choose another natural divisor, say 2. Then for every state, its population will be twice as large as the corresponding natural quota. This yields the following data presented in the table.

#	Population	Natural Quota	$\lfloor q \rfloor$ I. A.	Threshold Divisor	Rank	M. Quota (MD = 1.7)	$2 F. A.
1	0.34	0.17	0	0.34	5	0.2	0
2	2.42	1.21	1	1.21	4	1.4235	1
3	5.14	2.57	2	1.713$\overline{3}$	2$\prec$MD	3.0235	3
4	6.78	3.39	3	1.695	3	3.9882	3
5	5.32	2.66	2	1.773$\overline{3}$	1	3.1294	3
Total:		10	8				10

As it is evidenced in the table, the modified divisor is $1.7 = 2 \times 0.85$, which is twice as large as the corresponding divisor in the previous table. Choosing another natural divisor will lead to the same apportionment. Note that notation $2_{\prec MD}$ indicates that the modified divisor is smaller than the second largest threshold divisor.

Example 2.3.3. Jefferson's Method could be Inconclusive

Consider the results of the 2008/09 winner in the Northwest Division (NHL) Vancouver Canucks. The team won 45 games, lost 27 games, and lost 10 overtimes. In order to present the results as whole percent numbers, we make calculations to obtain the table

Game	#	Natural Quota	$\lfloor q \rfloor$ I. A.	Threshold Divisor	Rank	M. Quota (MD = 0.8)	\$2 F. A.
Win	45	54.878%	54	0.8181	1	0.2	55
Loss	27	32.926%	32	0.8181	$2_{\prec MD}$	1.4235	33
OTL	10	12.195%	12	0.7692	3	3.0235	12
Total:	82	100%	98				100

Since the sum of lower quotas is two seats short, we choose a modified divisor in the range between the second threshold divisor and the third one in the ordered list of all threshold divisors. Let us pick the modified divisor as **MD** = 0.8. After dividing the population of each state (Win, Loss, and OTL) by **MD**, we obtain the modified quotas that are rounded down to provide the final allocation. Note that we were lucky because of the tie between the threshold divisors for Win and Loss. If there were a tie between threshold divisors for Loss and OTL, Jefferson's method would be inconclusive because it is impossible to choose a modified divisor.

Now let us consider another hypothetical tournament with a team that finished with the following results:

Game	#	Natural Quota	$\lfloor q \rfloor$ I. A.	Threshold Divisor	Rank
Win	45	$\frac{45}{87} \times 100\% \approx 51.7241$	51	0.8653846	1
Loss	30	$\frac{30}{87} \times 100\% \approx 34.4827$	34	0.8571428	$2, 3_{\prec MD}$
OTL	12	$\frac{12}{87} \times 100\% \approx 13.7931$	13	0.8571428	2,3
Total:	87	100%	98		

There is a tie between two threshold divisors—the second and the third are the same. So we see that Jefferson's method is inconclusive because it is impossible to choose a modified divisor—any number will lead to any allocation but not the 100.

Example 2.3.4. Some States are Without a Seat

In the U.S., the state of Hawaii consists of many islands in the Pacific ocean. Travelers usually visit six of them: Hawaii's Big Island, Kauai, Oahu, Molokai, Lanai, and Maui. Using Jefferson's method, allocate 39 seats among the six islands based on their 2009 populations presented here.

Island	Hawaii	Kauai	Oahu	Molokai	Lanai	Maui
Population	175,784	63,689	905,601	7,404	3,193	143,574

We illustrate the details of how to use the algorithm shown on page 179 in the following table.

Island	Popu-lation	Natural Quota	$\lfloor q \rfloor$ I. A.	Threshold Divisor	Rank	M. Quota MD = 29300	\$2 F. A.
Hawaii	175,784	5.276584	5	29297.33333	3 $\prec$ MD	5.99	5
Kauai	63,689	1.911780	1	31844.5	2	2.17	2
Oahu	905,601	27.18381	27	32342.89286	1	30.9	30
Molokai	7,404	0.222249	0	7404	5	0.25	0
Lanai	3,193	0.095845	0	3193	6	0.10	0
Maui	143,574	4.309722	4	28714.8	4	4.90	4
Total:	1299245	39	37				41

Since one state, Oahu, gained more than one seat, we need to add two more threshold divisors for every additional seat that the state got:

$$\mathbf{TD_2} = \frac{905601}{29} \approx 31227.620, \quad \mathbf{TD_3} = \frac{905601}{30} = 30186.7.$$

Now we order all these threshold divisors; however, we only need the three largest ones because a modified divisor should be chosen between the second threshold divisor and the third one:

Island	Oahu[1]	Kauai	Oahu[2]	Oahu[3]	Others
TD	32342.892	31844.5	31227.620	30186.7	$\cdots$

Therefore, a modified divisor is a number from the interval $31227.62069 < \mathbf{MD} \leqslant 31844.5$. Choosing, for instance, $\mathbf{MD} = 31500$, we get the required allocation:

Island	Hawaii	Kauai	Oahu	Molokai	Lanai	Maui	Total
Modified Quota	5.58	2.02	28.74	0.23	0.10	4.55	
Final Allocation	5	2	28	0	0	4	39

Example 2.3.5. Violation of Quota Property

The state of Rhode Island consists of five counties: Bristol, Kent, Newport, Providence, and Washington. Suppose that the state wants to allocate 20 seats in the council between these 5 counties using Jefferson's method. Based on the 2000 census, we come up with the table, presented on the next page.

Following the general procedure, we pick up a modified divisor from the range $41772.5 < \mathbf{MD} \leqslant 42716.5$ because we need to add three seats to the initial allocation. If we choose, for instance, $\mathbf{MD} = 42000$, the final allocation will contain two extra seats. This means that the quota property is violated: the county of Providence gained three seats. Since all other counties did not gain seats more than the upper quota, we calculate two (others will be smaller, and remember that we need to add three seats) auxiliary threshold divisors:

County	Popu-lation	Natural Quota	I.A. $\lfloor q \rfloor$	Threshold Divisor	Rank	Mod. Quota	Final Alloc.
Bristol	50,648	0.9662	0	50648	2	1.2	1
Kent	167,090	3.1877	3	41772.5	4	3.9	3
Newport	85,433	1.6299	1	42716.5	3 $\prec$ MD	2.03	2
Providence	621,602	11.8590	11	51800.1$\overline{6}$	1	14.8	14
Washington	123,546	2.3570	2	41182	5	2.9	2
Total:	1048319	20	17				22

$$\mathbf{TD_2} = \frac{621,602}{13} \approx 47815.538, \quad \mathbf{TD_3} = \frac{621,602}{14} \approx 44400.14286.$$

Ranking all seven threshold divisors, we obtain the ordered list:

City	Providence[1]	Bristol	Providence[2]	Providence[3]	Others
TD	51800.1(6)	50648	47815.538	44400.14286	$\cdots$

Hence, a modified divisor should be chosen from the range $44400.14286 < \mathbf{MD} \leqslant 47815.\overline{538461}$. Let us pick it as $\mathbf{MD} = 45000$. Then the allocation will be as follows

City	Bristol	Kent	Newport	Providence	Washington	**Total**
MD	1.12	3.71	1.89	13.81	2.74	
F.A.	1	3	1	13	2	20

So we see that one county (Providence) gained two additional seats compared to the initial allocation, which is a violation of the quota property.

2.3.1 Adams' Method

As we learned from previous discussions and examples, Jefferson's method may violate the quota property (see Example 2.3.5) and is generous to large states at the expense of small states. Another method was proposed in 1832 by representative of Massachusetts and the former sixth president of the United States, John Quincy Adams (1767 – 1848). This method, also known as the method of smallest divisors, is a mirror image of Jefferson's method because it apportions seats starting with upper quotas. The method is called the **Adams method** and it favors small states. It has one main advantage compared to Jefferson's method—it always allocates at least one seat to every state. However, Adams' method suffers a similar flaw—it may violate the lower quota property. After considerable debate in 1832, the Senate rejected the Adams method claiming that a practical application of Adams' method would lead to the disintegration of parties and coalitions.

The implementation of the Adams method follows exactly the same steps as Jefferson's algorithm, page 179. Therefore, we will outline the differences in its application and illustrate the Adams method in the following examples.

The Adams method, which is a divisor method, starts with rounding up natural quotas. Since its initial allocation is larger than the house size (unless all natural quotas are integers), one needs to choose a modified divisor that is greater than the natural divisor. Thus, the next step is to calculate the state's threshold divisors:

$$\textbf{Threshold Divisor} = \frac{\text{state's population}}{\text{initial allocation} - 1} = \begin{cases} \infty, & \text{if } \lceil q_i \rceil = 1, \\ \frac{p_i}{\lceil q_i \rceil - 1}, & \text{if } \lceil q_i \rceil > 1. \end{cases} \qquad (2.3.4)$$

Then you need to repeat steps 5 through 7 in Jefferson's algorithm by rounding the modified quotas (2.3.2) up to obtain the final allocation. If the sum of these ceilings is less than the correct number of seats—which is a violation of the quota property—you should calculate additional threshold divisors similar to Eq. (2.3.3):

$$\textbf{TD}_1 = \frac{p_i}{\lceil q_i \rceil - 1}, \quad \textbf{TD}_2 = \frac{p_i}{\lceil q_i \rceil - 2}, \quad \ldots, \quad \textbf{TD}_s = \frac{p_i}{\lceil q_i \rceil - s}. \qquad (2.3.5)$$

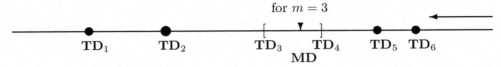

Finally, ranking all threshold divisors and choosing the appropriate range for picking a modified divisor, you should calculate the modified quotas, and round them up to obtain the final allocation. Note that when identifying the appropriate interval from which a modified divisor should be chosen, one needs to start counting from the smallest threshold divisor. If there are m surplus seats, choose the interval with endpoints of the mth and the $(m+1)$th smallest divisors from the ranking list. All details will become much clearer when we go through the examples.

Example 2.3.6. Adams' Method

In 2006, five states sent petitions where they solicited the government to provide the money needed to improve their economic situation. The committee decided to distribute 500 million dollars to these states using the Adams method because they wanted to use a method that benefited states with lower per capita incomes. Collecting the data, the committee came up with the following calculations:

State	Per Capita Income	Natural Quota	I.A. $\lceil q \rceil$	Threshold Divisor	Rank	Mod. Quota	Final Alloc.
Idaho	$29,920	84.9980	85	356.1904	1	84.460	85
Mississippi	$27,028	76.7823	77	355.6315	2	76.296	77
Nevada	$38,994	110.7758	111	354.4909	3	110.078	111
New Jersey	$46,763	132.8464	133	354.2651	4←MD	132.005	133
Oregon	$33,299	94.5972	95	354.2446	5	93.998	94
Total:	176,004	500	501				500

Let us examine the calculations for one state, for instance, Idaho. Its natural quota, $\frac{29,920}{176,004} \times 500 \approx 84.99806823$, is rounded up to get its initial allocation, 85. The threshold divisor for this state is $\dfrac{\text{Idaho's population}}{\text{lower quota}} = \dfrac{29,920}{84} \approx 356.1904762$, which is the largest among all threshold divisors. After ranking all threshold divisors, we see that the natural divisor, **ND** = 352.006, is smaller than any of the states' threshold divisors. Since we need to eliminate one seat from the initial allocation, we must choose a modified divisor in the range between the fifth threshold divisor, which is Oregon's, and the fourth divisor, which is New Jersey's. Therefore,

$$500178.8571 \approx \frac{33299}{94} \leqslant \textbf{Adams' Modified Divisor} < 354.26\overline{51},$$

and we pick **MD** = 354.25. Note that any other number within this range chosen as a modified divisor will lead to the same final allocation. Dividing the modified divisor into the populations of each state, we come up with modified quotas, presented in the second to the last column of the table, that are rounded up to provide the final allocation: all states but Oregon (having the largest per capita income) get their upper quotas.

Example 2.3.7. Violation of the Quota Property by Adams' Method
Each summer, all private schools in the Saint Paul area of Minneapolis choose new members for their educational committee for the next academic year. The committee consists of 58 people who are chosen from private schools based on students' enrollments. To apportion these seats, we use Adams' method, and summarize data in the following table.

School	Enroll-ment	Natural Quota	I.A. $\lceil q \rceil$	Threshold Divisor	Rank	Mod. Quota	Final Alloc.
Accell	16	0.2468	1	∞	1 ≺MD	0.22	1
Blake	1316	20.3	21	65.8	3	18.8	19
Totino	1112	17.1531	18	65.4117	4	15.88	16
Loyola	650	10.0265	11	65	5,6	9.28	10
Cotter	471	7.2654	8	67.2857	2	6.72	7
Groves	195	3.0079	4	65	5,6	2.78	3
Total:	3760	58	63				56

Since the initial allocation is larger than the house size, we need to eliminate five seats by choosing a modified divisor that is larger than the natural divisor, **ND** = total population/ house size ≈ 64.8275. We choose a modified divisor from the range with endpoints 67.2857 and ∞, which are threshold divisors in the first and second places in the ordered list. Note that we do not have an exact value of the upper bound for this range, so we can only guess its upper value. Picking the modified divisor to be 70, we come up with modified quotas presented in the seventh column of the table. Once we round them up, the final allocation becomes two seats short of the correct number. Hence, we observe a violation of the quota property: two states (Blake and Totino) get a number of seats less than their lower quotas.

In the case of such a violation, we need to calculate two more threshold divisors, one for the state *Blake* and one for *Totino*. Why only one divisor and why for two states? Because there are only two states that were initially allocated with the number of seats other than their lower or upper quotas. We only have to calculate one threshold divisor because their initial allocations differ by one seat from their lower quotas. So we find two more threshold divisors: $\mathbf{TD}_{\text{Blake}[2]} = \dfrac{1316}{19} \approx 69.2631$ and $\mathbf{TD}_{\text{Totino}[2]} = \dfrac{1112}{16} = 69.5$. Now we rank all threshold divisors

Rank	1	2	3	4	5	6	7
School	Accell	Totino[2]	Blake[2]	Cotter	Blake[1]	Totino[1]	Loyola/Groves
TD	∞	69.5	69.2631	67.2857	65.8	65.4117	65

Counting from the right (starting from the smallest threshold divisor) five times to eliminate five seats from the initial allocation, we conclude that a threshold divisor should be chosen in the range $67.\overline{285714} \leqslant \mathbf{MD} < 69.26315789$. For example, picking $\mathbf{MD}$ to be 69, we come up with the following allocation that gives the correct number of seats:

School	Accell	Blake	Totino	Loyola	Cotter	Groves	Total
M. Quota	0.23	19.35	16.35	9.55	6.92	2.86	
F.A. ($\lceil q \rceil$)	1	20	17	10	7	3	58

Example 2.3.8. Comparison of Four Apportionment Methods

The Hoover Dam, once known as the Boulder Dam, is a concrete arch-gravity dam in the Black Canyon of the Colorado River, in Nevada. When completed in 1936, it was the world's largest hydroelectric power generating station. The dam and the power plant are operated by the Bureau of Reclamation of the U.S. Department of the Interior. The Bureau of Reclamation reports that the energy generated in 2009 is allocated as follows:

Metropolitan Water District of Southern California – 28.5393%,

State of Nevada – 23.3706%,   State of Arizona – 18.9527%,

Los Angeles, California – 15.4229%,   Boulder City, Nevada – 1.7672%.

Southern California Edison Company – 5.5377%,

Seats on the 15-member board, which coordinates operations of the Bureau of Reclamation, are assigned according to the energy consumed. To apportion these seats in the board between six main consumers we will use four apportionment methods: Hamilton's, Lowndes', Jefferson's, and Adams' method.

We start with the first three methods because they all use the same initial allocation. Next, we need to calculate the natural quotas, `state's population * house size/total population`, and then round them down to obtain the tentative allocation:

State	Metro	NV	AZ	LA	Edison	Boulder
Natural Quota	$\dfrac{28.5393*15}{93.5904}$	$\dfrac{23.3706*15}{93.5904}$	$\dfrac{18.9527*15}{93.5904}$	$\dfrac{15.4229*15}{93.5904}$	$\dfrac{5.5377*15}{93.5904}$	$\dfrac{1.7672*15}{93.5904}$
Approx. of NQ	4.5740	3.7456	3.0376	2.4718	0.8875	0.2832
Initial allocation	4	3	3	2	0	0

The initial allocation is three seats short, so we need to distribute three surplus seats between these six customers. Using Hamilton's method, we distribute seats to states based on the values of their fractional parts in natural quotas. Then we get the Hamilton final allocation:

State	Metro	NV	AZ	LA	Edison	Boulder
Fractional part of NQ	0.5740	0.7456	0.0376	0.4718	0.8875	0.2832
Hamilton's final allocation	5	4	3	2	1	0

Lowndes' method uses the relative fractional parts (which are the ratios of the fractional parts over the corresponding integer parts) of natural quotas instead of their fractional parts. Hence, we get the following final allocation

State	Metro	NV	AZ	LA	Edison	Boulder
R.F.P.	$\frac{0.5740}{4}$	$\frac{0.7456}{3}$	$\frac{0.0376}{3}$	$\frac{0.4718}{2}$	∞	∞
Approximation of R.F.P.	.1435	.2485	.0125	.2359		
Lowndes' final allocation	4	4	3	2	1	1

For Jefferson's method, any number between the third largest threshold divisor and the fourth one may be chosen as the modified divisor. Choosing **MD** = 5.5 (or any number in the range $5.1409\overline{6} < \textbf{MD} \leqslant 5.5377$), we get

Customer	Energy consumed	Natural Quota	$\lfloor q \rfloor$ I. A.	Threshold Divisor	Rank	Modified Quota	$2 F. A.
Metro	28.5393	4.5740	4	5.707860	2	5.1	5
Nevada	23.3706	3.7456	3	5.842650	1	4.2	4
Arizona	18.9527	3.0376	3	4.738175	5	3.4	3
LA	15.4229	2.4718	2	**$5.1409\overline{6}$**	4	2.8	2
Edison	5.5377	0.8875	0	**5.5377**	3	1.0	1
Boulder	1.7672	0.2832	0	1.7672	6	0.3	0
Total	93.5904	15	12				15

The Adams method initially allocates upper quotas to every state:

State	South CA	NV	AZ	LA	Edison	Boulder	**Total**
Initial Allocation	5	4	4	3	1	1	18

This tentative allocation has three extra seats that must be removed. To determine the interval from which a modified divisor can be chosen, we calculate the threshold divisors according to the formula (2.3.4) for every state and then rank them.

State	Metro	NV	AZ	LA	Edison	Boulder
Threshold Divisor	$\frac{28.5393}{4}$	$\frac{23.3706}{3}$	$\frac{18.9527}{3}$	$\frac{15.4229}{2}$	∞	∞
Approximation of TD	7.134825	**7.7902**	6.31756	**7.71145**	∞	∞
Rank	5	3	6	4	1,2	1,2

Counting from the lowest threshold divisor three times (because we need to delete three seats from the initial Adams' allocation), we identify the upper bound for the modified divisors to be 7.7902; the next threshold divisor gives us the lower bound. Hence, a modified divisor should be chosen from the interval $7.71145 \leqslant \textbf{MD} < 7.7902$. Choosing **MD** = 7.75 and dividing it into every state's population, we obtain modified quotas that are rounded up to provide the final allocation:

State	Metro	NV	AZ	LA	Edison	Boulder
Modified Quota	$\frac{28.5393}{7.75}$	$\frac{23.3706}{7.75}$	$\frac{18.9527}{7.75}$	$\frac{15.4229}{7.75}$	$\frac{5.5377}{7.75}$	$\frac{1.7672}{7.75}$
Approximation of MQ	3.6824	3.0155	2.4455	1.9900	0.7145	0.2280
Adams' final allocation	4	4	3	2	1	1

Example 2.3.9. Adams' Method versus Jefferson's Method

We reconsider the previous example by allocating 58 seats to six schools using Jefferson's method. So we repeat almost all calculations to obtain the following data:

School	Enroll-ment	Natural Quota	I.A. $\lceil q \rceil$	Threshold Divisor	Rank	Mod. Quota	\$2 F. A.
Accell	16	0.2468	0	16	6	0.25	0
Blake	1316	20.3	20	$62.\overline{66}$	1 ≺MD	21.22	21
Totino	1112	17.1531	17	$61.\overline{77}$	2	17.93	17
Loyola	650	10.0265	10	$59.\overline{09}$	3	10.48	10
Cotter	471	7.2654	7	58.875	4	7.59	7
Groves	195	3.0079	3	48.75	5	3.14	3
Total:	3760	58	57				58

The initial allocation is only one seat short; hence, we choose a modified divisor from the interval with end points $62.\overline{66}$, which is the largest value in the ordered list, and $61.\overline{77}$, which is the second largest one in the ranked list of threshold divisors. For instance, choosing the modified divisor as **MD** $= 62$, we obtain the correct final allocation. This example shows that Jefferson's apportionment may be different from an allocation based on the Adams method.

Problems.

1. A labor council is being formed from the members of four unions. The electricians union has 33 members, the plumbers have 28 members, the painters have 23, and the carpenters have 36 members. Use the Jefferson method to apportion 10 representatives between these 4 unions.

2. Use Jefferson's method to apportion the 53-member committee between 4 states based on their populations:

State	A	B	C	D
Population	9,154	4,273	3,667	1,416

3. Use Jefferson's method to apportion the 70-member committee between 4 states based on their populations:

State	A	B	C	D
Population	18,156	41,723	13,766	12,465

4. In certain areas of the country, there are more people who speak a foreign language than in others. According to demographic surveys, there are high concentrations of non-English

speaking minorities in inner cities. Specifically, we look at cities closer to borders that have a high number of unemployed non-English speaking immigrants. For example, the city of Miami houses many Spanish speaking people because of its proximity to Cuba, and Cuba's high defection rate. Therefore, the Miami Police Department has a special need for Spanish speaking police officers because there is such a high frequency of Latino Americans in the city. Just as New York is divided into precincts, so is the city of Miami. Naturally there are certain areas of the city that are more prone to crime than others. There are also certain areas that would be considered very wealthy, meaning most of those residents have the ability to speak English. Again according to demographic research, the non-English speaking citizens tend to live in the lower class areas because of their inability to get a job without speaking English. For this fact, the Police Department will allocate their 150 Spanish speaking officers into areas that have a high concentration of Spanish speaking citizens, such as Olympia Heights, with a population of 14189; Butler Bay, with a population of 2770; West Miami, with a population of 5863; and Hallandale, with a population of 34282. Apportion 150 Spanish speaking officers using Jefferson's method.

5. The Nurek Dam, known simply as Nurek, is the tallest dam (its height is 300 m) in the world! Construction of the dam took 20 years to complete, it was opened in 1980, when Tajikistan was a republic within the Soviet Union. The reservoir on the Vakhsh River formed by the dam is the largest reservoir in Tajikistan near the border with Afghanistan with a capacity of 10.5 km^3. The reservoir is over 70 km (40 mi) in length, and has a surface area of 98 km^2 (38 sq mi). The reservoir fuels the hydroelectric plant located within the dam, and stored water is also used for irrigation of local agricultural land. Irrigation water is transported 14 kilometers through the Dangara irrigation tunnel and is used to irrigate about 700 km^2 (300 sq mi) of farmland.

There are four provinces in Tajikistan:

Province	Area in km^2	Population
Sughd	25,400	2,132,100
Republican Subordination	28,600	1,606,900
Khatlon	24,800	2,579,300
Gorno-Badakhshan	64,200	218,000

Use Jefferson's method to allocate a 15-member committee from each province based on the area and their 2008 populations. Are these two apportionments the same?

6. Use Jefferson's method to apportion 15 representatives among California, Kentucky, Hawaii, Rhode Island, and Texas. Their 2009 populations are presented here.

State	CA	KY	HI	RI	TX
Population	36,756,666	4,314,113	1,288,198	1,050,788	24,326,974

7. Fourteen fellowships are to be apportioned to students at a local university. The apportionment is to be based on the number of full-time graduate students enrolled in the writing program. There are 24 fiction majors, 12 poetry majors, 26 majoring in technical writing, and 31 majoring in writing for the media. Allocate these fellowships using the Jefferson method.

8. Using Jefferson's method, apportion 48 seats in a hypothetical legislature between 5 states based on their populations according to 2009 U.S. census.

State	Vermont	New Hampshire	Connecticut	Rhode Island	Maine
Population	621,270	1,315,809	3,501,252	1,050,788	1,316,456

9. Using Jefferson's method, apportion 37 seats in a hypothetical legislature between 5 states based on their 2009 populations.

State	Michigan	Minnesota	Montana	New York	Wyoming
Population	10,003,422	5,220,393	967,440	19,490,297	532,668

10. A health club instructor has a course load that allows her to teach 34 two-hour classes. A preregistration survey indicated the following interests:

    4 people want to take taekwondo;

    12 people want to take kick boxing;

    47 people want to take yoga; and

    41 people wants to take aerobics.

    Use the Jefferson method to apportion thirteen classes among four activities.

11. A community decided to set up a committee that will oversee the town's sport activities. The committee will consist of 20 people and will include representatives from each of the following sports. Apportion 20 people using the Jefferson method.

Sport	Basketball	Hockey	Football	Baseball	Golf
Number of teams	190	40	55	61	19

12. Uganda provides national elections for a president and a legislature. The National Assembly has 292 members. Of these, 214 members are elected without party labels directly in single seat constituencies, while 78 members are elected from so-called special interest groups. These include 53 district women representatives (one from each district), 10 army representatives, 5 youth representatives, 5 representatives from the disabled, and 5 from trade unions.

    The 2006 elections give the following results:

Party	Number of votes
National Resistance Movement	4,109,449
Forum for Democratic Change	2,592,954
Democratic Party	109,583
Independent	65,874
Uganda People's Congress	57,071
Total:	6,934,931

    Apportion 214 seats among these parties using Jefferson's method.

13. The city of Boston has asked its police chief to cut back on the use of motor vehicles in the summer to conserve money, The chief is charged with the task of re-allocating police cars throughout the city so that fewer are used, and only used when necessary. There are 11 active districts in the city of Boston and throughout those districts there are approximately 15 cars (including unmarked) per division. The chief will re-distribute the cars by general need. Instead of allocating all 165 cars, the BPD will only allocate 75 total cars over the 11 districts. Each district will have an increase of officers on foot and bike patrol in the spring and summer seasons. Districts E5, A, and E18 are the largest districts in the city and would require more cars because of their size. Each of these districts will get 10 cars. A15, A1, and D4 are the smallest and will only get 4 cars each. This leaves 33 cars remaining to be split between the 5 remaining districts in proportion to their populations. Apportion these cars using

    (a) Hamilton's method.   **(b)** Adams' method.   **(c)** Jefferson's method.

District	#1	#2	#3	#4	#5
Population	27,340	21,992	25,126	24,901	23,547

14. Suppose that a conference of Muslims from the seven countries of North Africa is being planned. There are 156 official delegates, allocated based on the populations (in thousands) of the countries in the year 2010:

County	Algeria	Egypt	Libya	Mauritania	Morocco	Tunisia	W. Sahara
Population	35,700	80,400	6,420	3,290	34,900	10,400	513

Apportion 156 seat between these seven countries using the Jefferson method.

15. There are 62 counties in the State of New York. The first 12 counties in New York were created immediately after the British annexation of the Dutch colony of New Amsterdam, although 2 of these counties have since been abolished. Use Jefferson's method to allocate 97 seats among 7 counties based on their 2009 populations.

County	Bronx	Erie	Kings	Nassau	Niagara	Oneida	Tioga
Population	1,332,650	950,265	2,465,326	1,334,544	219,846	235,469	51,784

16. There are 75 counties in the U.S. state of Arkansas. Allocate 36 seats between 7 counties based on their 2009 populations using
    (a) Hamilton's method.   (b) Adams' method.   (c) Jefferson's method.

County	Ashley	Benton	Calhoun	Desha	Fulton	Greene	Pulaski
Population	24,209	153,406	5,744	15,341	11,642	37,331	361,474

17. A summary of the 26 April 2009 Ecuadorian National Assembly election is presented here.

Party	Number of votes
PAIS Movement	1,655,785
Patriotic Society Party	538,769
Social Christian Party	491,033
Institutional Renewal Party of National Action	209,491
Ecuadorian Roldosist Party	148,846
Democratic People's Movement	146,594
United Left	68,153
National Democratic Coalition Movement	66,937
Municipalist Movement for National Integrity	66,619
Democratic Left	51,939
Pachakutik Plurinational Unity Movement, New Country	49,722
Total:	3,493,888

Apportion 124 seats among these 11 parties using
(a) Hamilton's method.   (b) Adams' method.   (c) Jefferson's method.

18. A legislative election in the Czech Republic took place on 28 May 2010. The Parliament has two chambers. The Chamber of Deputies has 200 members, elected for a four year term by proportional representation with a 5% election threshold.

Party	Number of votes
Czech Social Democratic Party	1,155,267
Civic Democratic Party	1,057,792
Tradition Responsibility Prosperity 09	873,833
Communist Party of Bohemia and Moravia	589,765
Public Affairs	569,127

Apportion 200 seats among these 5 parties using
(a) Hamilton's method.   (b) Adams' method.   (c) Jefferson's method.

19. Five European countries participate in building a new atomic plant that will supply electricity to these countries. Apportion 16-member committee among these countries using
    (a)   Hamilton's method.   (b)   Adams' method.   (c)   Jefferson's method.

County	Austria	Czech	Germany	Poland	Slovakia
2009 Population (in thousands)	8,357	10,507	81,880	38,164	5,380

20. During 2009, the following import cars were sold into the USA.

Maker	Porsche	Jaguar	Land Rover	Ferrari	Maserati	Bentley	Rolls Royce
Sales	3,176	1,575	3,599	143	155	162	66

Represent these numbers of cars sold in percentages using
(a)   Hamilton's method.   (b)   Lowndes' method.   (c)   Jefferson's method.

21. Using Jefferson's method, apportion 250 seats among 6 states based on their population (in thousands).

State	A	B	C	D	E	F
Population	1,646	6,936	154	2,091	685	988

22. The following data represent the sulfur emission of the following countries in 1000 metric tons in the year 1850.

State	Austria	Belgium	Denmark	France	Italy
Emission	7.386346	31.16598	1.572654	70.38246	1.180995

Represent these numbers as a percentage keeping one decimal place based on the Jefferson method.

23. For the new academic year, all fraternity and sorority members living on Greek Row were asked to vote on the new members for updating the Greek Student Council at their university. The distribution of seats depends on the number of votes each house receives:

Fraternity # 1 gets 213 (number of votes), Fraternity # 2 gets 178, Fraternity # 3 gets 40, Sorority # 1 gets 152, Sorority # 2 gets 195, Sorority # 3 gets 22.

Allocate 30 spots between 6 houses based on the Jefferson method.

24. The government wants to present the net costs of pursuing a master's degree in percentages based on the following data obtained in the USA during 2009. Using Jefferson's method, please help the government with its task based on the expenses in the table.

Major	Total Cost
Life science	$16,625
Math/Engineering	$17,082
Education	$14,518
Business	$19,853
Health	$20,073
Law	$29,252

25. The URI Fire Department is asked to re-do the smoke detector systems for the freshman dorms (Adams, Barlow, Bressler, Gorham, Merrow, and Tucker). The state fire marshal is only giving 120 smoke alarms to be allocated for these 6 buildings. Allocate smoke alarms based on the number of rooms in every dormitory using
    (a)   Hamilton's method.   (b)   Adams' method.   (c)   Jefferson's method.

Dormitory	Adams	Barlow	Bressler	Gorham	Merrow	Tucker
Number of Rooms	78	127	128	44	79	80

26. The European Parliament election of 2009 in Spain was intended to choose 50 delegates from Spain to the European Parliament. Allocate 50 seats between six parties based on the summary of results using
    (a) Hamilton's method.   (b) Adams' method.   (c) Jefferson's method.

Party	Votes
People's Party	6,615,015
Spanish Socialist Workers' Party	6,032,500
Coalition for Europe	802,225
The Left	583,708
Union, Progress and Democracy	449,499
The Greens	391,962

27. Solve Problem 27 from the previous section using Jefferson's method.

28. The government gave enough money to purchase 1000 new trees in order to restore fauna in five parks of New England. A committee decided that trees should be apportioned based on the number of visitors per year as shown in the following table. Use the Jefferson method to accomplish this goal.

Park	Number of visitors
Acadia National Park	2,602,227
Boston National Historical Park	2,330,011
Cape Cod National Seashore	4,915,414
Minute Man National Historical Park	869,884
Salem Maritime National Historical Site	762,869

29. Goddard Memorial State Park needs to plant new trees. So a committee decided to spend $10,000 according to the number of existing trees, presented in the following table. Use the Jefferson method to accomplish this goal.

Trees	Alder	Cedar	Hemlock	Sycamore	Tamaracks
Number of Trees	39	127	85	12	37

30. A school put up a sign up sheet for fall extra-curricular. The 78 students signed for soccer, 26 signed up for glee club, 107 signed up for chess, and 81 signed up for quidditch. There are 7 people available to supervise at those times. Using the Jefferson method, apportion the chaperones to the extra-curricular.

Sport	Popu-lation	Natural Quota	I.A. $10	Threshold Divisor	Rank	F.A. $2
Soccer	78					
Glee	26					
Chess	107					
Quidditch	81					
Total:	3770	7				7

31. A university college is setting up an organization that will oversee the school's sport events. The organization will consist of 30 people and will include representatives for each of the following sports. The number of students involved in each sport activity is presented in the table. Apportion the 30 people using the Jefferson method.

Sport	Basketball	Football	Tennis	Volleyball	Soccer
Number of Students	563	376	251	128	82

2.4   Webster's Method

Recall that a divisor apportionment method uses a number, d, called the modified divisor, instead of the natural divisor to calculate modified quotas for every state: $q_i = p_i/d$, where p_i is the population of state i. Then modified quotas are rounded according to a particular rule to obtain an allocation that preserves the total number of rounded modified quotas equal to the house size. A divisor method does not suffer from any paradoxes (see §2.1), but it may violate the quota rule. A divisor method favors larger states when the tentative apportionment is smaller than the house size, and it favors smaller states when the initial allocation is too large.

In this section, we consider another divisor method credited to Daniel Webster (1782 – 1852), a senator from Massachusetts (USA). The method, also known as the Webster-Willcox method as well as the method of major fractions, was proposed in 1832 as a compromise between the Jefferson method and the Adams method. The method was first applied in 1842 as a substitution for Jefferson's method in allocation of seats to the House of Representatives. After 10 years, it was replaced by the Hamilton-Vinton method, and reintroduced in 1901. It was an American statistician from Cornell University Walter Willcox (1861 – 1964) who strongly advocated in favor of Webster's method, and who had calculated apportionment for the House based on the 1900 census.

The **Webster method** uses the conventional rounding procedure that rounds natural quotas to nearest whole numbers: if the fractional part of a real number is greater than or equal to 0.5, then it rounds this number up to the next integer, if the fractional part is less than 0.5, it rounds this number down. In other words, Webster's method rounds natural quotas (or any positive real numbers) up or down depending on whether the number is smaller or not than the corresponding cutoff point: if the quota is less than this cutoff point, the method rounds it down to nearest integer, otherwise, the method rounds it up. The cutoff point in Webster's rule is the midpoint of two whole numbers surrounding the given quota (or number). From mathematical point of view, this midpoint is the average or **arithmetic mean** of its upper and lower quotas:

$$\frac{\lfloor q \rfloor + \lceil q \rceil}{2},$$

where q is the natural quota, its floor $\lfloor q \rfloor$ is the lower quota, and its ceiling $\lceil q \rceil$ is the upper quota.

According to Balinski and Young [2], Webster's method is the unique divisor method that is neutral to the sizes of states, namely, it shows no bias between large and small states. For instance, the small states were favored only nine times by Webster's method over the course of U.S. Congressional apportionments. A divisor method is biased in favor of large states when modified quotas are obtained by using a modified divisor that is smaller than the natural divisor. When a modified divisor is larger than the natural divisor, small states are favored. Since fractional parts of natural quotas are equally likely to be less than or greater than 0.5, we expect that on average the Webster method shows indifference to the sizes of states. It is the best method at minimizing the differences between representative shares.

Webster's method is the only divisor method that satisfies the quota rule when the number of states is either two or three, but it may violate the rule for a larger number of states. However, in real life problems, such violations are rare. Moreover, empirical observation makes clear that the event of a Webster apportionment not satisfying the quota rule is extremely unlikely. For example, had Webster's method been used for every apportionment in the U.S. House of Representatives, not a single violation of the quota rule would be observed. Despite all its advantages, Webster's method had a rather short tenure in the U.S. legislature: it was used only four times (1840, 1900, 1910, and 1930).

In practical applications, the Webster method has only one possible flaw: it can assign zero seats to one or more states. This is a violation of the U.S. Constitution requirement that each state receives a minimum of one seat. Other countries may have a different requirement; for instance, France ensures each of its departments with at least 2 seats; and the European Parliament has fixed minimum numbers of seats attached to each of the countries, ranging between 6 and 36.

Let us outline Webster's algorithm.

1. Calculate the natural quota for each state according to Eq. (2.1.2), page 150.

2. Initially assign a state its lower quota (floor of the natural quota) if the fractional part of its natural quota is less than 0.5. Tentatively assign a state its upper quota (ceiling of the natural quota) if the fractional part of its natural quota is greater than or equal to 0.5. In other words, the initial allocation is obtained by rounding natural quotas down or up to the nearest integers depending on whether they are smaller or not than the corresponding cutoff points—arithmetic means of upper and lower quotas.

3. Check whether the sum of initial allocations (the lower or upper quotas from step 2) is equal to the correct number of seats to be apportioned, called the house size. If the sum of the quotas in the initial allocation from step 2 is equal to the correct number of seats to be apportioned, then stick with this apportionment—you are done.

4. If the initial allocation is NOT equal to the house size, proceed as follows.

5. Calculate the difference, m, between the correct number of seats to be apportioned (house size) and the initial allocation. The number of surplus seats could be either positive if the initial allocation is less than the house size, or negative if the initial allocation is greater than the house size.

6. Calculate the natural divisor according to Eq. (2.1.1), page 150.

7. For each state i, calculate its **threshold divisor**:

$$\mathbf{TD} = \frac{\text{state's population}}{\text{lower quota} + 0.5} = \frac{p_i}{\lfloor q_i \rfloor + 0.5}, \tag{2.4.1}$$

where p_i is the population of state i, $q_i = p_i * h/p$ is the natural quota, h is the house size, p is the total population.

8. Rank all threshold divisors from largest to smallest. Such ordering breaks the axis of real numbers into subintervals having endpoints as threshold divisors.

9. Identify the position of the natural divisor among all ordered threshold divisors; that is, find the smallest interval that contains the natural divisor. Note that its position can be above all of them (if the natural divisor is larger than any threshold divisor), it can be below all of them (if the natural divisor is smaller than any threshold divisor), or it can be somewhere in between.

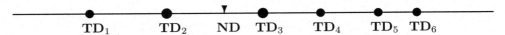

If the difference, m, between the house size and the initial allocation is positive—the initial allocation is m seats less than required—move the position of the natural divisor m steps down. Then choose a modified divisor as a number between these modified divisors that serve as the boundary of this position.

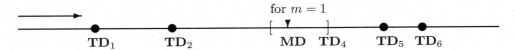

If the difference, m, between the house size and the initial allocation is negative—the initial allocation is m seats more than the house size—move the position of the natural divisor m steps up. Then choose a modified divisor as a number between these modified divisors.

10. For each state, compute the modified quota (the result of division of each state's population by the modified divisor instead of the natural divisor):

$$\mathbf{Modified\ Quota} = \frac{\text{state's population}}{\text{modified divisor}} = \frac{p_i}{\mathbf{MD}}. \tag{2.4.2}$$

By rounding modified quotas up or down in the conventional way to the nearest integer (ceilings of modified quotas if their fractional parts are greater than or equal to 0.5, or floors if fractional parts of modified quotas are less than 0.5), obtain the final allocation. If the sum of seats in the final allocation is the exact number of seats to be apportioned, you are done.

11. If the sum of the rounded in the conventional way modified quotas is not equal to the house size, this indicates that either the quota property is violated or the wrong number of states gain or lose seats. In this case, calculate the threshold divisors for each additional seat each state gains/loses. Thus, if a state gains s ($s \geqslant 1$) additional seats compared to the upper quota, calculate the threshold divisors:

$$\mathbf{TD}_1 = \frac{p_i}{\lfloor q_i \rfloor + 0.5}, \quad \mathbf{TD}_2 = \frac{p_i}{\lfloor q_i \rfloor + 1.5}, \quad \ldots, \quad \mathbf{TD}_s = \frac{p_i}{\lfloor q_i \rfloor + s - 0.5}. \quad (2.4.3)$$

If a state loses s ($s \geqslant 1$) seats compared to the lower quota, calculate the threshold divisors:

$$\mathbf{TD}_1 = \frac{p_i}{\lfloor q_i \rfloor - 0.5}, \quad \mathbf{TD}_2 = \frac{p_i}{\lfloor q_i \rfloor - 1.5}, \quad \ldots, \quad \mathbf{TD}_s = \frac{p_i}{\lfloor q_i \rfloor - s + 0.5}. \quad (2.4.4)$$

12. Rank all threshold divisors (from parts 7 and 11). Identify the position of the natural divisor in this ordered list of threshold divisors. Move the natural divisor m positions up or down depending on whether you need to add seats or eliminate surplus seats. If you need to eliminate m seats, move m positions up. If the initial allocation is m seats short, move the position of the natural divisor down m places. The two threshold divisors that serve as the boundary of this new position are the upper and lower bounds for your choice of a new modified divisor.

13. With a modified divisor chosen in the previous step, repeat step 10 of this algorithm to obtain the required final allocation. Namely, use a modified divisor instead of the natural divisor to calculate the modified quotas, Eq. (2.4.2). Round modified quotas up or down to the nearest whole number, then the sum of all the rounded modified quotas is the house size. Apportion each state with its modified rounded quota. ∎

Now we are going to demonstrate the application of Webster's algorithm in numerous examples.

Example 2.4.1. 2008 Senate Elections
In the United States, elections to Congress take place every two years. Congress has two chambers: the Senate and the House of Representatives. The Senate has hundred members, elected for a six year term in dual-seat constituencies (two from each state) with one-third being renewed every two years. The results of 2008 Senate elections are presented here.

Party	Number of votes
Democratic Party	34,276,327
Republican Party	29,729,539
Libertarian Party	670,231
Independence Party	437,505
Green Party	427,418
Constitution Party	240,726
Total:	65781746

It is common to present the results of the election in percentage form, keeping one decimal place. This means that the house size is $h = 1000$; for instance, the Democratic Party had 52.1% of votes. Since there are six parties, the number of states is six, their populations are given in the right column. Calculating the natural quotas and rounding them to the nearest integer, we get

Party	DP	RP	LP	IP	GP	CP	Total
Natural Quota	521.0613	451.9420	10.1887	6.6508	6.4975	3.6594	1000
Allocation	521	452	10	7	6	4	1000

So the initial apportionment is exactly what we need. Does it mean that Webster's method is always correct? We make an experiment with calculations of the initial allocations for different values of the house size. Adding or subtracting 1 from the previous house size within a loop, we get the correct allocation for 18 values of the house size: $h = 990, 991, \ldots, 1008$, but for $h = 899, 1009$, and 1011 the initial apportionments do not equal to the required house sizes. For instance, the initial allocation for $h = 1009$ has a surplus "seat:"

Party	DP	RP	LP	IP	GP	CP	Total
Natural Quota	525.7509	456.0095	10.2804	6.7107	6.5559	3.6924	1009
Allocation	526	456	10	7	7	4	1010

Based on this numerical experiment, we can conclude that most of the time the Webster method provides the required allocation without extra work—the initial allocation gives the correct apportionment—though sometimes it fails. The outcome of the application of Webster's method is much better than any of four apportionment methods we discussed so far: almost always the tentative allocation of Hamilton's, Lowndes', Jefferson's, or Adams' is not equal to the house size.

Example 2.4.2. Airplane Allocation

In 2009, six airlines ordered twenty one new commercial airplanes that should be allocated between these companies based on their share in the domestic market:

Airline	American	United	Delta	US Airways	Northwest	Alaska
Share	13.9%	10.5%	10.4%	8.0%	6.2%	3.1%

Once the natural quotas have been calculated according to the formula (2.1.2), page 150, it becomes clear that the initial allocation is one seat short against the house size. This forces us to compute the threshold divisors for every state. The results are summarized in the following table:

Airline	Popu-lation	Natural Quota	Webster I. A.	Threshold Divisor	Rank	Modif. Quota	Web. F. A.
American	13.9	5.602687140	6	2.527272727	1◄ND	5.79	6
United	10.5	4.232245681	4	2.333333333	3	4.37	4
Delta	10.4	4.191938580	4	2.311111111	4	4.33	4
US Airways	8.0	3.224568138	3	2.285714286	5	3.33	3
Northwest	6.2	2.499040307	2	2.48	2◄MD	2.58	3
Alaska	3.1	1.249520154	1	2.066666667	6	1.29	1
Total:	52.1	21	20				21

The natural divisor, $\mathbf{ND} = \dfrac{\text{total population}}{\text{house size}} = \dfrac{52.1}{21} \approx 2.480952381$ is less than the threshold divisor of American, but larger than Northwest. To add one seat to the initial allocation we should choose a modified divisor between the second largest threshold divisor and the third one:

$$2\tfrac{1}{3} = 2.333\overline{3} < \textbf{Webster's Modified Divisor} \leqslant 2.48.$$

By choosing the modified divisor to be $\mathbf{MD} = 2.4$, we obtain the modified quotas (ratios of states' populations over the modified divisor) for each state presented in the table. By rounding these modified quotas in the conventional way, we get the required allocation, as shown in the last column. Note that the final allocation of airplanes obtained does not depend exactly on your choice of the modified divisor, but only on the range from which this divisor was picked. For example, if we choose another modified divisor, say, $\mathbf{MD} = 2.34$, modified quotas will be as follows

Airline	American	United	Delta	US Airways	Northwest	Alaska
Modified Quota	5.94	4.48	4.44	3.41	2.64	1.32
Final Allocation	6	4	4	3	3	1

Hence, the final allocation is exactly the same as the one obtained with the modified divisor of 2.4. So we see that the final allocation does not depend on a particular choice for the modified divisor—there could be infinitely many of them—but on the interval from which this divisor was chosen.

Example 2.4.3. Family Apportionment

The Johnson family has a large yard that needs constant maintenance with different fruits and vegetables growing there. The Johnsons would like to have their kids help out in the yard and would also like to teach them about the value of work. So, they decide to distribute 52 chocolate chip cookies evenly among four kids if they help their parents. However, seeing that their children were not motivated enough, the parents decided to give cookies in proportion to the amount of time each child spends working in the yard. To do this fairly, the Johnsons create a table with the exact minutes that each child had worked.

Names	Anna	Deanna	Frank	Tony	Total
Time	125	146	82	223	576 (or 9.6 hours)

Based on this information, the parents calculated the natural quota for each child using Eq. (2.1.2), and made an initial allocation using conventional rounding to obtain the following table.

Child	Anna	Deanna	Frank	Tony	**Total**
Natural Quota, $q_i = \frac{p_i * 52}{576}$	$\frac{125*52}{576}$	$\frac{146*52}{576}$	$\frac{82*52}{576}$	$\frac{223*52}{576}$	52
Approximation of NQ	11.2847	13.1805̄	7.4027̄	20.1319	
Initial Allocation	11	13	7	20	51

Since the initial distribution is short by one cookie, the father suggests keeping this initial allocation, and to give the surplus cookie to himself because he worked so hard. However, mom vetoes his amendment, and wants to proceed with Webster's allocation. It took a minute for her to calculate threshold divisors.

Child	Anna	Deanna	Frank	Tony
Modified Divisor, $\frac{p_i}{\lfloor q_i \rfloor + 0.5}$	$\frac{125}{11.5}$	$\frac{146}{13.5}$	$\frac{82}{7.5}$	$\frac{223}{20.5}$
Approximation of MD	10.8695	10.8148	10.9333̄	10.8780
Rank	3	4	1	2

Since the natural quota is larger than any of threshold divisors, a modified divisor should be chosen between the largest threshold divisor, which is Frank's, and the second largest divisor, which is Tony's:

$$10.87804878 < \textbf{Webster's Modified Divisor} \leq 10.9333\overline{3}.$$

By choosing the modified divisor to be **MD** = 10.9, the family gets the final distribution summarized in the following table.

Name	Number of Minutes	Natural Quota	Webster I. A.	Threshold Divisor	Rank	Modif. Quota	Webster F. A.
Anna	125	11.2847	11	10.86956522	3	11.4678	11
Deanna	146	13.1805̄	13	10.81481481	4	13.3944	13
Frank	82	7.4027̄	7	10.93333333	1 ≺MD	7.5229	8
Tony	223	20.1319	2	10.87804878	2	20.4587	20
Total:	576	52	51				52

The final decision left only one person unhappy—the father—who disagreed with this allocation of cookies, which, in his opinion, is unfair: Frank worked less than anybody, but he was finally awarded with the surplus cookie.

Example 2.4.4. Mexico-USA Border Crossing

In 2009, the number of private visitors entering the USA from Mexico is presented in the table.

Check Point	Nogales	Sasabe	Calexico	Pinecreek	Hidalgo	Roma
Number of Visits	1,758,494	14,625	2,917,035	2,900	3,492,088	480,631

The government wants to assign 278 police officers to these six check points based on the number of visitors crossing the USA-Mexico border. Using the Webster method, the government comes to the following data that indicate that the initial allocation is one seat less than the house size. Recall that the initial allocation of a state (in our case, it is a check point) is obtained by rounding the corresponding natural quota up or down depending on whether its fractional part is less than 0.5 or not.

Check Point	Number of Visits	Natural Quota	Web. I. A.	Threshold Divisor	Rank	Modif. Quota	Web. F. A.
Nogales,AZ	1758494	56.4128938	56	31123.7876	2	56.54	57
Sasabe,AZ	14625	0.46917337	0	29250	5	0.47	0
Calexico,CA	2917035	93.5791567	94	31198.2352	1 <ND	93.79	94
Pinecreek,NM	2900	0.09303267	0	5800	6	0.09	0
Hidalgo,TX	3492088	112.026989	112	31040.7822	3	112.28	112
Roma,TX	480631	15.4187535	15	31008.4516	4	15.45	15
Total:	8665773	278	277				278

To achieve the required allocation, we need to calculate threshold divisors for every state:

$$\textbf{TD}_{\text{Nogales}} = \frac{1758494}{56.5} \approx 31123.78761, \qquad \textbf{TD}_{\text{Sasabe}} = \frac{14625}{0.5} = 29250,$$

$$\textbf{TD}_{\text{Calexico}} = \frac{2917035}{93.5} \approx 31198.23529, \qquad \textbf{TD}_{\text{Pinecreek}} = \frac{2900}{0.5} = 5800,$$

$$\textbf{TD}_{\text{Hidalgo}} = \frac{3492088}{112.5} \approx 31040.78222, \qquad \textbf{TD}_{\text{Roma}} = \frac{480631}{15.5} \approx 31008.45161.$$

Ordering these threshold divisors, from largest to smallest, we identify the position of the natural divisor $\textbf{ND} = 31171.84532$ to be between the largest threshold divisor, which is Calexico's check point and the second one, which is Nogales' check point. Since we need to add one seat to the initial allocation, we pick up a modified divisor from the range

$$31040.78222 < \textbf{Webster's Modified Divisor} \leqslant \tfrac{1758494}{56.5} \approx 31123.78761.$$

Choosing a modified divisor as $\textbf{MD} = 31100$, we get the modified quotas that, after rounding to the nearest integers, provide the required allocation:

Check Point	Nogales	Sasabe	Calexico	Pinecreek	Hidalgo	Roma
Modified Divisor	$\frac{1{,}758{,}494}{31100}$	$\frac{14{,}625}{31100}$	$\frac{2{,}917{,}035}{31100}$	$\frac{2{,}900}{31100}$	$\frac{3{,}492{,}088}{31100}$	$\frac{480{,}631}{31100}$
Approximation	56.54	0.47	93.79	0.09	112.28	15.45
Final Allocation	57	0	94	0	112	15

As we see from this table, two check points (Sasabe and Pinecreek) are without seats. Obviously, the government cannot allow this allocation, so the Webster method should be replaced with another one that has no such flaw (see Example 2.5.9, page 232).

Example 2.4.5. Canada-USA Border Crossing
In 2009, the number of trucks entering the USA from Canada is presented in the table.

State	Alaska	Idaho	Maine	Michigan	Minnesota
Number of Trucks	11,406	49,794	477,906	2,625,761	109,728

The government wants to assign 7,400 border officers to each of the 5 states based on the number of trucks to be checked. Using the Webster method, the government comes up with the following results presented in the table.

State	Popula-tion	Nat. Quota	Web. I. A.	Threshold Divisor	Rank	Modified Quota	Web. F. A.
AK	11406	25.7755	26	447.29411	1	25.7	26
ID	49794	112.5256	113	442.61$\overline{3}$	4	112.5	113
ME	477906	1079.9828	1080	442.71051	3	1079.7	1080
MI	2625761	5933.7546	5934	442.5315$_{\prec}$ND	5	5932.5	5933
MN	109728	247.9613	248	443.33737	2	247.9	248
Total:	3274595	7400	7401				7400

As we can see, the initial allocation consists of upper quotas because all fractional parts are greater than 0.5. The natural divisor, $\mathbf{ND} = \frac{2625761}{7400} \approx 442.5128378$ is less than any threshold divisor. Since we need to eliminate one seat from the initial allocation, we must choose a modified divisor as a number from the interval bounded by the lowest modified divisor, which is Michigan, and the second lowest one, which has to be determined. The main candidate for this role is Idaho's threshold divisor. However, when the discrepancy in the sizes of states' populations is significant, the allocations under Webster's rule may experience violation of the quota property. In our case, the population of Michigan is more than 230 times larger than Alaska's population. It makes sense to calculate Michigan's threshold divisor that corresponds to 5933 seats: $\mathbf{TD}_{\text{Michigan}[0]} = \frac{2625761}{5932.5} \approx 442.6061$, and to 5935 seats: $\mathbf{TD}_{\text{Michigan}[2]} = \frac{2625761}{5934.5} = 442.4569888$. Ranking all threshold divisors

Ranking	2	3	4	5	$\leftarrow$	6
State	$\cdots$	**ID**	**MI[0]**	**MI[1]**	**ND**	**MI[2]**
Threshold Divisor		442.61$\overline{3}$	442.6061525	442.5315581	442.51	442.457

we see that the natural divisor, $\mathbf{ND}$, is between two threshold divisors for the state of Michigan: $\frac{2625761}{5934.5} = 442.4569888 < \mathbf{ND} < \frac{2625761}{5933.5} = 442.5315581$. Therefore, a modified divisor for our problem should be chosen from the interval

$$442.5315581 < \textbf{Webster's Modified Divisor} \leqslant \tfrac{2625761}{5932.5} \approx 442.6061525.$$

Choosing a modified divisor as, say, $\mathbf{MD} = 442.6$, we compute the modified quotas for every state: `state's population/modified divisor` to obtain values presented in the next to the last column. Rounding all modified quotas up, we obtain the required final allocation.

Example 2.4.6. Continuation of Example 2.4.1
Let us return to Example 2.4.1 where we found that for the house size $h = 1009$ the tentative allocation has one surplus seat. What should we do in such cases when the initial

allocation differs from the house size? The answer is simple: follow the Webster algorithm. To eliminate one surplus seat, we calculate the threshold divisors for every party to obtain

Party	DP	RP	LP	IP	GP	CP
Natural Quota	525.75	456	10.28	6.7	6.555	3.69
Initial Allocation	526	456	10	7	7	4
Threshold Divisor, $\frac{p_i}{\lfloor q_i \rfloor + 0.5}$	$\frac{34{,}276{,}327}{525.5}$	$\frac{29{,}729{,}539}{456.5}$	$\frac{670{,}231}{10.5}$	$\frac{437{,}505}{6.5}$	$\frac{427{,}418}{6.5}$	$\frac{240{,}726}{3.5}$
Approximation of TD	65226	65124	63831	67308	65756	68778
Rank	4	5	6	2	3	1

The natural divisor, $\mathbf{ND} = 65194.99108$, is larger than the fifth divisor, which is the Republican Party, but smaller than the fourth divisor, which is the Democratic Party. To eliminate one seat from the initial allocation, we have to choose a divisor from the interval bounded by the third and the fourth largest threshold divisor:

$$65226.12179 < \textbf{Modified Divisor} < 65756.61538.$$

You will find it much to your surprise that the divisor $\mathbf{MD} = 65700$ does not work:

Party	DP	RP	LP	IP	GP	CP	Total
M. Quota, $\frac{p_i}{\text{MD}}$	$\frac{34{,}276{,}327}{65700}$	$\frac{29{,}729{,}539}{65700}$	$\frac{670{,}231}{65700}$	$\frac{437{,}505}{65700}$	$\frac{427{,}418}{65700}$	$\frac{240{,}726}{65700}$	
M. Quota, Approx.	521.70	452.50	10.20	6.65	6.50	3.66	
Final Allocation	522	453	10	7	7	4	1003

Analyzing the table, we see that two main parties (DP and RP) lost four and three seats, respectively. Therefore, we need to calculate the threshold divisors for each additional seat they lost (see part 11 in Webster's algorithm). However, since only one seat should be dropped from the initial allocation, we evaluate two additional threshold divisors (others will be larger than these two):

$$\textbf{TD}_{\text{DP}[-1]} = \frac{34{,}276{,}327}{524.5} \approx 65350.48, \qquad \textbf{TD}_{\text{RP}[-1]} = \frac{29{,}729{,}539}{455.5} \approx 65267.92316.$$

Since the first threshold divisor that is larger than $\mathbf{MD}_{\text{DP}} = 65226$ is $\mathbf{MD}_{\text{RP}[-1]} = 65267.92$ we conclude that the modified divisor for this problem should be chosen from the interval

$$65226.12179 < \textbf{Webster's Modified Divisor} \leqslant \tfrac{34276327}{525.5} \approx 65267.92316.$$

Indeed, choosing a modified divisor, say $\mathbf{MD} = 65260$, we get the correct allocation

Party	DP	RP	LP	IP	GP	CP	Total
M. Quota, $\frac{p_i}{\text{MD}}$	$\frac{34{,}276{,}327}{65260}$	$\frac{29{,}729{,}539}{65260}$	$\frac{670{,}231}{65260}$	$\frac{437{,}505}{65260}$	$\frac{427{,}418}{65260}$	$\frac{240{,}726}{65260}$	
M. Quota, Approx.	525.22	455.55	10.27	6.7	6.54	3.68	
Final Allocation	525	456	10	7	7	4	1009

Example 2.4.7. Deuce
Four friends have bought a collection of 32 high-quality pearls for $248,000 at an auction.

Anna contributed \$26,000, Boris contributed \$90,000, and the remaining funds were supplied by Chris and Donna, with \$66,000 each. Before the auction, the friends decided to use Webster's method to distribute the pearls, assuming that twins, Chris and Donna, would get an equal number pearls. After the auction, the friends made preliminary calculations and found that the initial distribution has one surplus pearl.

Friends	Anna	Boris	Chris	Donna	Total
Contribution	26,000	90,000	66,000	66,000	
Natural Quota, $q_i = \frac{p_i*32}{248}$	$\frac{26*32}{248}$	$\frac{90*32}{248}$	$\frac{66*32}{248}$	$\frac{66*32}{248}$	32
Approximation of NQ	3.3548	11.6129	8.5161	8.5161	
Initial Allocation	3	12	9	9	33

To determine who out of the four friends should sacrifice one pearl, the threshold divisors were calculated.

Friends	Anna	Boris	Chris	Donna
Threshold Divisor, $\frac{p_i}{\lfloor q_i \rfloor + 0.5}$	$\frac{26000}{3.5}$	$\frac{90000}{12.5}$	$\frac{66000}{9.5}$	$\frac{66000}{9.5}$
Approximation of TD	7428.57	7826.08	7764.70	7764.70
Rank	$4 \prec$**ND**	1	2,3	2,3

The natural divisor, $\mathbf{ND} = \frac{248,000}{32} = 7750$, is larger than the smallest threshold divisor of Anna, but is less than the threshold divisor of the twins; in this case, the Webster algorithm suggests choosing a modified divisor between the second and the third largest divisor. However, it is impossible because the twins have the same threshold divisor, and the friends cannot fairly separate pearls between them.

Example 2.4.8. Webster's Method versus Jefferson's and Adams' Methods

The seats in the South African National Assembly are allocated by proportional representation. The results of the 2004 elections are as follows.

Party	Number of Votes
African National Congress (ANC)	10,880,915
Democratic Alliance (DA)	1,931,201
Inkatha Freedom Party (IFP)	1,088,664
United Democratic Movement (UDM)	355,717
Independent Democrats (ID)	269,765
Total:	14,526,262

Suppose that we want to allocate 356 seats for these 5 parties using the *Webster* method. The initial allocation has one extra seat.

Party	ANC	DA	IFP	UDM	ID
N. Q., $q_i = \frac{p_i*356}{14526262}$	$\frac{10880915*356}{14526262}$	$\frac{1931201*356}{14526262}$	$\frac{1088664*156}{14526262}$	$\frac{355717*356}{14526262}$	$\frac{269765*356}{14526262}$
N. Quota, Approx.	266.6622	47.3285	26.6802	8.7176	6.6112
Initial Allocation	267	47	27	9	7

Therefore, we have to calculate threshold divisors for every party.

Party	ANC	DA	IFP	UDM	ID
T. Divisor, $\frac{p_i}{\lfloor q_i \rfloor + 0.5}$	$\frac{10880915}{266.5}$	$\frac{1931201}{47.5}$	$\frac{1088664}{26.5}$	$\frac{355717}{8.5}$	$\frac{269765}{6.5}$
Approximation of TD	40828.94	40656.86	41081.66	41849.05	41502.30
Ordering	4 ◁ND	5	3	1	2

The natural divisor, $\mathbf{ND} = 40804.10674$, is less than the fourth threshold divisor (ANC), but greater than the smallest one (DA), so to determine the range for a modified divisor we have to take the interval bounded by the fourth largest modified divisor (ANC) and the third one (IFP):

$$40828.94934 < \mathbf{Modified\ Divisor} < 41081.66038.$$

Choosing the modified divisor to be $\mathbf{MD} = 41081$, we will get a wrong allocation of seats.

Party	ANC	DA	IFP	UDM	ID	Total
Modified Quota, $\frac{p_i}{\mathbf{MD}}$	$\frac{10880915}{41081}$	$\frac{1931201}{41081}$	$\frac{1088664}{41081}$	$\frac{355717}{41081}$	$\frac{269765}{41081}$	
M. Quota, Approximation	264.8649	47.0095	26.5004	8.6589	6.5666	
Allocation	265	47	27	9	7	355

As we see, the party ANC lost two seats, which is a violation of the quota rule. The Webster algorithm suggests calculating one more threshold divisor that corresponds to zero surplus allocation for ANC:

$$\mathbf{TD}_{\text{ANC}[0]} = \frac{10880915}{265.5} \approx 40982.7307.$$

Hence, the correct interval for a modified divisor would be

$$40828.94934 < \mathbf{Webster's\ Modified\ Divisor} \leqslant \tfrac{10880915}{265.5} \approx 40982.7307.$$

Indeed, choosing $\mathbf{MD} = 40980$, we get the correct allocation.

Party	ANC	DA	IFP	UDM	ID	Total
Modified Quota, $\frac{p_i}{\mathbf{MD}}$	$\frac{10880915}{40980}$	$\frac{1931201}{40980}$	$\frac{1088664}{40980}$	$\frac{355717}{40980}$	$\frac{269765}{40980}$	
M. Quota, Approximation	265.5176	47.1254	26.5657	8.6802	6.5828	
Webster Final Allocation	266	47	27	9	7	356

Now we apply the *Jefferson* method to obtain another initial allocation (based on the natural quotas calculated previously):

Party	ANC	DA	IFP	UDM	ID	Total
Natural Quota	266.6622	47.3285	26.6802	8.7176	6.6112	356
$2 Initial Allocation	266	47	26	8	6	353

So the Jefferson initial allocation is three seats short. Calculating the the threshold divisors according to the formula (2.3.1) and then ordering them, we get

Party	ANC	DA	IFP	UDM	ID
Threshold Divisor, $\frac{p_i}{\lceil q_i \rceil}$	$\frac{10880915}{267}$	$\frac{1931201}{48}$	$\frac{1088664}{27}$	$\frac{355717}{9}$	$\frac{269765}{7}$
Approximation of TD	40752.49	40233.35	40320.88	39524.11	38537.85
Ordering	1	3	2	4	5

According to Jefferson's algorithm, we choose a modified divisor from the interval bounded by the third largest threshold divisor, which is DA, and the fourth one, which is UDM:

$$39524.1\bar{1} < \textbf{Modified Divisor} < 40233.35417.$$

Taking $\textbf{MD} = 40000$, we will obtain a *wrong* allocation:

Party	ANC	DA	IFP	UDM	ID	Total
Modified Quota, $\frac{p_i}{\text{MD}}$	$\frac{10880915}{40000}$	$\frac{1931201}{40000}$	$\frac{1088664}{40000}$	$\frac{355717}{40000}$	$\frac{269765}{40000}$	
M. Quota, Approximation	272.02	48.28	27.21	8.89	6.74	
Allocation	272	48	27	8	6	361

which has five surplus seats. This means that we need to calculate additional threshold divisors:

$$\textbf{TD}_{\text{ANC}[2]} = \frac{10880915}{268} \approx 40600.42, \qquad \textbf{TD}_{\text{ANC}[3]} = \frac{10880915}{269} \approx 40449.49,$$

$$\textbf{TD}_{\text{ANC}[4]} = \frac{10880915}{270} \approx 40299.68, \qquad \textbf{TD}_{\text{ANC}[5]} = \frac{10880915}{271} \approx 40150.97.$$

Hence, the correct interval for Jefferson's modified divisor is

$$40299.68519 < \textbf{Jefferson's Modified Divisor} \leqslant \tfrac{10880915}{269} \approx 40449.49814.$$

Taking $\textbf{MD} = 40400$, we obtain the *correct* final allocation:

Party	ANC	DA	IFP	UDM	ID	Total
Modified Quota, $\frac{p_i}{\text{MD}}$	$\frac{10880915}{40400}$	$\frac{1931201}{40400}$	$\frac{1088664}{40400}$	$\frac{355717}{40400}$	$\frac{269765}{40400}$	
M. Quota, Approximation	269.32	47.80	26.94	8.80	6.67	
$2 Final Allocation	269	47	26	8	6	356

This allocation gives another example that Jefferson's method violates the quota rule. Finally, we turn our attention to *Adams'* method. Its initial allocation

Party	ANC	DA	IFP	UDM	ID
Natural Quota, Approximation	266.6622	47.3285	26.6802	8.7176	6.6112
Adams' Initial Allocation	267	48	27	9	7

has two surplus seats. The table of Adams' threshold divisors

Party	ANC	DA	IFP	UDM	ID
Threshold Divisor, $\frac{p_i}{\lfloor q_i \rfloor}$	$\frac{10880915}{266}$	$\frac{1931201}{47}$	$\frac{1088664}{26}$	$\frac{355717}{8}$	$\frac{269765}{6}$
Approximation of TD	40905.69	41089.38	41871.69	44464.62	44960.83
Ordering	5	4	3	2	1

suggests that we can choose a modified divisor from the interval formed by the fourth largest threshold divisor, which is DA, and the third one, which is IFP (because we have to start counting from the smallest threshold divisor):

$$41089.38 < \textbf{Modified Divisor} < 41871.69.$$

This leads to a *wrong* allocation if we take **MD** = 41100:

Party	ANC	DA	IFP	UDM	ID	Total
Modified Quota, $\frac{p_i}{\textbf{MD}}$	$\frac{10880915}{41100}$	$\frac{1931201}{41100}$	$\frac{1088664}{41100}$	$\frac{355717}{41100}$	$\frac{269765}{41100}$	
M. Quota, Approximation	264.74	46.98	26.48	8.65	6.56	
Allocation	265	47	27	9	7	355

Since this apportionment is one seat short, we calculate an additional threshold divisor for "state" ANC because it is the only party that lost more than one seat in comparison to the initial allocation (which is a violation of the quota rule):

$$\textbf{TD}_{\text{ANC}[2]} = \frac{10880915}{265} \approx 41060.05660.$$

Adding this threshold divisor to the set of all previously calculated threshold divisors and ranking them, we conclude that the correct interval for Adams' modified divisors is

$$41060.05660 \approx \tfrac{10880915}{265} \leqslant \textbf{Adams' Modified Divisor} < 41089.38298.$$

Choosing a modified divisor, for instance, to be **MD** = 41070, we get the *correct* Adams' apportionment

Party	ANC	DA	IFP	UDM	ID	Total
Modified Quota, $\frac{p_i}{\textbf{MD}}$	$\frac{10880915}{41070}$	$\frac{1931201}{41070}$	$\frac{1088664}{41070}$	$\frac{355717}{41070}$	$\frac{269765}{41070}$	
M. Quota, Approximation	269.32	47.80	26.94	8.80	6.67	
Adams' Final Allocation	265	48	27	9	7	356

Comparing final allocations under these three apportionment methods, we observe that they all are different!

Example 2.4.9. Violation of the Quota Rule

As it was emphasized at the beginning of this section, Webster's method may violate the quota rule but only in rare cases. So it is not easy to find real world data where such a violation occurs. In this example, we use hypothetical populations of six states, which we would like to represent in percentage form carried out to one decimal place. So the house size in our example is 1000, and the states' populations are presented here.

State	A	B	C	D	E	F	Total
Population	101	10	9	55	3674	291	4140

The initial allocation under Webster's rule is

State	A	B	C	D	E	F	Total
N. Q., $\quad q_i = \frac{p_i * 1000}{4140}$	$\frac{101*100}{414}$	$\frac{10*100}{414}$	$\frac{9*100}{414}$	$\frac{55*100}{414}$	$\frac{3674*100}{414}$	$\frac{291*100}{414}$	1000
N. Quota, Approx.	24.39	2.41	2.17	13.28	887.43	70.28	
Initial Allocation	24	2	2	13	887	70	998

To find a correct allocation, we calculate Webster's threshold divisors, and then rank them to obtain

State	A	B	C	D	E	F
Threshold Divisor, $\frac{p_i}{\lfloor q_i \rfloor + .5}$	$\frac{101}{24.5}$	$\frac{10}{2.5}$	$\frac{9}{2.5}$	$\frac{55}{13.5}$	$\frac{3674}{887.5}$	$\frac{291}{70.5}$
Approximation of TD	4.12244	4.0	3.6	4.07407	4.13971	4.12765
Ordering	3	5	6	4	1	2

Since the natural divisor, $\mathbf{ND} = 4.14$, is larger than any of these six threshold divisors, we choose an interval for a modified divisor with bounds of the second largest threshold divisor, which is state **F**, and the third one, which is state **A**:

$$4.12244 < \textbf{Modified Divisor} < 4.12765.$$

This leads to a *wrong* allocation if we take $\mathbf{MD} = 4.123$:

State	A	B	C	D	E	F	Total
Modified Quota, $\frac{p_i}{\mathbf{MD}}$	$\frac{101}{4.123}$	$\frac{10}{4.123}$	$\frac{9}{4.123}$	$\frac{55}{4.123}$	$\frac{3674}{4.123}$	$\frac{291}{4.123}$	
M. Quota, Approximation	24.49	2.42	2.18	13.33	891.09	70.57	
Allocation	24	2	2	13	891	71	1003

because it has three surplus seats. According to part 11 of Webster's algorithm, our next step is to calculate threshold divisors for every additional seat state **E** gained:

$$\mathbf{TD_{E[2]}} = \frac{3674}{888.5} \approx 4.13505, \qquad \mathbf{TD_{E[3]}} = \frac{3674}{889.5} \approx 4.13041,$$

$$\mathbf{TD_{E[4]}} = \frac{3674}{890.5} \approx 4.12577, \qquad \mathbf{TD_{E[5]}} = \frac{3674}{891.5} \approx 4.12114.$$

Adding these four threshold divisors to the list of previously calculated divisors, and ordering them, we conclude that the interval to choose a modified divisor is

$$4.130410343 < \textbf{Webster's Modified Divisor} \leqslant \frac{3674}{888.5} \approx 4.135059088.$$

Indeed, taking a modified divisor, say, $\mathbf{MD} = 4.131$, we obtain the correct allocation

State	A	B	C	D	E	F	Total
Modified Quota, $\frac{p_i}{\mathbf{MD}}$	$\frac{101}{4.131}$	$\frac{10}{4.131}$	$\frac{9}{4.131}$	$\frac{55}{4.131}$	$\frac{3674}{4.131}$	$\frac{291}{4.131}$	
M. Quota, Approximation	24.44	2.42	2.17	13.31	891.37	70.44	
Webster Final Allocation	24	2	2	13	889	70	1000

Since state **E** gained two seats compared to its initial allotment of 887 seats, it is a violation of the quota rule.

Example 2.4.10. Natural Quota Equals a Threshold Divisor

The following example demonstrates how to handle a case where the natural divisor coincides with one of the threshold divisors. This case may happen in practice.

Four local elementary schools in Brooklyn, New York, are reapportioning their 20 elementary teachers according to the following enrollments: Beikvei, 225 students; Cheder, 440 students; Hanson, 150 students; and Nefesh, 185 students. Using Webster's rule, the administration found that their initial allocation has one surplus "seat:"

School	Beikvei	Cheder	Hanson	Nefesh	Total
Enrollment	225	440	150	185	1000
Natural Quota	4.5	8.8	3	3.7	20
Webster Initial Allocation	5	9	3	4	21

Calculating their threshold divisors and ordering them, we obtain

School	Beikvei	Cheder	Hanson	Nefesh
Threshold Divisor, $\frac{p_i}{\lfloor q_i\rfloor+.5}$	$\frac{225}{4.5}$	$\frac{440}{8.5}$	$\frac{150}{3.5}$	$\frac{185}{3.5}$
Approximation of TD	50.0	51.76470588	42.85714286	52.85714286
Ordering	3	2	4	1

From the table shown above we see that the natural divisor, $\mathbf{ND} = 50.0$, is the same as the third largest threshold divisor corresponding to Beikvei's school. Hence, a modified divisor should be taken from the interval formed by the third largest threshold divisor and the second divisor:

$$50.0 < \textbf{Webster's Modified Divisor} \leqslant \tfrac{440}{8.5} \approx 51.76470588.$$

If we choose, for instance, $\mathbf{MD} = 50.01$, we get the correct allocation

School	Beikvei	Cheder	Hanson	Nefesh	Total
Modified Quota, $\frac{p_i}{\text{MD}}$	$\frac{225}{50.01}$	$\frac{440}{50.01}$	$\frac{150}{50.01}$	$\frac{185}{50.01}$	
M. Quota, Approximation	4.4991	8.7982	2.9994	3.699	
Webster Final Allocation	4	9	3	4	20

Problems.

1. The 2,438-member carpenters' union, the 1,784-member plumbers' union, and the 984-member painters' union are supposed to choose 10 representatives to negotiate a contract with the city of Boston. Apportion representatives between unions using Webster's rule.

2. Six seats on a state performing arts board are to be apportioned. In the state, there are 204 participating in theater, 187 in dance, and 43 in chorus. Use the Webster method to apportion six seats to this board.

3. A family health services agency has 4 offices and 36 doctors. The agency decides to apportion the doctors based on the average weekly patient load for each clinic, given in the following table. Use Webster's method to apportion the 36 doctors.

Clinic	A	B	C	D
Average Weekly Patient Load	2191	4798	5703	6953

4. A labor council is being formed from the members of four unions. The electricians' union has 78 members, plumbers have 26 members, the carpenters have 107 members, and the general workers have 81 members. The council will have nine representatives, with each union having at least one representative. Use the Webster method to make the apportionment.

5. A condominium association board consists of seven members. There are four buildings A, B, C, and D. There are 32 units in building A, 48 units in building B, 26 units in building C, and 14 units in building D. Use the Webster method to apportion seven seats between buildings.

6. The U.S. state of West Virginia comprises 55 counties. The population of five counties is given in the following table. Use Webster's method to apportion the 128 congressional seats between these 5 counties.

County	Boone	Cabell	Gilmer	Hardy	Kanawha
Population	25,535	96,784	7,160	12,669	200,073

7. Twelve fellowships are to be apportioned among the schools of humanities, science, engineering, and business at a small university. The apportionment is based on the number of full-time graduate students in each school, which is 12 for humanities, 32 for science, 40 for engineering, and 36 for business. Use the Webster method to carry out the apportionment.

8. Five people pool their money to buy a hundred shares of stock. The amount that each person contributes is shown in the following table. Apportion these shares of stock between five people using

   (a)   the Hamilton method,   (b)   the Lowndes method,
   (c)   the Jefferson method,   (d)   the Webster method.

Person	A	B	C	D	E
Contribution	$3260	$4170	$5920	$6990	$7260

9. Suppose that the four Scandinavian countries want to form an economic council with 21 members; apportion the council based on the 2008 estimates of their populations given here, using

   (a)   the Hamilton method,   (b)   the Lowndes method,
   (c)   the Jefferson method,   (d)   the Webster method.

Country	Denmark	Norway	Sweden	Finland
Population	5,493,621	4,768,212	9,219,637	5,313,399

10. The Transit Authority governs the five bridges connecting the New York City boroughs of Bronx, Brooklyn, Manhattan, Queens, and Staten Island. Assume that representation on the nine-member authority is allocated in proportion with the 2008 population of each of the five boroughs presented here.

Bronx	Manhattan	Brooklyn	Queens	Staten Island
1,391,903	1,634,795	2,556,598	2,293,007	487,407

Apportion the nine authority seats using

   (a)   the Hamilton method,   (b)   the Lowndes method,
   (c)   the Jefferson method,   (d)   the Webster method.

11. Apportion 18 patrol officers among 5 regions, having the following number of incidents per region using

    (a) the Hamilton method, (b) the Lowndes method,
    (c) the Jefferson method, (d) the Webster method.

Region	A	B	C	D	E
Incidents	39	48	87	21	42

12. Brandeis University is an American private research university with a liberal arts focus. It is located in the southwestern corner of Waltham, Massachusetts, nine miles (14 km) west of Boston. The university has an enrollment of approximately 3,200 undergraduate and 2,100 graduate students. The schools of the university include:

College	Number of Faculty Members
The College of Arts and Sciences	793
The Graduate School of Arts and Sciences	49
The Heller School for Social Policy and Management	100
Rabb School of Summer and Continuing Studies	62
Brandeis International Business School	51

    Apportion the representatives to the 16-member council based on the number of faculty in each department using

    (a) the Hamilton method, (b) the Lowndes method,
    (c) the Jefferson method, (d) the Webster method.

13. A company has five divisions: Assembling (A), Business products (B), Computers (C), Human resources (H), and Networking (N). Division A has 112 employees, B has 34, C has 81, H has 4, and N has 36. Assume that a 12-member quality control council consists of people from each division in proportion to the number of employees. Apportion the representatives to the council using

    (a) the Hamilton method, (b) the Lowndes method,
    (c) the Jefferson method, (d) the Webster method.

14. Suppose that a town must apportion its polling places among six districts. Apportion 23 polling places to the districts based on the number of eligible voters in each district using

    (a) Jefferson's method, (b) Lowndes' method, (c) Webster's method.

District	1	2	3	4	5	6
Number of Eligible Voters	2908	3157	4284	5422	6013	7316

15. The state of Maine consists of 16 counties. The 2009 populations of six of them are presented in the following table.

State	Aroostook	Franklin	Knox	Lincoln	Oxford	Waldo
Population	71488	29735	40801	34576	56244	38287

    Apportion 71 seats in a hypothetical legislature using

    (a) Jefferson's method, (b) Lowndes' method, (c) Webster's method.

16. There are fourteen counties in the U.S. state of Vermont. The 2009 populations of six of them are presented in the following table.

State	Addison	Chittenden	Essex	Lamoille	Rutland	Windsor
Population	35,974	146,571	6,459	23,233	63,400	57,418

    Apportion 83 seats in a hypothetical legislature using

    (a) Jefferson's method, (b) Lowndes' method, (c) Webster's method.

17. A cafeteria is being renovated to further extend the original area in use. A manager figures out that 88 new tiles are needed. He wants to order tiles in proportion to the existing colors of tiles. Use Webster's method to apportion colored tiles based on the current number of colored tiles presented here.

Color	Blue	Gold	Green	White	Black
Number of tiles	418	450	87	55	494

18. On the national level, Argentina elects a head of state (the president) and a legislature. Voting is mandatory for citizens between 18 and 70 years of age, with some exceptions. The National Congress (Congreso Nacional) has two chambers. The Chamber of Deputies of the Nation has 257 members, elected for a four-year term in each electoral district (23 provinces and the federal capital) by proportional representation using the d'Hondt method, with half of the seats renewed every two years in all districts. Allocate 127 seats in the National Congress based on the 2009 election as shown in the table using

    (a) Jefferson's method   (b) Webster's method.

Party	Votes
Justicialist Party	2,778,326
Front for Victory	1,679,084
Justicialist Front	415,404
Civic and Social Agreement	3,794,853
Radical Civic Union	639,818
Front for Everyone	381,067

19. Regional elections were held in Belgium on June 7, 2009 to choose representatives in the regional councils of Flanders, Wallonia, Brussels, and the German-speaking community of Belgium. These elections were held on the same day as the European elections, and their results are presented below. Compare results of allocations under Jefferson's method and Webster's rule.

    (a) Distribute 124 seats in the **Flemish Parliament** according to the number of votes received.

Party	Votes
Christian Democratic and Flemish	939,873
Flemish Interest	628,564
Socialist Party Different	627,852
Open Flemish Liberals and Democrats	616,610
New-Flemish Alliance	537,040
List Dedecker	313,176
Green	278,211
Union of Francophones	47,319

    (b) Allocate 75 seats in the **Walloon Parliament** according to the number of votes received.

Party	Votes
Socialist Party	657,803
Reformist Movement	469,792
Ecolo	372,067
Humanist Democratic Centre	323,952
National Front	57,374

    (c) Apportion 72 seats in the **Brussels Regional Parliament** (French language group) according to the number of votes received.

Party	Votes
Reformist Movement	121,905
Socialist Party	107,303
Ecolo	82,663
Humanist Democratic Centre	60,527
National Front	7,803
Pro Bruxsel	6,840

(d) Apportion 17 seats in the **Brussels Regional Parliament** (Dutch language group) according to the number of votes received.

Party	Votes
Open Flemish Liberals and Democrats	11,957
Socialist Party Different	10,085
Flemish Interest	9,072
Christian Democratic and Flemish	7,696
Green	5,806
New-Flemish Alliance	2,586
List Dedecker	1,957

(e) Assign 25 seats in the **Parliament of the German-speaking Community** according to the number of votes received.

Party	Votes
Christian Social Party	10,122
Socialist Party	7,231
Party for Freedom and Progress	6,562
ProDG	6,553
Ecolo	4,310
Vivant	2,684

20. India held general elections to the 15th Lok Sabha (lower house of the parliament of India) in five phases between April 16, 2009 and May 13, 2009. With an electorate of 714 million (larger than the European Union and United States combined), it was the largest democratic election in the world to date. Apportion 543 seats in the Lok Sabha according to the number of votes received using

   (a) Jefferson's method.   **(b)** Adams' method.   **(c)** Webster's method.

Party	Votes
Indian National Congress	119,110,776
All India Trinamool Congress	13,355,986
Dravida Munnetra Kazhagam	7,625,397
Indian Union Muslim League	877,503
National Conference	498,374
Sikkim Democratic Front	159,351

21. The Hillside Family and Community Center (HFCC) operates 6 clinics with a total staff of 27 doctors. Each doctor works five days a week, except for three doctors, who can only work four days. Each doctor can work at only one location during a day. Therefore, the HFCC has $5 \times 24 + 3 \times 4 = 132$ doctor-days available to apportion among clinics. The average weekly patient load shown in the following table:

Clinic	A	B	C	D	E	F
Patient Loads	178	237	251	269	284	301

Allocate 132 doctor-days using
(a) Jefferson's method,   (b) Adams' method,   (c) Webster's method.

22. Five clubs are sending reps to a student council. The populations are: Computer club with 53 members, Dance club with 47 members, Drama club with 16 members, Garden club with 29 members, and Fitness club with 81 members. There are 13 seats available on the council. How do we divide the seats using one of the known apportionment methods?

23. In order to reduce the federal deficit, the government analyzed U.S. import spending in 2004.

Commodity	Airplane	Coffee	Gold	Sugar	TV	Vehicles
Spending (in millions of dollars)	11389	1868	3996	516	87885	187723

Represent the import spending in percentages keeping one decimal place using
(a) Jefferson's method,   (b) Adams' method,   (c) Webster's method.

24. Six countries want to form a 44-member committee to develop a project to construct a highway around the Black Sea. Apportion seats in the committee using Webster's method based on the population of the states in 2008.

Country	Russia	Ukraine	Romania	Bulgaria	Greece	Turkey
Population	141,950,000	46,258,200	21,513,622	7,623,395	11,237,094	73,914,260

25. In the city of New York there are 76 police precincts collectively throughout the 5 boroughs (Manhattan, Queens, Bronx, Brooklyn, and Staten Island). The city of New York Police Department separates their crime statistics by groupings: Manhattan North, Manhattan South, Brooklyn North, Brooklyn South, Queens North, Queens South, Staten Island, and Bronx. Crime statistics are based upon violent and serious crimes (murder, rape, larceny, assault, and burglary) and they are compiled weekly. The highest crime grouping in New York is the Bronx. During the first half of 2010, there have been 346 reported violent crimes in the Bronx district. Manhattan North has reported 214 violent crimes, while Manhattan South has reported 298. Brooklyn North also has reported 298 and Brooklyn South 289. Queens North is reporting 205 to date, while Queens South is showing 203. Staten Island has reported 49 violent crimes to date, making them the lowest crime area in the 5 boroughs. The NYPD must allocate its 2763 police officers according to needs within each borough. According to this basic synopsis the most police support must be stationed in the Bronx district, while the lowest would be in Staten Island. Apportion these officers using
(a) Jefferson's method,   (b) Adams' method,   (c) Webster's method.

26. Suppose that 6 Caribbean countries decide to convene 116 tourist companies for a conference to make recommendations on coordination and promotion of their business. Apportion 116 seats between these countries based on their 2010 populations using
(a) Jefferson's method.   (b) Adams' method.   (c) Webster's method.

Country	Bahamas	Cuba	Haiti	Dominican	Puerto Rico	Jamaica
Population	307451	11451652	10033000	9760000	3996381	2825928

27. A store makes an order for 134 sneakers. The weekly average number of sneakers the store has sold is presented in the table below.

Maker	Adidas	Air Jordan	Converse	Nike	Puma	Reebok	Fila
Sales	476	273	239	361	409	527	197

Determine how many sneakers of each brand the store should order using
(a) Jefferson's method.   (b) Adams' method.   (c) Webster's method.

28. Apportion the 201 meals that a local food store makes during rush hours using the numbers in the table based on
    (a) Jefferson's method. (b) Adams' method. (c) Webster's method.

Meal	Burgers	McFlurry	Salads	Fries	Nuggets	Nachos
Sales	384	189	41	299	316	471

29. A TV company wants to show 237 sport games based on the audience watching competitions, presented in the table.

Sport	NBA	NFL	NHL	MLB	MLS	Tennis
Audience	633	501	487	368	295	186

Apportion sport activities to be shown on TV using
    (a) Jefferson's method. (b) Adams' method. (c) Webster's method.

30. The Knesset is the legislature of Israel, located in Givat Ram, Jerusalem. The Knesset enacts laws, elects the president and prime minister (although he or she is ceremonially appointed by the President), supervises the work of the government, reserves the power to remove the President of the State and the State Comptroller from office and to dissolve itself and call new elections. Summary of the 10 February 2009 Israeli Knesset election results are presented in the following table.

Party	Votes	Party	Votes
Kadima	758,032	Likud	729,054
Yisrael Beiteinu	394,577	Labor	334,900
Shas	286,300	United Torah Judaism	147,954
The Jewish Home	96,765	The Greens	12,378

Distribute 120 seats between these eight parties using
    (a) Jefferson's method. (b) Adams' method. (c) Webster's method.

31. China's Three Gorges Dam is known as the world's largest hydropower project and most notorious dam. It is 185 m high and 2,309 m long, on the Chang (Yangtze) River, in the central Hubei province, 48 km west of Yichang. The dam holds a reservoir 660 km long and contains as much water as Lake Superior. Seats on the 22-member board, which coordinates operations of the power plant, are assigned according to the populations of 7 provinces surrounding the dam.

State	Anhui	Chongqing	Henan	Hubei	Hunan	Jiangxi	Shaanxi
Population	61,350	31,442	98,690	60,160	66,980	44,000	37,050

Distribute 22 seats between these seven provinces using
    (a) Jefferson's method. (b) Adams' method. (c) Webster's method.

32. Show that the apportionment produced by Webster's method violates the quota rule for a 110-member house with five states, whose populations are given in the following table.

State	A	B	C	D	E
Population	10100	430	332	148	46

33. Brazil elects on a national level the president and a legislature. The president is elected for a four-year term by the people. The National Congress has two chambers. The Chamber of Deputies has 513 members, elected for a 4-year term by proportional representation. The Federal Senate has eighty one members, elected for an eight-year term, with elections every four years for alternately one-third and two-third of the seats. Brazil was the first country in

the world to have the fully electronic elections. Based on the summary of the 2006 elections, allocate 513 seats between 7 main parties using

(a)   Jefferson's method.   (b)   Adams' method.   (c)   Webster's method.

Party	Votes
Workers' Party	13,989,859
Brazilian Democratic Movement Party	13,580,517
Brazilian Social Democratic Party	12,691,043
Democrats	10,182,308
Progressive Party	6,662,309
Brazilian Socialist Party	5,732,464
Democratic Labour Party	4,854,017

34. Italy elects, on a national level, a parliament consisting of two houses, the Chamber of Deputies (630 members) and the Senate of the Republic (315 elected members, plus a few senators for life). The president of the republic is elected for a seven-year term by the parliament houses in joint session. In Italy, there are two main political coalitions: Silvio Berlusconi leads the center-right opposition coalition and Walter Veltroni leads the Democratic Party. However, historically there are many parties, including the left-wing coalition, the right-wing one, and some regional parties. The results of the 2008 elections are presented here. Allocate 630 seats between 7 parties using

(a)   Jefferson's method.   (b)   Adams' method.   (c)   Webster's method.

Party	Berlusconi	Veltroni	Center	Left	Right	Socialist	People
Votes	17,064,314	13,686,501	2,050,309	1,124,428	885,226	355,575	147,666

35. Solve Problem 27 from §2.2 using Webster's method.

36. The 2010 Philippine House of Representatives elections were held on May 10, to elect 286 new members. The results of this election are summarized in the following table. Distribute these seats between parties using the Webster method.

Party	Lakas	Liberal	Nacionalista	PMP	NPC	Independents
Votes	13,242,191	6,923,162	3,995,334	971,026	5,227,075	2,437,480

37. The House of Representatives in the Kingdom of Thailand is the lower house of the National Assembly of Thailand, the legislative branch of the Thai Government. The House of Representatives has 480 members: 400 of them are democratically elected through single constituency elections, while the other 80 are appointed through proportional representation. Summary of the 2007 general election is presented here. Apportion 398 seats between parties using the Webster method.

Party	People	Democrat	Thai	Motherland	United	Neutral
Votes	26,293,456	21,745,696	6,363,475	6,599,422	3,395,197	3,844,673

Compare your distribution with allocation obtained with the aid of the Jefferson method and the Adams method.

38. A country comprises 6 states. Distribute 94 seats between these states using the Webster method according to their populations (in thousand) presented in the following table.

State	A	B	C	D	E	F
Population	6256	2569	3477	2422	1978	843

Compare your distribution with allocation obtained with the aid of the Jefferson method and the Adams method.

2.5 The Hill-Huntington Method

The divisor method that has been used to distribute seats in the U.S. House of Representatives since 1941 is called the Hill-Huntington method, which is also known as the method of equal proportions. It was developed around 1911 by Joseph Adna Hill (1860 – 1938), chief statistician of the Bureau of the Census. Later, in the 1920s, a Professor from Harvard University, Edward Vermilye Huntington (1874 – 1952), revised his ideas to obtain a rigorous apportionment method. The method assigns seats between states to achieve the smallest relative unfairness, Eq. (2.1.6) on page 156, among all states. We illustrate it first in the following example, then go to the general case, and at the end provide a streamline of the method followed by numerous examples.

The Hill-Huntington method is a divisor method, which means that it may violate the quota rule, but it is immune from the three paradoxes discussed in §2.1. In contrast to Jefferson's method, the Hill-Huntington method rarely fails to satisfy the quota rule. The method is neutral to the populations of states, with a slight bias to smaller states. It favored small states 13 times over the course of U.S. Congressional apportionments. The method never assigns any state an apportionment of zero seats. The Hill-Huntington algorithm is the best apportionment method that minimizes the relative unfairness.

Example 2.5.1. Using the Apportionment Criterion
Suppose that in a legislature two parties, call them A and B, compete for seats. Party A gained support of 14,475 voters and 5 representatives and party B had support of 8,682 and two representatives. To apply the apportionment criterion, we need to calculate the relative unfairness for each state, Eq. (2.1.6) on page 156.

To determine which of the two states deserves a surplus seat, we first give the representative to A instead of B. So A will have six representatives and B will have two. Their average constituencies are as follows:

$$\text{A's average constituency} = \frac{\text{A's population}}{\text{A's representatives}} = \frac{14475}{6} = 2412.5,$$

$$\text{B's average constituency} = \frac{\text{B's population}}{\text{B's representatives}} = \frac{8682}{2} = 4341.$$

Because state A has the smaller average constituency, the relative unfairness of this apportionment becomes

$$\text{relative unfairness} = \frac{4341 - 2412.5}{2412.5} = \frac{1928.5}{2412.5} \approx 0.799378.$$

Assume that we give the representative to party B instead of A. Hence, A will have five representatives and B will have three. Their average constituencies are as follows:

$$\text{A's average constituency} = \frac{\text{A's population}}{\text{A's representatives}} = \frac{14475}{5} = 2895,$$

$$\text{B's average constituency} = \frac{\text{B's population}}{\text{B's representatives}} = \frac{8682}{3} = 2894.$$

Now B has the smaller average constituency; therefore, the relative unfairness of this apportionment becomes

$$\text{relative unfairness} = \frac{2895 - 2894}{2894} = \frac{1}{2894} \approx 0.0003.$$

So we see that we should give B a third representative before A receives a bonus seat since it will result in a smaller relative unfairness.   □

Now we consider a general case. Suppose that we have two states, call them A and B, with populations a and b, respectively. Suppose that initially state A was allotted with x seats and state B was apportioned with y seats. The averages of constituencies of these states are $\frac{a}{x}$ and $\frac{b}{y}$, respectively. If the latter average is the smallest, then we reward state B with an extra seat and the relative unfairness would be

$$\text{relative unfairness} = \frac{\frac{b}{y+1} - \frac{a}{x}}{a/x} = \frac{b\,x}{(y+1)\,a} - 1.$$

If state B has the smallest average, we give state A an extra seat and the relative unfairness would be

$$\text{relative unfairness} = \frac{\frac{a}{x+1} - \frac{b}{y}}{b/y} = \frac{a\,y}{(x+1)\,b} - 1.$$

Whichever of these two quantities is bigger, the corresponding state gets a seat. So we compare these numbers:

$$\frac{b\,x}{(y+1)\,a} - 1 \overset{?}{<} \frac{a\,y}{(x+1)\,b} - 1 \qquad \text{or} \qquad \frac{b\,x}{(y+1)\,a} \overset{?}{<} \frac{a\,y}{(x+1)\,b}.$$

Multiplying both sides by $\frac{ab}{xy}$, we get

$$\frac{b^2}{y(y+1)} \overset{?}{<} \frac{a^2}{x(x+1)} \qquad \text{or} \qquad \frac{b}{\sqrt{y(y+1)}} \overset{?}{<} \frac{a}{\sqrt{x(x+1)}}.$$

Therefore, the determination of which state deserves a surplus seat depends on whichever of these two numbers is larger. If the number of states is greater than two, the distribution of seats between states becomes tedious and time consuming because it involves the calculation of roots for possible seat allocations using these quantities for comparisons. Now we are going to present a streamline of this method, starting with some definitions, demonstrating the application of the method in the U.S. apportionment of the House, and then formulate the Hill-Huntington algorithm, followed by various examples.

The **geometric mean** of two real numbers a and b is defined to be the root $\sqrt{a \cdot b}$. It is the edge of the square with the same area as the rectangle with edges a and b.

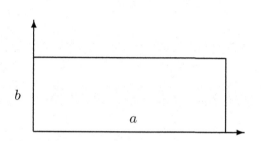
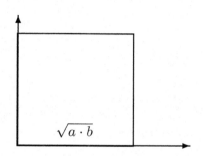

Example 2.5.2. Allocation of Seats in the U.S. House of Representatives

Since 1941, the 435 seats in the U.S. House of Representatives have been apportioned according to the Hill-Huntington method. As a first step, each of the 50 states is initially guaranteed at least 1 seat in the House of Representatives, leaving the remaining 385 seats to be assigned.

The remaining seats are allocated, one at a time, to the state that deserves the next assigned seat the most. Thus, the 51st seat would go to the most populous state (currently California). The measure of how much a state deserves the next allocatable seat is determined by a priority formula that is mathematically computed to be the ratio of the state population to the geometric mean of the number of seats it currently holds in the assignment process, n (initially 1), and the number of seats it would hold if the seat were assigned to it, $n + 1$.

The formula for determining the priority of a state to be apportioned the next available seat defined by the method of equal proportions is

$$A_n = \frac{p}{\sqrt{n(n+1)}},$$

where p is the population of the state, and n is the number of seats it currently holds before the possible allocation of the next seat. An equivalent, recursive definition is

$$A_{n+1} = \sqrt{\frac{n}{n+2}}\, A_n,$$

where n is still the number of seats the state has before allocation of the next, and for $n = 1$, the initial A_1 is explicitly defined as

$$A_1 = \frac{p}{\sqrt{2}}.$$

Consider the reapportionment following the 2010 U.S. Census. Beginning with all states initially being allocated one seat, the largest value of $A_1 = \frac{38,292,687}{\sqrt{2}} \approx 27077018$ corresponds to the largest state, California, which is allocated seat 51. But after being allocated its second seat, its priority value decreases to its $A_2 = \frac{38,292,687}{\sqrt{2 \cdot 3}} \approx 15632924$ value, which is reordered to a position back in line. The 52nd seat goes to Texas, the second largest state, because its $A_1 = \frac{24,873,773}{\sqrt{2}} \approx 17588413$ priority value is larger than the A_n of any other state. However, the 53rd goes back to California because its A_2 priority value is larger than the A_n

of any other state. The 54th seat goes to New York because its $A_1 = \frac{19,541,453}{\sqrt{2}} \approx 13817893$ priority value is larger than the A_n of any other state at this point. This process continues until all 435 seats have been assigned. Each time a state is assigned a seat, n is incremented by 1, causing its priority value to be reduced and reordered among the states which normally results in another state rising to the top of the list.   □

Let q be the natural quota of a state with population p. Recall that the root $\sqrt{q(q+1)}$ is called the **geometric mean** of numbers q and $q+1$. The quantity

$$\frac{p}{\sqrt{\lfloor q \rfloor \cdot \lceil q \rceil}} = \frac{\text{state's population}}{\sqrt{(\lfloor \text{State's Natural Quota} \rfloor)(\lfloor \text{State's Natural Quota} \rfloor + 1)}} \qquad (2.5.1)$$

is referred to as the (Hill-Huntington) **threshold divisor** of the state if its lower quota is not zero. We assign ∞ to be the threshold divisor if its lower quota is 0. Here $\lfloor q \rfloor$ denotes the floor of the quota (real number) q, and $\lceil q \rceil$ denotes its ceiling.

The Hill-Huntington rounding procedure is based on comparison of the natural quota with the geometric mean of its upper and lower quotas, called the **cutoff point**. For every rational number, q, it rounds the number to the lower quota, $\lfloor q \rfloor$, if it is less than the geometric mean of the two whole numbers that the number q is immediately between (for example, 3.71 is immediately between 3 and 4). Otherwise, the rounding procedure assigns q its upper quota, $\lceil q \rceil$. In other words,

$$\text{Hill-Huntington's procedure rounds } q \text{ to } \begin{cases} \lfloor q \rfloor, & \text{if } q < \sqrt{\lfloor q \rfloor \cdot \lceil q \rceil}, \\ \lceil q \rceil, & \text{if } q > \sqrt{\lfloor q \rfloor \cdot \lceil q \rceil}. \end{cases} \qquad (2.5.2)$$

For instance, if the natural (standard) quota is 8.49, its upper quota is 9, and the lower quota is 8. The geometric mean of numbers 8 and 9 is

$$\sqrt{8 \cdot 9} = \sqrt{72} \approx 8.485281374.$$

Therefore, the natural quota is larger than the mean, and we round the quota up. On the other hand, the number 8.47 is rounded down, to 8, because it is less than the geometric mean of 8 and 9.

Note that the natural quota of state i, which we denote by q_i, is a rational number because it is the ratio of two integers where the numerator is the state's population times the house size and the denominator is the total population, see Eq. (2.1.1) on page 150. On the other hand, the geometric mean of two sequential integers is always an irrational number. Hence these two numbers, the natural quota, q_i, and the geometric mean of its upper and lower quotas, $\sqrt{\lfloor q_i \rfloor \cdot \lceil q_i \rceil}$, are never equal. So their comparison always leads to a conclusion that one of these numbers is larger than the other. Moreover, the square root of two sequential integers has a fractional part that always starts with the digit 4. Therefore, the comparison of the rational number q_i with the irrational one, $\sqrt{\lfloor q_i \rfloor \cdot \lceil q_i \rceil}$, can be done without actual calculation of the geometric mean if the fractional part of the quota, q_i, starts with any digit other than 4.

Actually, the Hill-Huntington method is very close to Webster's algorithm: they differ only in utilizing the cutoff points—the points for rounded quotas. When the house size is large, these two cutoff points become very close to each other. Since the geometric mean of two positive real numbers is always less than or equal to the arithmetic mean ($\sqrt{a \cdot b} < (a+b)/2$, $a \neq b$), the Hill-Huntington cutoff point is strictly less than the Webster cutoff point, approaching it when the house size grows.

You might be wondering as to why we should learn and use such a complicated cousin to Webster's method when these two algorithms usually give indistinguishable allocations. But the answer is simple: they round quotas similarly only when their integer parts are not zero. If the lower quota is zero, then the geometric mean of 0 and 1 is $\sqrt{0 \cdot 1} = 0$, while the arithmetic mean is $(0 + 1)/2 = 0.5$. So the cutoff point for Hill-Huntington method is 0 whereas for Webster's rule it is 0.5. Since the natural quota is a positive number, it is always larger than zero; therefore, the Hill-Huntington method never leaves a state without a seat, while the Webster rule could assign no seat to a state when its modified quota is less than 0.5.

Now we are ready to present the Hill-Huntington algorithm for allocation of seats between states.

1. Calculate each state's natural (standard) quota according to Eq. (2.1.2), page 150.

2. Initially assign a state its lower quota (floor of the natural quota) if the natural quota is less than the geometric mean of its lower and upper quotas; assign a state its upper quota (ceiling of the natural quota) if the standard quota is larger than the geometric mean of its lower and upper quotas. In other words, if q_i is the natural quota of the state i, then the tentative allocation of the state is obtained by rounding q_i down if $q_i < \sqrt{\lfloor q_i \rfloor \cdot \lceil q_i \rceil}$, or rounding up if $q_i > \sqrt{\lfloor q_i \rfloor \cdot \lceil q_i \rceil}$, see Eq. (2.5.2).

3. Check whether the sum of initial allocations (the lower or upper quotas from step 2) is equal to the correct number of seats to be apportioned, called the house size. If the sum of the quotas in the initial allocation from step 2 is equal to the correct number of seats to be apportioned, then stick with this apportionment—no other work is required.

4. If the initial allocation is NOT equal to the house size, proceed as follows.

5. Calculate the difference, m, between the correct number of seats to be apportioned (house size) and the initial allocation. The number of surplus seats could be either positive if the tentative allocation is less than the house size, or negative if the initial allocation is greater than the house size.

6. Calculate the natural (standard) divisor according to Eq. (2.1.1), page 150.

7. For each state i, calculate its **threshold divisor**:

$$\mathbf{TD} = \frac{\text{state's population}}{\text{geometric mean of lower quota and upper quota}} = \frac{p_i}{\sqrt{\lfloor q_i \rfloor \cdot \lceil q_i \rceil}}, \quad (2.5.3)$$

where p_i is the population of state i, $q_i = p_i * h/p$ is the natural quota, h is the house size, and p is the total population. The threshold divisor is assigned ∞ if the lower quota is zero, $\lfloor q_i \rfloor = 0$.

8. Rank all threshold divisors from largest to smallest. Such an ordering breaks the axis of real numbers into subintervals having endpoints as threshold divisors.

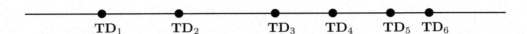

9. Identify the position of the natural divisor among all ordered threshold divisors; that is, find the smallest interval that contains the natural divisor. Note that its position can be above all of them (if the natural divisor is larger than any threshold divisor), it can be below all of them (if the standard divisor is smaller than any threshold divisor), or it can be somewhere in between.

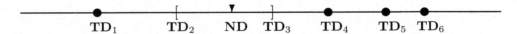

If the difference, m, between the house size and the initial allocation is positive—the initial allocation is m seats less than required—move the position of the natural divisor m steps down. Then choose a modified divisor as a number between these modified divisors that serve as the boundary of this position.

If the difference, m, between the house size and the initial allocation is negative—the initial allocation is m seats more than required—move the position of the natural divisor m steps up. Then choose a modified divisor as a number between these two modified divisors that bound its position.

10. For each state, compute the modified quota (by dividing each state's population by the modified divisor instead of the natural divisor):

$$\mathbf{Modified\ Quota} = \frac{\text{state's population}}{\text{modified divisor}} = \frac{p_i}{\mathbf{MD}}. \quad (2.5.4)$$

By rounding modified quotas up or down to the nearest integers (ceilings of modified quotas if the modified quota is larger than the geometric mean of upper and lower quotas, or floors if the modified quotas are less than the cutoff points), obtain the final allocation. If the sum of seats in the final allocation is the exact number of seats to be apportioned, you are done!

11. If the sum of rounded modified quotas in the previous step is not equal to the house size, this indicates that either the quota property is violated or a wrong number of states gain or lose seats. In this case, calculate the threshold divisors for each additional seat each state gains/loses. If a state gains s ($s \geqslant 1$) additional seats compared to the upper quota, calculate the threshold divisors:

$$\mathbf{TD}_1 = \frac{p_i}{\sqrt{\lfloor q_i \rfloor \cdot \lceil q_i \rceil}}, \quad \mathbf{TD}_2 = \frac{p_i}{\sqrt{\lfloor q_i + 1 \rfloor \cdot \lceil q_i + 1 \rceil}}, \quad \ldots, \quad \mathbf{TD}_s = \frac{p_i}{\sqrt{\lfloor q_i + s \rfloor \cdot \lceil q_i + s \rceil}}.$$
$$(2.5.5)$$

If a state loses s ($s \geqslant 1$) seats compared to the lower quota, calculate the threshold divisors:

$$\mathbf{TD}_1 = \frac{p_i}{\sqrt{\lfloor q_i \rfloor \cdot \lceil q_i \rceil}}, \quad \mathbf{TD}_2 = \frac{p_i}{\sqrt{\lfloor q_i - 1 \rfloor \cdot \lceil q_i - 1 \rceil}}, \quad \ldots, \quad \mathbf{TD}_s = \frac{p_i}{\sqrt{\lfloor q_i - s \rfloor \cdot \lceil q_i - s \rceil}}.$$
$$(2.5.6)$$

12. Rank all threshold divisors (from parts 7 and 11). Identify the position of the natural divisor in this ordered list of threshold divisors. Move the natural divisor m positions up or down depending on whether you need to add seats or eliminate surplus seats. If you need to eliminate m seats from the initial allocation, move the natural quota m positions up. If your allocation is m seats short, move the position of the natural divisor down m places. The threshold divisors that serve as the boundary of this new position are the upper and lower bounds for your choice of a new modified divisor.

13. With a modified divisor chosen in the previous step, repeat step 10 of this algorithm to obtain the required final allocation. Namely, use a modified divisor instead of the natural divisor to calculate the modified quotas, Eq. (2.5.4). Round modified quotas up or down based on comparison with the geometric mean, Eq. (2.5.2), then the sum of all the rounded modified quotas will be the house size. Apportion each state its modified rounded quota. ∎

The Hill-Huntington apportionment procedure is demonstrated in the following examples.

Example 2.5.3. Hill-Huntington's Apportionment

Consider five local libraries in the city of Boston, Massachusetts, United States. A committee of 30 supervises their operations and distribution of new books. It is filled by representatives from each library based on the average number of visitors per week presented in the table.

Library	Public	Charlestown	Eddy	Science	French	Total
Number of visitors	10495	8343	5487	3812	1293	29430

Based on their natural quotas, we find the tentative allocation to be

Library	Public	Charles	Eddy	Science	French	Total
N. Quota, $\frac{p_i * 30}{29430} = \frac{p_i}{981}$	$\frac{10495}{981}$	$\frac{8343}{981}$	$\frac{5487}{981}$	$\frac{3812}{981}$	$\frac{1293}{981}$	
NQ Approximation	10.6982	8.5045	5.5932	3.8858	1.3180	
Initial allocation	11	9	6	4	1	31

To determine state's initial allocation in this example, we do not need to apply the Hill-Huntington rounding procedure (2.5.2) and calculate geometric means of upper and lower natural quotas for every state ($\sqrt{\text{lower quota} \cdot \text{upper quota}}$):

Library	Public	Charlestown	Eddy	Science	French
$\sqrt{\lfloor q_i \rfloor \cdot \lceil q_i \rceil}$	$\sqrt{10 \cdot 11}$	$\sqrt{8 \cdot 9}$	$\sqrt{5 \cdot 6}$	$\sqrt{3 \cdot 4}$	$\sqrt{1 \cdot 2}$
Geometric mean	10.488088	8.485281	5.477225	3.464101	1.414213

As we know, the first digit after the period in a geometric mean of any two sequential integers is always 4; however, in our case, all fractional parts of natural quotas but the last one are larger than 0.4. The last natural quota for the French library has the fractional part less than 0.4. Hence, we have to round natural quotas for the first four libraries up, while we have to round the last quota down. This gives us the initial allocation that exceeds the house size by one seat.

In this case, the algorithm asks to find threshold divisors for every state and identify the position of the natural divisor, which is $\mathbf{ND} = \frac{29430}{30} = 981$, in this ranked list of divisors. So we use the formula (2.5.3) to calculate the threshold divisors and then rank them.

Library	Public	Charlestown	Eddy	Science	French
Threshold Divisor	$\frac{10495}{\sqrt{10 \cdot 11}}$	$\frac{8343}{\sqrt{8 \cdot 9}}$	$\frac{5487}{\sqrt{5 \cdot 6}}$	$\frac{3812}{\sqrt{3 \cdot 4}}$	$\frac{1293}{\sqrt{1 \cdot 2}}$
Approximation of TD	1000.65	983.23	1001.78	1100.42	914.28
Ranking	3	4	2	1	5

Since the natural divisor, $\mathbf{ND} = 981$, is less than Charlestown's threshold divisor and larger than French's threshold divisor, we must choose a modified divisor as a number from the next interval bounded by the fourth largest modified divisor, which is Charlestown, and the third largest one, which is Public:

$$983.2319792 < \textbf{H.H. Modified Divisor} < 1000.658987.$$

Choosing a modified divisor as, say, $\mathbf{MD} = 1000$, we compute the modified quotas for every state: `state's population/modified divisor` to obtain values presented in the table. Rounding all modified quotas according to the Hill-Huntington rounding procedure (2.5.2), we obtain the required final allocation.

Library	Public	Charlestown	Eddy	Science	French	Total
Modified quota, $\frac{p_i}{1000}$	10.495	8.343	5.487	3.812	1.293	
Final allocation	11	8	6	4	1	30

Example 2.5.4. Natural Divisor is the Lowest

The Republic of Banana is a small country consisting of five states (A, B, C, D, and E). The total population of Banana is 12 million. Allocate 120 seats in the legislature using the Hill-Huntington method if the natural quota of each state is given in the following table:

States:	A	B	C	D	E
Natural Quota:	40.50	29.70	23.65	14.60	11.55

Since the populations of the states are not provided, we have to find their values. Using Eq. (2.1.5) on page 155, we obtain the states' populations by multiplying the natural quotas by their **Natural Divisor** $= \dfrac{\text{total population}}{\text{house size}} = \dfrac{12.0}{120} = 0.1$. The populations in millions are given:

States:	A	B	C	D	E	Total
Population:	4.05	2.97	2.365	1.460	1.155	12

The initial allocation is obtained by rounding all natural quotas up, to their upper quotas, because all their fractional parts are larger than 0.5. Hence this allocation has two surplus seats that should be eliminated. To identify which of the states's allocations must be lowered, we have to calculate the threshold divisors:

State	A	B	C	D	E
Threshold Divisor	$\dfrac{4.05}{\sqrt{40\cdot41}}$	$\dfrac{2.97}{\sqrt{29\cdot30}}$	$\dfrac{2.365}{\sqrt{23\cdot24}}$	$\dfrac{1.46}{\sqrt{14\cdot15}}$	$\dfrac{1.155}{\sqrt{11\cdot12}}$
Approximation of TD	0.100007	0.100692	0.100661	0.100749	0.100529
Ranking	5	2	3	1	4

The natural divisor, $\mathbf{ND} = 0.1$, is smaller than any of these threshold divisors, so we need to pick up a modified divisor from the interval bounded by the second smallest threshold divisor, which corresponds to state E, and the third one, which corresponds to state C:

$$0.1005298463 < \textbf{H.H. Modified Divisor} < 0.1006610847.$$

For instance, we choose a modified divisor as $\mathbf{MD} = 0.1006$. Then we divide the population of each state by this modified divisor to obtain the modified quota. We compare modified quotas with the values of roots $\sqrt{n(n+1)}$, where n is the integer part of the quota: if the root is larger, then round the corresponding modified quota down; otherwise, take a ceiling. This yields the final allocations that have the correct sum value—house size. We summarize the calculations in the following table:

State	Natural Quota	$\sqrt{n(n+1)}$	H.-H. I. A.	Threshold Divisor	Rank	Modif. Quota	H.-H. F. A.
A	40.50	40.496913	41	0.1000076216	5	40.25	40
B	29.70	29.495762	30	0.1006924303	2	29.52	30
C	23.65	23.494680	24	0.1006610847	3	23.50	24
D	14.60	14.491376	15	0.1007495717	1	14.51	15
E	11.55	11.489125	12	0.1005298463	4	11.481	11
Total:			122				120

Example 2.5.5. Natural Divisor is the Largest

Consider a typical apportionment problem of rounding rational numbers in a sum to integers:

$$1.4 + 2.44 + 3.45 + 4.47 + 5.24 = 17.$$

In this sum, we can consider each summand as a state's natural quota, so $q_1 = 1.4$, $q_2 = 2.44$, $q_3 = 3.45$, $q_4 = 4.47$, and $q_5 = 5.24$ are natural quotas, and $h = 17$ is the house size. The populations of each state are not provided; however, we do not need their actual values to round quotas. According to Eq. (2.1.5) on page 155, a state's population is proportional to its natural quota, and the coefficient of proportionality is the natural divisor, which can be chosen arbitrarily because it does not affect the final allocation. If we take the natural divisor to be 1, we assign the state's populations to be their natural quotas.

To obtain the tentative allocation, the Hill-Huntington rounding procedure compares the natural quota, q_i, with the cutoff point—the root $\sqrt{\lfloor q_i \rfloor \cdot \lceil q_i \rceil}$. If the quota is less than the geometric mean of its ceiling and floor, then the quota is rounded down (called the lower quota). If the natural quota is larger than the cutoff point, the quota is rounded up (called the upper quota). In such a way, we obtain the following tentative allocation:

Number	#1	#2	#3	#4	#5	Total
Natural Quota	1.4	2.44	3.45	4.47	5.24	17
Mean, $\sqrt{\lfloor q_i \rfloor \cdot \lceil q_i \rceil}$	$\sqrt{1 \cdot 2}$	$\sqrt{2 \cdot 3}$	$\sqrt{3 \cdot 4}$	$\sqrt{4 \cdot 5}$	$\sqrt{5 \cdot 6}$	
Mean Approximation	1.4142	2.4494	3.4641	4.4721	5.4772	
Initial allocation	1	2	3	4	5	15

Since the house is not full, we need to calculate the threshold divisors for every number in the sum.

Number	#1	#2	#3	#4	#5
Modified Divisor, $\dfrac{p_i}{\sqrt{\lfloor q_i \rfloor \cdot \lceil q_i \rceil}}$	$\dfrac{1.4}{\sqrt{1 \cdot 2}}$	$\dfrac{2.44}{\sqrt{2 \cdot 3}}$	$\dfrac{3.45}{\sqrt{3 \cdot 4}}$	$\dfrac{4.47}{\sqrt{4 \cdot 5}}$	$\dfrac{5.24}{\sqrt{5 \cdot 6}}$
MD, Approximation	.989949	.996125	.995929	.999522	.956688
Ordering	4	2	3	1	5

The natural divisor, **ND** $= 1.0$ is larger than any of these threshold divisors. Because we need to add two surplus seats to the initial allocation, we have to choose a modified divisor in the range between the second largest threshold divisor and the third one:

$$0.9959292146 < \textbf{H.H. Modified Divisor} < 0.9961258289.$$

If we choose the modified divisor to be **MD** $= 0.996$, we will get the required allocation:

Number	#1	#2	#3	#4	#5
Modified Quota, $\dfrac{p_i}{0.996}$	$\dfrac{1.4}{0.996}$	$\dfrac{2.44}{0.996}$	$\dfrac{3.45}{0.996}$	$\dfrac{4.47}{0.996}$	$\dfrac{5.24}{0.996}$
M. Quota, Approximation	≈ 1.4056	≈ 2.4497	≈ 3.4638	≈ 4.4879	≈ 5.2610
Mean, $\sqrt{\lfloor q_i \rfloor \cdot \lceil q_i \rceil}$	1.4142	2.4494	3.4641	4.4721	5.4772
Final Allocation	1	3	3	5	5

Example 2.5.6. U.S. Contributions to International Organizations

The government wants to represent U.S contributions to international organizations in 2004 in a percentage way. The main contributions were made to UNESCO, \$70 million; World Trade Organization, \$21 million; North Atlantic Treaty Organization, \$62 million; Organization of American States, \$57 million; Asia-Pacific Economic Cooperation, \$1 million; and

North Atlantic Assembly, \$1 million, with total of \$212 million. Using the Hill-Huntington method with the house size $h = 100$, the initial allocation becomes

Organization	UNESCO	WTO	NATO	OAS	APEC	NAA
Natural Quota, $\frac{p_i*100}{212}$	$\frac{70*100}{212}$	$\frac{21*100}{212}$	$\frac{62*100}{212}$	$\frac{57*100}{212}$	$\frac{1*100}{212}$	$\frac{1*100}{212}$
N. Quota, Approximation	33.0188	9.9056	29.2452	26.8867	0.4716	0.4716
Initial Allocation	33	10	29	27	1	1

Since the tentative allocation has one surplus seat, we need to calculate threshold divisors, and then rank them.

Organization	UNESCO	WTO	NATO	OAS	APEC	NAA
T. Divisor, $\frac{p_i}{\sqrt{\lfloor q_i \rfloor \cdot \lceil q_i \rceil}}$	$\frac{70}{\sqrt{33\cdot34}}$	$\frac{21}{\sqrt{9\cdot10}}$	$\frac{62}{\sqrt{29\cdot30}}$	$\frac{57}{\sqrt{26\cdot27}}$	∞	∞
TD, Approximation	2.08978	2.21359	2.10199	2.15132		
Ordering	6	3	5	4	1,2	1,2

The natural divisor, **ND** $= 2.12$, belongs to the interval bounded with endpoints of the fourth threshold divisor, which is OAS, and the fifth one, which is NATO. Since we need to eliminate one surplus seat from the initial allocation, the Hill-Huntington algorithm suggests choosing a modified divisor from the interval bounded by the fourth largest threshold divisor and the third one:

$$2.151326365 < \textbf{Modified Divisor} < 2.213594362.$$

Choosing the modified divisor as **MD** $= 2.2$, you will be surprised to get an incorrect allocation.

Organization	UNESCO	WTO	NATO	OAS	APEC	NAA	Total
Modified Quota, $\frac{p_i}{MD}$	$\frac{70}{2.2}$	$\frac{21}{2.2}$	$\frac{62}{2.2}$	$\frac{57}{2.2}$	$\frac{1}{2.2}$	$\frac{1}{2.2}$	
MQ, Approximation	$31.\overline{81}$	$9.\overline{54}$	$28.\overline{18}$	$25.\overline{90}$	$0.\overline{45}$	$0.\overline{45}$	
Allocation	32	10	28	26	1	1	98

This indicates that our choice was too large to get the correct allocation—it causes the loss of a seat not only in OAS, but also in two other organizations: UNESCO and NATO. So we have to add two more threshold divisors for these states:

$$\textbf{TD}_{\text{UNESCO}} = \frac{70}{\sqrt{32\cdot33}} \approx 2.154101092, \qquad \textbf{TD}_{\text{NATO}} = \frac{62}{\sqrt{28\cdot29}} \approx 2.17577346.$$

Now we see that the correct interval for choosing a modified divisor is

$$2.151326365 < \textbf{H.H. Modified Divisor} < 2.154101092.$$

Indeed, choosing the modified divisor to be **MD** $= 2.153$, we get the required allocation.

Organization	UNESCO	WTO	NATO	OAS	APEC	NAA	Total
M. Quota, $\frac{p_i}{MD}$	$\frac{70}{2.153}$	$\frac{21}{2.153}$	$\frac{62}{2.153}$	$\frac{57}{2.153}$	$\frac{1}{2.153}$	$\frac{1}{2.153}$	
M.Q., Approx.	32.5127	9.7538	28.7970	26.4746	0.4644	0.4644	
F. Allocation	33	10	29	26	1	1	100

Example 2.5.7. Violation of the Quota Rule

We are going to show that allocation of 1000 seats to 6 states considered in Example 2.4.9, page 209, also violates the quota property under the Hill-Huntington algorithm. Its initial allocation is exactly the same as apportionment under the Webster rule, which is two seats short. The calculation of threshold divisors is straightforward:

State	A	B	C	D	E	F
T. Divisor, $\dfrac{p_i}{\sqrt{\lfloor q_i \rfloor \cdot \lceil q_i \rceil}}$	$\dfrac{101}{\sqrt{24\cdot25}}$	$\dfrac{10}{\sqrt{2\cdot3}}$	$\dfrac{9}{\sqrt{2\cdot3}}$	$\dfrac{55}{\sqrt{13\cdot14}}$	$\dfrac{3674}{\sqrt{887\cdot888}}$	$\dfrac{291}{\sqrt{70\cdot71}}$
Approximation of TD	4.12330	4.08248	3.67423	4.07687	4.13971	4.12776
Ordering	3	4	6	6	1	2

Part 9 of the Hill-Huntington algorithm suggests choosing a modified divisor within the interval:

$$4.123307733 < \textbf{Modified Divisor} < 4.127763388.$$

This leads to a *wrong* allocation if we take **MD** = 4.124 within this interval:

State	A	B	C	D	E	F	Total
Modified Quota, $\dfrac{p_i}{\textbf{MD}}$	$\dfrac{101}{4.124}$	$\dfrac{10}{4.124}$	$\dfrac{9}{4.124}$	$\dfrac{55}{4.124}$	$\dfrac{3674}{4.124}$	$\dfrac{291}{4.124}$	
M. Quota, Approximation	24.49	2.42	2.18	13.33	890.88	70.56	
Allocation	24	2	2	13	891	71	1003

Since this allocation has three surplus seats, we calculate additional modified divisors for state **E** because this state allocation exceeds the initial one by 4 seats.

$$\textbf{TD}_{\textbf{E}[2]} = \frac{3674}{\sqrt{888 \cdot 889}} \approx 4.13505, \qquad \textbf{TD}_{\textbf{E}[3]} = \frac{3674}{\sqrt{889 \cdot 890}} \approx 4.13041,$$

$$\textbf{TD}_{\textbf{E}[4]} = \frac{3674}{\sqrt{890 \cdot 891}} \approx 4.12577, \qquad \textbf{TD}_{\textbf{E}[5]} = \frac{3674}{\sqrt{891 \cdot 892}} \approx 4.12114.$$

Adding these four threshold divisors to the list of previously calculated divisors, and ordering them, we conclude that the interval from which to choose a modified divisor is

$$4.130410996 < \textbf{H.H. Modified Divisor} < 4.135059744.$$

For instance, choosing the same modified divisor as we did in Example 2.4.9, **MD** = 4.131, we obtain the same final allocation, where state **E** violates the quota rule.

Example 2.5.8. Hill-Huntington versus Jefferson

In January 2010, airborne time (in minutes) of six U.S. carriers is presented in the following table.

U.S. Carriers:	Southwest	JetBlue	Virgin	Gulfstream	Vision	Tanana
Time (minutes):	8,191,330	2,395,346	572,910	146,543	15,178	2,585

Suppose that 2,434 flight attendants are available to be distributed among these six air carriers. We are going to compare their allocations under Jefferson's method and the Hill-Huntington one. The first step in all methods is determination of the tentative allocation.

Hence, we start with calculations of the carriers' natural quotas that are used in rounding to obtain the initial allocation.

Carriers:	Southwest	JetBlue	Virgin	Gulf	Vision	Tanana	Total
N. Quota	1760.67	514.86	123.14	31.4985	3.26	0.55	2434
Jeff. I.A.	1760	514	123	31	3	0	2431
$\sqrt{n(n+1)}$	1760.4999	514.4997	123.4989	31.4960	3.4641	0	
H.-H. I.A.	1761	515	123	32	3	1	2435

So we see that Jefferson's initial allocation is three seats short, while the Hill-Huntington initial allocation has one surplus seat. Our next step is to find threshold divisors for every state. Recall that Jefferson's threshold divisor is determined according to Eq. (2.3.2):

Carriers:	Southwest	JetBlue	Virgin	Gulf	Vision	Tanana
Formula	$\frac{8191330}{1761}$	$\frac{2395346}{515}$	$\frac{572910}{124}$	$\frac{146543}{32}$	$\frac{15178}{4}$	$\frac{2585}{1}$
Jeff. TD	4651.521863	4651.15728	4620.241935	4579.46875	3794.5	2585
Rank	1	2	3	4	5	6

Since we need to add three seats to the initial allocation, we pick up a modified divisor from the interval bounded by the third largest threshold divisor, which is Virgin's divisor, and the fourth one, which is Gulfstream's:

$$4579.468750 < \textbf{Modified Divisor} < 4620.241935.$$

By choosing $\textbf{MD} = 4600$, we get the following allocation (which is not correct):

Carriers:	Southwest	JetBlue	Virgin	Gulf	Vision	Tanana	Total
M. Quota	1780.7	520.7	124.5	31.4985	3.29	0.56	
Jeff. F.A.	1780	520	124	31	3	0	2458

This table shows that two U.S. carriers (Southwest and JetBlue) violate the quota rule. Hence, we need to calculate additional threshold divisors for each additional seat these two states gain. Remember that we need to add exactly three seats to the initial allocation. So we can calculate only two additional threshold divisors for Southwest airlines:

$$\textbf{TD}_2 = \frac{8191330}{1762} \approx 4648.881952, \quad \textbf{TD}_3 = \frac{8191330}{1763} \approx 4646.245037$$

because all others are smaller than $\textbf{TD}_3$. Similarly, calculations for JetBlue carrier show that its second threshold divisor is less than $\textbf{TD}_3$. Then we order the threshold divisors:

Carriers:	SW	JetBlue	SW [2]	SW [3]	SW [4]	JetBlue [2]
Ranked TDs	4651.52	4651.15	**4648.88**	**4646.24**	4643.61	4642.14

Now it is clear that a modified threshold divisor should be chosen from the interval

$$4646.245037 < \textbf{Jefferson's Modified Divisor} \leqslant \tfrac{8191330}{1762} \approx 4648.881952.$$

Picking the modified divisor as $\textbf{MD} = 4647$, we obtain the correct final allocation:

Carriers:	Southwest	JetBlue	Virgin	Gulf	Vision	Tanana	Total
M. Quota	1762.7	515.4	123.2	31.5	3.2	0.55	
Jeff. F.A.	1762	515	123	31	3	0	2434

The Hill-Huntington threshold divisors are determined according to the formula (2.5.3). This leads to the following ranking of these threshold divisors

Carriers:	Southwest	JetBlue	Virgin	Gulf	Vision	Tanana
Formula	$\dfrac{8191330}{\sqrt{1760\cdot1761}}$	$\dfrac{2395346}{\sqrt{514\cdot515}}$	$\dfrac{572910}{\sqrt{123\cdot124}}$	$\dfrac{146543}{\sqrt{31\cdot32}}$	$\dfrac{15178}{\sqrt{3\cdot4}}$	
H.-H. TD	4652.8431	4655.6795	4638.9853	4652.7449	4381.5111	∞
Rank	3	2	5	4	6	1

The Hill-Huntington natural divisor, which is the ratio of the total population over the house size, $\mathbf{ND} = \dfrac{11323892}{2434} \approx 4652.379622$, is less than the fourth threshold divisor, which is Gulfstream's, but larger than the fifth threshold divisor, which is Virgin's. Since we need to eliminate one seat from the tentative allocation, we choose a modified divisor from the interval bounded by the third threshold divisor, which is Southwest's, and the fourth threshold divisor, which is Gulfstream's:

$$4652.744902 < \mathbf{Hill - Huntington's\ Modified\ Divisor} < 4652.843131.$$

For instance, if we choose a modified divisor to be $\mathbf{MD} = 4652.8$, we obtain the following final allocation.

Carriers:	Southwest	JetBlue	Virgin	Gulf	Vision	Tanana	Total
M. Quota	1760.5162	514.8181	123.1323	31.4956	3.2621	0.5555	
H.-H. F.A.	1761	515	123	31	3	1	2434

Example 2.5.9. Continuation of Example 2.4.4
In Example 2.4.4, page 202, the final allocation under Webster's rule left two states without a seat. We reconsider this example to show that the Hill-Huntington algorithm, in contrast to the Webster rule, never behaves like this. Since all preliminary work has already been done, we can start by figuring out the initial allocation under the Hill-Huntington method. Its apportionment is based on comparison of the natural quota with the corresponding cutoff point—the geometric mean of upper and lower quotas ($\sqrt{\lfloor q_i \rfloor \cdot \lceil q_i \rceil}$).

Check Point	Nogales	Sasabe	Calexico	Pinecreek	Hidalgo	Roma	Total
N. Quota	56.4128	0.4691	93.5791	0.0930	112.0269	15.4187	278
$\sqrt{n(n+1)}$	56.4977	0	93.4986	0	112.4988	15.4919	
I. A.	56	1	94	1	112	15	279

Hence, the initial allocation has one surplus seat that should be eliminated. According to the main algorithm, our next step is to find the threshold divisors:

$$\mathbf{TD}_{\text{Nogales}} = \frac{1758494}{\sqrt{56 \cdot 57}} \approx 31125.00641, \qquad\qquad \mathbf{TD}_{\text{Sasabe}} = \infty,$$

$$\mathbf{TD}_{\text{Calexico}} = \frac{2917035}{\sqrt{93 \cdot 94}} \approx 31198.68139, \qquad \mathbf{TD}_{\text{Pinecreek}} = \infty,$$

$$\mathbf{TD}_{\text{Hidalgo}} = \frac{3492088}{\sqrt{112 \cdot 113}} \approx 31041.0888, \qquad \mathbf{TD}_{\text{Roma}} = \frac{480631}{\sqrt{15 \cdot 16}} \approx 31024.59764.$$

By ranking them, from largest to smallest, we see that the natural divisor $\mathbf{ND} = 31171.84532$ is between the third largest threshold divisor, which is Calexico's check point and the fourth one, which is Nogales' check point.

Rank	1,2	3	ND	4	5	$\cdots$
Threshold Divisor	∞	31198.68	31171.84	31125.00	31041.08	
Check Point		Calexico		Nogales	Hidalgo	

Since we need to eliminate one seat from the initial allocation, we should pick a modified divisor from the range bounded by the second and the third threshold divisor. However, the second threshold divisor is unbounded, and we cannot determine the upper value. In this case, we need to use a trial and error approach (which is the least wanted resource, but we have no choice), and pick up a number that exceeds Calexico's threshold divisor. For instance, we may try the threshold divisor $\mathbf{MD} = 31200$, which leads to the following allocation

Check Point	Nogales	Sasabe	Calexico	Pinecreek	Hidalgo	Roma
Modified Divisor	$\frac{1,758,494}{31200}$	$\frac{14,625}{31200}$	$\frac{2,917,035}{31200}$	$\frac{2,900}{31200}$	$\frac{3,492,088}{31200}$	$\frac{480,631}{31200}$
Approximation	56.36	0.46	93.494	0.09	111.92	15.40
Final Allocation	56	1	93	1	112	15

We summarize our calculations in the following table:

Check Point	Number of visits	Natural Quota	H.-H. I. A.	Threshold Divisor	Rank	Modif. Quota	H.-H. F.A.
Nogales,AZ	1758494	56.4128938	56	31125.0064	4	56.54	56
Sasabe,AZ	14625	0.46917337	1	∞	1,2	0.47	1
Calexico,CA	2917035	93.5791567	94	31198.6813	3≺ND	93.79	93
Pinecreek,NM	2900	0.09303267	1	∞	1,2	0.09	1
Hidalgo,TX	3492088	112.026989	112	31041.0888	5	112.28	112
Roma,TX	480631	15.4187535	15	31024.5976	6	15.45	15
Total:	8665773		279				278

So we are done. But wait a minute, suppose we were not so lucky with the trial and error approach, what would we do? Since we need to eliminate only one seat from the initial allocation, we calculate four auxiliary threshold divisors (for Sasabe and Pinecreek check points they are not needed):

$$\mathbf{TD}_{\text{Nogales}} = \frac{1758494}{\sqrt{55 \cdot 56}} \approx 31685.86245, \qquad \mathbf{TD}_{\text{Calexico}} = \frac{2917035}{\sqrt{92 \cdot 93}} \approx 31535.97423,$$

$$\mathbf{TD}_{\text{Hidalgo}} = \frac{3492088}{\sqrt{111 \cdot 112}} \approx 31319.48979, \qquad \mathbf{TD}_{\text{Roma}} = \frac{480631}{\sqrt{14 \cdot 15}} \approx 33166.68999.$$

Ranking all previous threshold divisors together with these four ones, we get

Ranked **TD**	$\cdots$	31535.97	31319.48	31198.68	31171.84	31125.00
Check Point		Calexico $[-1]$	Hidalgo $[-1]$	Calexico	**ND**	Nogales

We observe that the new threshold divisor for Hidalgo, $\mathbf{TD}_{\text{Hidalgo}[-1]} = 31198.6813$, is the first threshold divisor that is larger than the previously calculated one for Calexico. Therefore, a threshold divisor that will provide us the right final Hill-Huntington allocation should be chosen from the interval

$$31198.68139 < \textbf{H.H. Modified Divisor} < 31319.48979.$$

These inequalities explain why our previous guess for **MD** = 31200 was the right choice. So any other number from this interval will serve as a correct threshold divisor. For instance, **MD** = 31300 would also work, although **MD** = 31400 would not work.

Example 2.5.10. Hill-Huntington versus Webster
In March 2010, the state of Rhode Island suffered from heavy rain and flood, damaging a lot of real estate property. The federal government gave \$14 million to help six cities to overcome the disaster. This money should be distributed among these six cities based on the number of houses damaged by flood.

City	Scituate	Foster	Warwick	Kingston	Cranston	Providence
# of houses	355	615	830	295	835	570

Using the house size $h = 14$, we find the initial allocations under the Hill-Huntington and Webster methods to be

City	Scituate	Foster	Warwick	Kingston	Cranston	Prov.	**Total**
Natural Quota	1.42	2.46	3.32	1.18	3.34	2.28	14
H.H. I.A.	2	3	3	1	3	2	14
Webster I.A.	1	2	3	1	3	2	12

The initial allocation under the Hill-Huntington algorithm is right because its initial allocation provides what we need—the total number of seats equals the house size. Therefore, the natural divisor, $\mathbf{ND} = \texttt{total population / house size} = \frac{3500}{14} = 250$, works well for this method; however, the Webster initial allocation is two seats short. For Webster's rule, the natural divisor is too large, and we need to use a smaller modified divisor to obtain the correct allocation. So we calculate the threshold divisors:

City	Scituate	Foster	Warwick	Kingston	Cranston	Prov.
Threshold Divisor	$\frac{355}{1.5}$	$\frac{615}{2.5}$	$\frac{830}{3.5}$	$\frac{295}{1.5}$	$\frac{835}{3.5}$	$\frac{570}{2.5}$
Approximation	$236.\overline{6}$	246	237.1428	$196.\overline{6}$	238.5714	228
Rank	4	1	**3**	6	**2**	5

Since the natural divisor is larger than any of these threshold divisors, the interval for a modified divisor is determined by the second largest threshold divisor, which is Cranston's, and the third largest one, which is Warwick's:

$$237.1428571 < \textbf{Webster's Modified Divisor} \leqslant \frac{835}{3.5} \approx 238.5714286.$$

Choosing a modified divisor to be, say, **MD = 238**, we get the correct allocation

City	Scituate	Foster	Warwick	Kingston	Cranston	Prov.
Modified Divisor	$\frac{355}{238}$	$\frac{615}{238}$	$\frac{830}{238}$	$\frac{295}{238}$	$\frac{835}{238}$	$\frac{570}{238}$
Approximation	1.49	2.58	3.48	1.23	3.50	2.39
Webster F.A.	1	3	3	1	4	2

Hence, we see that these two methods allocate seats differently despite the fact that most of the time they produce the same results.

Example 2.5.11. Webster versus Hill-Huntington

Consider U.S. government expenditure (in millions of dollars) to seven states during the fiscal year 2006, presented in the table.

State	CA	TX	GA	CT	RI	Virgin Islands	Palau
Expenditure	253906	166618	64551	30617	8766	621	37

To represent the government expenditure in percentages keeping one decimal place, we set the house size to be 1000; this leads to the following allocations.

State	Expen- diture	Natural Quota	Web. I. A.	H.H. I. A.	Threshold Divisor	Rank	Modif. Quota	H.H. F.A.
California	253906	483.5236	484	484	525.1419	4 ‹ND	482.7	483
Texas	166618	317.2975	317	317	524.7817	5	316.7	317
Georgia	64551	122.9271	123	123	526.9513	3 ‹MD	122.7	123
Connecticut	30617	58.3052	58	58	523.3866	6	58.2	58
Rhode Island	8766	16.6934	17	17	531.5168	2	16.6	17
Virgin Islands	621	1.1825	1	1	439.1133	7	1.1	1
Palau	37	0.07046	0	1	∞	1	0.07	1
Total:	525116		1000	1001				1000

The fourth column tells us that the natural divisor, **ND = 525.116**, is the right choice for the Webster rule, but is too small for Hill-Huntington's method. This is the case when Webster's method is easier than Hill-Huntington's one because its initial allocation is correct.

To find the apportionment under Hill-Huntington's algorithm, we need to calculate the threshold divisors, presented in the sixth column. The natural divisor is between the fifth largest threshold divisor, which is TX, and the fourth one, which is CA. So the interval that contains a modified divisor is bounded by the third largest threshold divisor and the fourth one:

$$525.1419561 < \textbf{H.H. Modified Divisor} < 526.9513282.$$

Choosing a modified divisor to be, say, **MD = 526**, we get the correct allocation

State		CA	TX	GA	CT	RI	Virgin	Palau
Modified Quota,	$\frac{p_i}{\text{MD}}$	$\frac{253906}{526}$	$\frac{166618}{526}$	$\frac{64551}{526}$	$\frac{30617}{526}$	$\frac{8766}{526}$	$\frac{621}{526}$	$\frac{37}{526}$

Writing their approximations into the eighth column and rounding them, we obtain the final Hill-Huntington allocation (given in the last column).

Example 2.5.12. Hill-Huntington and Webster versus Jefferson and Adams

The following table represents the U.S. auto sales in April 2010.

Cars	Midsize	Small	Pickup	Luxury	Minivan	Large	Total
Sales	254,214	167,040	129,603	75,759	65,235	8,585	700,436

Suppose we need to represent the sales in percentages, carrying out one decimal place. Then the house size would be 1000, and the natural quotas are as follows.

Cars	Midsize	Small	Pickup	Luxury	Minivan	Large
Natural Quota	362.9367	238.4800	185.0318	108.1597	93.1348	12.2566

Since none of the methods provide the required apportionment, we need to apply each method in order to allocate 1000 "seats." Let us start with Hill-Huntington's method. Since its initial allocation is one seat short, we calculate the threshold divisors:

Cars	Midsize	Small	Pickup	Luxury	Minivan	Large
Formula for TD	$\frac{254214}{\sqrt{362\cdot363}}$	$\frac{167040}{\sqrt{238\cdot239}}$	$\frac{129603}{\sqrt{185\cdot186}}$	$\frac{75759}{\sqrt{108\cdot109}}$	$\frac{65235}{\sqrt{93\cdot94}}$	$\frac{8585}{\sqrt{12\cdot13}}$
H.-H. TD	701.28	700.37	698.67	698.24	697.71	687.35
Rank	1	2	3	4	5	6

The natural divisor, $\mathbf{ND} = 700.436$, is smaller than the largest threshold divisor, which corresponds to the "state" *Midsize*, but is larger than the second divisor, which is *Small*, so we pick up a modified divisor from the interval bounded by the second largest threshold divisor and the third one:

$$698.6710014 < \mathbf{Modified\ Divisor} < 700.3788977.$$

For instance, choosing a modified divisor to be $\mathbf{MD} = 699$, we get a *wrong* allocation

Cars	Midsize	Small	Pickup	Luxury	Minivan	Large	Total
Modified Quota	$\frac{254214}{699}$	$\frac{167040}{699}$	$\frac{129603}{699}$	$\frac{75759}{699}$	$\frac{65235}{699}$	$\frac{8585}{699}$	
H.-H. M. Quota	363.68	238.96	185.41	108.38	93.32	12.28	
Allocation	364	239	185	108	93	12	1001

because each of two "states," *Midsize* and *Small* gained a seat (but we were supposed to add only one seat to the initial allocation). Following part 11 of Hill-Huntington's algorithm, we add one more threshold divisor to the list of previously calculated divisors:

$$\mathbf{MD}_{Midsize[1]} = \frac{254214}{\sqrt{363\cdot364}} \approx 699.3514.$$

Since this divisor is the third largest one, it moves the threshold divisor of "state" *Small* in the ranked list of divisors to the fourth position. Therefore, the correct interval for choosing a modified divisor becomes

$$699.3514181 < \mathbf{H.H.\ Modified\ Divisor} < 700.3788977.$$

This explains why our previous choice for the modified divisor did not work; now we pick up a correct one, $\mathbf{MD} = 700$, that leads to the correct final allocation:

Cars	Midsize	Small	Pickup	Luxury	Minivan	Large	Total
Modified Quota	$\frac{254214}{700}$	$\frac{167040}{700}$	$\frac{129603}{700}$	$\frac{75759}{700}$	$\frac{65235}{700}$	$\frac{8585}{700}$	
H.-H. M. Quota	363.16	238.62	185.14	108.22	93.19	12.26	
H.-H. F. A.	363	239	185	108	93	12	1000

Because we are done with the Hill-Huntington apportionment, we go to allocation under the Webster rule. Since its initial allocation is identical to Hill-Huntington's one (being one seat short), we calculate the Webster threshold divisors:

Cars	Midsize	Small	Pickup	Luxury	Minivan	Large
Formula	$\frac{254214}{362.5}$	$\frac{167040}{238.5}$	$\frac{129603}{185.5}$	$\frac{75759}{108.5}$	$\frac{65235}{93.5}$	$\frac{8585}{12.5}$
Web. TD	701.28	700.37	698.66	698.23	697.70	686.8
Rank	1	2	3	4	5	6

Trying to avoid a wrong choice for a modified divisor that we learned from Hill-Huntington's algorithm, we calculate the second threshold divisor for *Midsize*: $\mathbf{MD}_{\text{Midsize}[1]} = \frac{254214}{363.5} \approx 699.3507565$, which is the third largest divisor. Therefore, we get the Webster interval for modified divisors:

$$699.3507565 < \textbf{Webster's Modified Divisor} \leqslant \tfrac{167040}{238.5} \approx 700.3773585.$$

This tells us that we can choose the same modified divisor as we did in the previous case when we considered allocation under the Hill-Huntington method. This will yield the same final apportionment—these two methods usually produce the same allocations.

Now we turn our attention to the other twins of apportionment methods—Jefferson's and Adams' rules. Based on natural quotas calculated previously, we obtain their initial allocations:

Cars	Midsize	Small	Pickup	Luxury	Minivan	Large	Total
$2 Allocation	362	238	185	108	93	12	998
Adams Allocation	363	239	186	109	94	13	1004

Since these two methods—Jefferson's and Adams'—never produce a correct initial allocation (unless all natural quotas are integers), we are forced to calculate the threshold divisors:

Cars	Midsize	Small	Pickup	Luxury	Minivan	Large
$2 Formula	$\frac{254214}{363}$	$\frac{167040}{239}$	$\frac{129603}{186}$	$\frac{75759}{109}$	$\frac{65235}{94}$	$\frac{8585}{13}$
$2 TD	700.31	698.91	696.79	695.03	693.98	660.38
Rank	1	2	3	4	5	6

Hence, we need to add two seats to the Jefferson initial allocation, the interval for choosing a modified divisor is bounded by the second largest threshold divisor, which is *Small*, and the third one, which is *Pickup*:

$$696.7903226 < \textbf{Modified Divisor} < 698.9121339.$$

It is no surprise that the final allocation is not correct when we set **MD** = 697:

Cars	Midsize	Small	Pickup	Luxury	Minivan	Large	**Total**
\$2 M. Quota	$\frac{254214}{697}$	$\frac{167040}{697}$	$\frac{129603}{697}$	$\frac{75759}{697}$	$\frac{65235}{697}$	$\frac{8585}{697}$	
\$2 M.Q., Approx.	364.72	239.65	185.94	108.69	93.59	12.31	
\$2 Allocation	364	239	185	108	93	12	1001

Adding to the list of threshold divisors another one that corresponds to two surplus seats of "state" *Midsize*, we get $\mathbf{MD}_{\text{Midsize}[2]} = \frac{254214}{364} \approx 698.39$, which is the third largest divisor. Therefore, we get the interval for Jefferson's modified divisors:

$$698.3901099 < \textbf{Jefferson's Modified Divisor} \leqslant \tfrac{167040}{699} \approx 698.9121339.$$

For instance, if we choose a modified divisor to be **MD** = 698.5, we will get the correct allocation

Cars	Midsize	Small	Pickup	Luxury	Minivan	Large	**Total**
\$2 Modified Quota	$\frac{254214}{698.5}$	$\frac{167040}{698.5}$	$\frac{129603}{698.5}$	$\frac{75759}{698.5}$	$\frac{65235}{698.5}$	$\frac{8585}{698.5}$	
\$2 M.Q., Approx.	363.94	239.14	185.54	108.45	93.39	12.29	
\$2 Final Allocation	363	239	185	108	93	12	1000

The threshold divisors under Adams' method are as follows:

Cars	Midsize	Small	Pickup	Luxury	Minivan	Large
Adams' Formula	$\frac{254214}{362}$	$\frac{167040}{238}$	$\frac{129603}{185}$	$\frac{75759}{108}$	$\frac{65235}{93}$	$\frac{8585}{12}$
Adams' TD	702.24	701.84	700.55	701.47	701.45	715.41
Rank	2	3	6	4	5	1

Counting from the back four times (because we need to eliminate four surplus seats from the initial allocation), we conclude that the Adams interval should be bounded by the third largest threshold divisor, which is *Small*, and the second one, which is *Midsize*:

$$701.8487395 \approx \tfrac{167040}{238} \leqslant \textbf{Adams' Modified Divisor} < 702.2486188.$$

Choosing **MD** = 702, we come to the required allocation

Cars	Midsize	Small	Pickup	Luxury	Minivan	Large	**Total**
Adams' M. Quota	$\frac{254214}{702}$	$\frac{167040}{702}$	$\frac{129603}{702}$	$\frac{75759}{702}$	$\frac{65235}{702}$	$\frac{8585}{702}$	
Adams' M.Q., App.	362.12	237.94	184.61	107.91	92.92	12.22	
Adams' Final Allo.	363	238	185	108	93	13	1000

Problems.

1. Use the apportionment criterion to decide which state is more deserving of an additional representative in the House. The 2009 populations of the states are provided by U.S. Bureau of Census:

   (a) Alabama with population 4,708,708 and 7 representatives; Arkansas with population 2,889,450 and 4 representatives.

(b) Colorado with population 5,024,748 and 7 representatives; Pennsylvania with population 12,604,767 and 19 representatives.

(c) Georgia with population 9,829,211 and 13 representatives; Delaware with population 885,122 and 1 representative.

(d) Florida with population 18,537,969 and 25 representatives; Iowa with population 3,007,856 and 5 representatives.

2. Use the Hill-Huntington method to apportion 27 representatives among Kansas, Nebraska, and Missouri. The 2009 populations of the states were as follows: Kansas had population 2,818,747; Nebraska had population 1,796,619; and Missouri had population 5,987,580.

3. Use the Hill-Huntington method to apportion 23 representatives among Maine, Massachusetts, and Vermont. The 2009 populations of the states were as follows: Vermont had population 621,760; Massachusetts had population 6,593,587; and Maine had population 1,318,301.

4. Use the Hill-Huntington method to apportion 17 representatives among Washington, Oregon, and Idaho. The 2009 populations of the states were as follows: Washington had population 6,664,195; Oregon had population 3,825,657; and Idaho had population 1,545,801.

5. Use the Hill-Huntington method to apportion 24 representatives among Arizona, New Mexico, and Texas. The 2009 populations of the states were as follows: Arizona had population 6,595,778; New Mexico had population 2,009,671; and Texas had population 24,782,302.

6. Suppose that Rhode Island, Connecticut, and Massachusetts are cooperating to build a power plant to provide electricity for three states. Allocate a 11-member authority to these 3 states using the Hill-Huntington method. There are about 67,000 customers in Connecticut, 78,000 customers in Massachusetts, and 11,000 customers in Rhode Island.

7. A hospital administrator wants to assign eight emergency medical teams to four community outreach centers. The number of patients treated last week at each center is listed in the following table:

Center	A	B	C	D
Patients	96	87	74	23

Use the Hill-Huntington method to decide how the administrator should assign the medical teams to these four centers.

8. A health club instructor has a course load that allows her to teach twelve two-hour classes. A preregistration survey indicated the following interests:
15 want to take taekwondo;
12 want to take kick boxing;
48 want to take yoga; and
54 wants to take aerobics.

Use the Hill-Huntington method to apportion twelve classes among four activities.

9. Statistics are showing that the city of Buffalo is experiencing a downward trend in the overall crime rate, though there was a jump in violent crimes in 2008. Using Hill-Huntington's method, allocate 127 police officers specialized in the following categories of crime:

Crimes	Murder	Rape	Robbery	Assault
Numbers of Cases	51	212	1485	2056

10. Four food chain stores decide to merge. They want to apportion the new 15 store managers to run their company in proportion to the net worth of the chains. Use the Hill-Huntington method to determine how many managers will come from each store.

Store	Net Worth (in millions of dollars)
Good Burger	132
Eat Well	74
Love Pizza	241
Hungry Dad	133

11. Suppose that four main parties compete for 80 seats in the legislature. According to the constitution, parties that have less than 5% votes are eliminated from the legislature. Apportion the 80 representatives to German's parliament using the Hill-Huntington method based on the number of votes each party has after recent elections.

Party	Social Democrats	Christian Democrats	Greens	Free Democrats	Total
Votes	39.8%	36.4%	8.1%	5.7%	90%

12. Oroville Dam is on the Feather River above the city of Oroville in Butte County, California, United States. Oroville is the tallest dam (with height 234.696 m) in the United States and is ranked among the top twenty dams in the world for its height as well as volume of dam materials. It creates Lake Oroville, generates electricity, and provides drinking and irrigation water for Central and Southern California. The dam, lake, and other facilities are owned and operated by the State of California Department of Water Resources and are part of the California State Water Project. It supplies electricity to Sacramento, 20.8%; to Reno, 18.7%; to Sonoma, 12.6%; to Arnold, 9.3%; and to Chico, 8.6%. Using Hill-Huntington's method, allocate 14 members to these cities in proportion to the electricity supply.

13. The city of Los Angeles spans over 498.3 square miles making it one of the largest cities in the world. With a city this large comes the burden of allocating the number of fire stations located within the city and where they go. There are approximately 114 fire houses in the city of L.A. and 18 battalions. While the allocation of fire house jurisdiction is laid out fairly well in Los Angeles, the allocation of how many firemen need to be on call at each station at a certain time is important to reflect on. Fire statistics are based upon calls, and what those calls entail (fires, explosions, medical, hazmat, service, and false calls). Battalion 13 is by far the busiest, receiving over 24,000 calls in 2009. A vast majority of these calls are medical calls. Battalion 13 reported almost 12,000 medical emergencies which is roughly half of that battalion's calls in a year. This is a common trend in not only the busiest fire houses but also in the quietest ones. Battalion 13 reported under 800 fire calls last year—proving that fire is a secondary function of a fire house. Consequently, Battalion 13 or any other for that matter would want to allocate more EMTs and medical workers than certified fire fighters to each house. Use the Hill-Huntington method to allocate 302 medical workers to six battalions based on the statistics presented.

Battalion	9	10	11	12	13	14
Number of Calls	18,771	16,799	21,315	13,530	24,541	6,228

14. A law firm is hiring 14 new lawyers specialized in divorce issues. However, they are specialized in different cases based on the age of the clients. There are five basic groups of divorces:

Age	Under 20 years	20 to 24	25 to 29	30 to 34	35 to 39 years old
Percent	27.6%	36.6%	16.4%	8.5%	5.1%

How many lawyers specialized in the five groups are needed based on the Hill-Huntington method?

15. Drunk driving accidents claim lives at the rate of 1 every 39 minutes in the United States, based on the most recent data available (13,470 deaths during 2006 in crashes caused by DUI drivers with a BAC—Blood Alcohol Content—at or above 0.08 percent). The New England administration has ordered 403 breathalyzers that can detect and measure current alcohol levels. Distribute these breathalyzers between six states using the Hill-Huntington method, based on the number of drunk driving fatalities (DDF) presented here.

State	ME	NH	VT	MA	RI	CT
DDFs	66	34	22	146	25	101

16. The Congress of the Republic of Peru or the National Congress of Peru is the unicameral body that assumes legislative power in Peru. Congress consists of 120 members of congress, who are elected for five-year periods in office on a proportional representation basis. The last congressional election was held on April 9, 2006, which gives the following results:

Party	Number of Votes
Union for Peru	3,758,258
Peruvian Aprista Party	2,985,858
National Unity	2,923,280
Alliance for the Future	912,420
Center Front	706,156
National Restoration	537,564
Decentralization Coalition	76,105

Use the Hill-Huntington algorithm to allocate 120 seats between these 7 parties.

17. Georgetown University is a private Jesuit university located in northwest Washington, DC. Father John Carroll founded the school in 1789, though its roots extend back to 1634. The student body is noted for its pluralism and political activism, as well as its sizable international contingent. The university has an 17-member faculty council. Apportion the representatives of this council based on the 2010 number of faculty in each college of the university using
    (a) Hamilton's method.   (b) Lowndes' method.   (c) Jefferson's method.
    (d) Webster's method.   (e) Hill-Huntington's method.

College	Number of Faculty Members
Graduate School of Arts and Sciences	123
Georgetown College	556
School of Business	131
Law Center	177
School of Nursing and Health Studies	129
Edmund Walsh School of Foreign Service	137
Total:	1253

18. The French legislative elections took place in June 2007 to elect the 13th National Assembly of the Fifth Republic, a few weeks after the French presidential election. There were 7,639 candidates competing for 577 seats,

The procedure by which deputies are elected is a mixture of plurality and run-off systems. A candidate must take an absolute majority (more than 50%) in their constituency to win in the first round, and receive the support of at least 25% of all registered voters. Otherwise, if they get at least 12.5% of the votes of all registered voters in the first round, or are one of the top two candidates remaining, they go through to the second round, where only a simple plurality is needed to win. The results of the elections are summarized in the following table.

Party	Number of Votes
Union for a Popular Movement	10,289,737
Socialist Party	6,436,520
Democratic Movement	1,981,107
National Front	1,116,136
French Communist Party	1,115,663
Other far-left	888,250

Use the Hill-Huntington algorithm to allocate 577 seats between these 6 parties.

19. A company has purchased 55 lottery tickets to award its best employees. It decides to allocate tickets to its various divisions in proportion to the number of employees in the divisions as listed below.

Division	Number of Employees
Accounting	37
Customer service	52
Human resources	5
Public relations	16
Repair	43
Research	28

Distribute tickets among divisions using
(a) Hamilton's method. (b) Lowndes' method. (c) Jefferson's method.
(d) Webster's method. (e) Hill-Huntington's method.

20. Hungarian parliamentary elections were held in 2006. Allocate 386 seats in the Hungarian National Assembly using the Hill-Huntington system based on the following results.

Party	Number of Votes
Hungarian Socialist Party	2,336,705
Fidesz-KDNP	2,272,979
Hungarian Democratic Forum	272,831
Third Way Alliance of Parties	119,007

21. In order to reduce the federal deficit, the government analyzes the U.S. export earning in 2004. Using the Hill-Huntington method, make a table showing the percentage of each commodity exported, rounded to the nearest tenth of a percent (to one decimal place).

Commodity	Earning (in millions of dollars)
Airplane	24493
Cigarettes	1294
Diamonds	939
Machinery	34824
Spacecraft	467
Vehicles	65217

22. September 2008 is the month that automobile makers in the United States would like to forget. Using the Hill-Huntington method, represent the following sales of seven car makers in terms of percentage.

Car Maker	GM	Honda	BMW	Subaru	Volvo	Porsche	Isuzu
Sales	282806	96626	18583	14491	4054	1458	258

23. Four friends have bought thirty three identical diamonds at a cost of $44,400. Alex invested $15,700. Beth contributed $19,500, Charles gave $7,880, and Doris invested $1,320. They

want to divide the diamonds according to the proportion of their investment. Determine the number of diamonds that should go to each friend if the apportionment method used is the

(a) Hamilton's method.   (b) Lowndes' method.   (c) Jefferson's method.
(d) Adams' method.   (e) Webster's method.   (f) The Hill-Huntington method.

24. A catering company has to deliver 234 meals on a particular day. The average number of meals per week is presented here.

Meal	Pizza	Spaghetti	Burgers	Lasagna	Steak
Average	457	426	360	394	603

Apportion the 234 meals the catering business has to deliver based on the numbers in the table using

(a) Hamilton's method.   (b) Lowndes' method.   (c) Jefferson's method.
(d) Adams' method.   (e) Webster's method.   (f) The Hill-Huntington method.

25. A television program is conducting a poll to determine a committee that will represent the biggest sports cities in the United States. The committee will be made up of 2,500 people proportionally to the 2007 populations of cities.

City	New York	Los Angeles	Chicago	Houston	Philadelphia	Phoenix
Population	8274527	3834340	2836658	2208180	1449634	1552259

Apportion the 2,500 people using

(a) Hamilton's method.   (b) Lowndes' method.   (c) Jefferson's method.
(d) Adams' method.   (e) Webster's method.   (f) The Hill-Huntington method.

26. A summer camp can accommodate only 512 students. The education committee wants to fairly distribute these 512 seats among students of different races. The data collected in 2009 show that in Boston area there are 20,096 white students, 1,819 African-American students, 1,745 Hispanic students, 533 Asians, and 54 Indians. Distribute 512 seats using

(a) Hamilton's method.   (b) Lowndes' method.   (c) Jefferson's method.
(d) Adams' method.   (e) Webster's method.   (f) The Hill-Huntington method.

27. The education department in Massachusetts wanted to allocate 144 available tutors for students in the Danvers public school system. Based on the number of students in different grades

Grade	First	Second	Third	Fourth	Fifth
Number of Students	1860	1748	1738	1750	1769

distribute 144 tutors using

(a) Hamilton's method.   (b) Lowndes' method.   (c) Jefferson's method.
(d) Adams' method.   (e) Webster's method.   (f) The Hill-Huntington method.

28. Distribute 965 computers between five private schools in Boston based on their enrollments

School	Latin	Trinity	School of Boston	Cathedral	Donald McKay
Enrollment	2427	210	300	204	675

using

(a) Hamilton's method.   (b) Lowndes' method.   (c) Jefferson's method.
(d) Adams' method.   (e) Webster's method.   (f) The Hill-Huntington method.

Chapter 2 Review

Important Terms, Symbols, and Concepts

2.1 Introduction

For any positive real number q, let $\lfloor q \rfloor$ be the integer part of the number, called its floor or lower quota. When a real number is rounded up to the nearest integer, this integer is called the ceiling or upper quota and is denoted by $\lceil q \rceil$. Note that the floor and ceiling of a whole number give the same value. The fractional part of a real positive number is the difference, $q - \lfloor q \rfloor$. For instance, the number $\dfrac{5}{3} = 1.666\ldots = 1.\overline{6}$ has floor 1 and ceiling 2; its fractional part is $0.666\ldots = 0.\overline{6}$.

In practical applications, we usually have n states, with populations $p_1, p_2, \ldots, p_n$, and a certain number of seats in the legislature. The number of seats, h, is called the house size. The sum of states' populations is called the total population: $p = p_1 + p_2 + p_3 + \ldots + p_n$. The natural (standard) divisor is the ratio: $\mathbf{ND} = \dfrac{\text{total population}}{\text{house size}} = \dfrac{p}{h}$. The value of the expression $q_i = p_i * h/p$ is called the natural (standard) quota for state i, with population p_i. Usually, natural quotas are not integers; however, their sum is an integer—the house size: $q_1 + q_2 + \cdots + q_n = h$. The rule(s) for rounding natural quotas so that their sum is maintained at its original value is called an **apportionment method**:

$$q_1 + q_2 + \cdots + q_n = \tilde{q}_1 + \tilde{q}_2 + \cdots + \tilde{q}_n = \text{house size},$$

where $\tilde{q}_1, \tilde{q}_2, \ldots, \tilde{q}_n$ are integers.

The **constituency size** (also called average constituency or district population) of a state is the quotient:

$$\textbf{Average constituency} = \frac{\text{population of the state}}{\text{number of representatives from the state}},$$

if the denominator is not zero. When the number of a state's representatives is zero, we assign ∞ to be its constituency size. Its reciprocal is called the **representative share**:

$$\textbf{Representative share} = \frac{\text{number of representatives from the state}}{\text{population of the state}}.$$

So it represents the share of a congressional seat given to each citizen of the state. On the other hand, the constituency size of a state is defined to be the number of persons per representative in this state. Comparing the representations of two states A and B, we say that state A is **more poorly represented** than state B if the constituency size of A is greater than B's constituency size. In other words, state A is more poorly represented than state B if its representative share is less than the representative share of state B.

2.2 Hamilton's and Lowndes' Methods

Hamilton's method is also known as the method of largest remainders and sometimes as Vinton's method. It initially assigns to each state its lower quota. If the initial allocation is less than the required number of seats, called the house size, give surplus seats, one at a time, to states in descending order of the fractional parts of their natural quota.

Lowndes' method works exactly as the Hamilton method, but it assigns surplus seats, one at a time, to states in descending order of the relative fractional parts of their natural quota. The relative fractional part of a positive real number q is

$$\begin{cases} \frac{q_i - \lfloor q_i \rfloor}{\lfloor q_i \rfloor}, & \text{if } \lfloor q_i \rfloor \neq 0, \\ \infty, & \text{if } \lfloor q_i \rfloor = 0. \end{cases}$$

So Lowndes' method always assigns at least one seat to every state. These two apportionment methods—Hamilton and Lowndes'—satisfy the **quota rule** or property:
Every state's apportionment is equal to either its lower quota or its upper quota.

These two apportionment methods suffer from three paradoxes:
Alabama paradox: An increase in the total number of seats to be apportioned causes a state to lose a seat. Such a property is referred to as non-monotonic.

Population paradox: When state A's population is growing faster than state B's population, yet state A loses a representative to state B. It is assumed that the total number of representatives remains the same.

New state paradox: When a new state is added and its share of seats is added to the legislature, it causes a change in allocation of seats previously given to other states.

2.3 Jefferson's Method

A **divisor method** is the method that uses a modified divisor instead of the natural divisor to calculate modified quotas:

$$\text{modified quota of the state} = \frac{\text{state's population}}{\text{modified divisor}}.$$

Based on modified quotas, every divisor method rounds modified quotas to integers according to some rule so that their sum is equal to the house size.

Divisor methods never have the Alabama paradox, the population paradox, nor the new state paradox. However, divisor methods do not always satisfy the quota condition.

The Jefferson method is a divisor method that rounds quotas down. Thus, if q_i is a modified quota of state i, the state's apportionment is $\lfloor q_i \rfloor$. It always favors larger states.

Adams' method is a twin brother of Jefferson's method—it rounds quotas up to nearest integer (called ceiling or upper quota). The method favors small states at the expense of large states.

2.4 Webster's Method

The Webster method is a divisor method of apportionment that rounds quotas to the nearest integer. Thus, if q_i is a quota of state i, Webster's algorithm rounds it down if q_i is less than the cutoff point, otherwise it rounds it up. The cutoff point is the arithmetic mean of its lower quota and its upper quota, that is, $(\lfloor q_i \rfloor + \lceil q_i \rceil)/2$. The method is neutral to the size of the states.

2.5 The Hill-Huntington Method

The Hill-Huntington method is a divisor method of apportionment that rounds quotas by comparing them with the corresponding cutoff points. Thus, if q_i is a quota of state i, Hill-Huntington's algorithm rounds it down if q_i is less than the cutoff point, otherwise it rounds it up. The cutoff point is the geometric mean of its lower quota and its upper quota, that is, $\sqrt{\lfloor q_i \rfloor \cdot \lceil q_i \rceil}$. The Hill-Huntington method slightly favors smaller states. The method never assigns any state an apportionment of zero.

Review Exercises.

1. Which state is more poorly represented in 2009: the state of Maine with a population of 1,316,456 and 2 representatives, or Oklahoma with a population of 3,642,361 and 5 representatives? What is the absolute unfairness of this apportionment? What is the relative unfairness of this apportionment?

2. According to U.S. Bureau of the Census estimates, in 2010, Oregon's population was approximately 3,825,657 people, while Idaho had 1,545,801 people. Oregon was allotted five seats in the House, and Idaho was allotted two. Calculate the absolute and relative unfairness of this apportionment.

3. In the sum
$$1.47 + 2.38 + 3.25 + 4.41 + 4.56 = 16,$$
round numbers either up (ceiling) or down (floor) to whole numbers preserving the total of 16. Find the apportionment that minimizes average constituencies, and compare it with an allocation that minimizes representative shares.

4. In the sum
$$0.08 + 0.73 + 1.41 + 2.54 + 3.24 = 8,$$
round numbers either up (ceiling) or down (floor) to whole numbers preserving the total of 8. Find the apportionment that minimizes average constituencies, and compare it with an allocation that minimizes representative shares.

5. Use the Hamilton method to round each of the following numbers in the sum to a whole number preserving the total of 10.
$$0.28 + 2.37 + 1.87 + 2.29 + 3.19 = 10.$$

6. Repeat the previous exercise, but now use Lowndes' method.

7. Use the Hamilton method to round each of the following numbers in the sum to a whole number preserving the total of 15.
$$0.17 + 1.23 + 2.45 + 3.67 + 4.89 + 2.59 = 15.$$

8. Repeat the previous exercise, but now use Lowndes' method.

9. The employees of a big company are negotiating a new contract. There are 47 cashiers, 237 workers, 88 cooks, and 33 janitors. The eight-member negotiations committee consists of persons in proportion to the number of employees in each of the four groups. Assign members to the negotiations committee using
   (a) Hamilton's method; (b) Lowndes' method.

10. Portugal elects on a national level the president and the national parliament, the Assembly of the Republic. The president is elected for a five year term by the people while the parliament has 230 members, elected for a four-year term by proportional representation in the districts. Also, Portugal elects 24 members of the European Parliament. The results of the Portuguese legislative election, 2009 (September 27) are presented in the following table.

Party	Number of Votes
Socialist Party (PS)	2,077,238
Social Democratic Party (PSD)	1,653,665
Democratic and Social Center – People's Party (CDS/PP)	592,778
Left Bloc (BE)	557,306
Democratic Unity Coalition (CDU)	446,279
Communist Party (PCTP-MRPP)	52,761

   Apportion 230 seats in the Portuguese parliament using
   (a) Hamilton's method; (b) Lowndes' method.

11. There are 87 local police, 15 federal agents, and 38 state police involved with drug enforcement in a city. A special 42-person task force will be formed to investigate a particular case. Officers will be assigned to this task force with membership proportional to the number of the three types of law enforcement officers.

   (a) Allocate officers using Hamilton's method.

   (b) Allocate officers using Lowndes' method.

   (c) Now increase the task force's size to 43, and redo the apportionment using Hamilton's method. Do you observe the Alabama paradox? What group loses an officer when the size of the task force increases?

12. Suppose a company has four divisions: Acquisition (A), Business Relations (B), Computers (C), and Digital (D). Division A has 41 employees, Division B has 152 employees, C has 78, and D has 34. Assume that a 41-member quality improvement council has membership on the council proportional to the number of employees in the four divisions.

   (a) Apportion this council using Hamilton's method.

   (b) Apportion this council using Lowndes' method.

   (c) Now increase the council's size to 42, under what method does an Alabama paradox occur?

13. A hospital has four emergency clinics throughout the city. Statistics have been gathered regarding the number of patients seen on the night shift for each clinic.

Clinic	South Side	North Side	West Side	East Side
Patients	141	59	253	367

(a) If there are 14 emergency physicians available, how should they be apportioned using Hamilton's rule?

(b) Increase the number of physicians until an Alabama paradox occurs. At what increase does the paradox occur and what clinic loses a doctor?

14. Show that Hamilton's method leads to the Alabama paradox as the house size increases from 36 to 37 seats, with populations given by the following table.

State	A	B	C	D
Population	37	191	224	158

15. Show that Hamilton's method leads to the Alabama paradox as the house size increases from 32 to 33 seats, with populations given by the following table. Show that Lowndes' method leads to the Alabama paradox as the house size increases from 23 to 24 seats, with the same populations.

State	A	B	C	D
Population	2300	324	1236	456

16. Show that Hamilton's method leads to the Alabama paradox as the house size increases from 36 to 37 seats, with populations given by the following table. Show that Lowndes' method leads to the Alabama paradox as the house size increases from 35 to 36 seats, with the same populations.

State	A	B	C	D
Population	3456	456	2340	650

17. Show that Hamilton's method leads to the Alabama paradox as the house size increases from 14 to 15 seats, with populations given by the following table.

State	A	B	C	D	E
Population	374	581	764	1406	1406

18. Show that Hamilton's method leads to the Alabama paradox as the house size increases from 68 to 69 seats, with populations given by the following table.

State	A	B	C	D	E
Population	661	413	325	785	1297

19. Show that Hamilton's method leads to the Alabama paradox as the house size increases from 31 to 32 seats, with populations given by the following table.

State	A	B	C	D	E
Population	807	470	280	99	83

20. Show that Hamilton's method is inconclusive to allocate 44 seats, with populations given by the following table.

State	A	B	C	D	E
Population	3844	3367	2772	903	903

21. Show that Hamilton' method is inconclusive to allocate 44 seats, with populations given by the following table.

State	A	B	C	D	E
Population	875	738	691	289	289

22. A college has five branch campuses with student enrollment in media shown below. Currently the college has 76 computer labs, but with its new electronic media concentration, the administration wants to increase the number of labs.

Campus	Middletown	Newport	Fall River	Kingston	Warwick
Students	328	571	309	673	112

    (a) How should the 76 labs be apportioned using Lowndes' method?

    (b) Increase the number of labs until an Alabama paradox occurs. At what increase does the paradox occur and what branch loses a lab?

23. Show that Lowndes' method leads to the Alabama paradox as the house size increases from 47 to 48 seats, with populations given by the following table.

State	A	B	C	D	E
Population	864	751	642	319	188

24. Show that Lowndes' method leads to the Alabama paradox as the house size increases from 76 to 77 seats, with populations given by the following table.

State	A	B	C	D	E
Population	93	150	450	221	657

25. Show that Lowndes' method leads to the Alabama paradox as the house size increases from 56 to 57 seats, with populations given by the following table.

State	A	B	C	D	E
Population	284	678	244	603	327

26. Show that Lowndes' method leads to the Alabama paradox as the house size increases from 36 to 37 seats, with populations given by the following table.

State	A	B	C	D	E
Population	341	519	742	1506	2393

27. Show that Lowndes' method leads to the Alabama paradox as the house size increases from 58 to 59 seats, with populations given by the following table.

State	A	B	C	D	E
Population	1185	1270	323	2177	392

28. Show that Hamilton's method leads to the population paradox for a 10-member house, with populations given by the following table. Show that Lowndes' method leads to the population paradox for a 23-member house, with the same populations.

State	A	B	C
Old Population	22537	60744	157021
New Population	23381	62676	157214

29. Show that Hamilton's method leads to the population paradox for an 18-member house, with populations given by the following table. Show that Lowndes' method leads to the population paradox for a 32-member house, with the same populations.

State	A	B	C
Old Population	3457	12785	36018
New Population	3592	13167	36639

30. Show that Lowndes' method leads to the population paradox for a 81-member house, with populations given by the following table.

State	A	B	C	D
Old Population	1,396	912	568	322
New Population	1,473	963	571	326

31. Use Jefferson's method to apportion the 35-member committee between 4 states based on their populations:

State	A	B	C	D
Population	35,491	81,819	25,207	44,075

32. Using Jefferson's method, apportion 68 seats in a hypothetical legislature between 5 states based on their populations according to the 2009 U.S. census.

State	Ohio	Michigan	Wisconsin	Illinois	Tennessee
Population	11,542,645	9,969,727	5,654,774	12,910,409	6,296,254

33. Using Jefferson's method, apportion 59 seats in a hypothetical legislature between 6 states based on their 2009 populations.

State	N. Dakota	S. Dakota	Nebraska	Kansas	Oklahoma	Iowa
Population	646,844	812,383	1,796,619	2,818,747	3,687,050	3,007,856

34. Slovakia elects on a national level a head of state—the president—and a legislature. The president is elected for a five-year term by the people. The National Council has 150 members, elected for a four year term by proportional representation. Based on the results of the 2006 parliamentary elections presented here, allocate 150 seats between 6 parties using Jefferson's method.

Party	Number of Votes
Direction Social Democracy	671,185
Slovak Democratic and Christian Union – Democratic Party	422,815
Slovak National Party	270,230
Party of the Hungarian Coalition	269,111
People's Party – Movement for a Democratic Slovakia	202,540
Christian Democratic Movement	191,443

35. There are 82 counties in the U.S. state of Mississippi. The 2009 populations of six counties are presented in the following table. Apportion 85 seats between these counties using Jefferson's method.

County	Attala	Benton	Carroll	DeSoto	Forrest	Greene
Population	19,661	8,026	10,769	107,199	72,604	13,299

36. There are 114 counties and 1 independent city in the U.S. state of Missouri. Most of the counties in Missouri are named after politicians; other counties are named after war heroes, natural resources, explorers, and former U.S. territories. The city of St. Louis is an independent city, and is not within the limits of a county. Populations of six counties are based on the 2008 United States Census estimates. Apportion 78 seats in a hypothetical legislature using Jefferson's method.

State	Adair	Barton	Clay	Dade	Franklin	Gentry
Population	24,943	12,531	215,707	7,418	100,898	6,185

37. Show that the apportionment produced by Jefferson's method violates the quota rule for a 100-member house and 6 states, with populations given by the following table.

State	A	B	C	D	E	F
Population	17,273	13,180	10,657	5,978	4,031	1,533

38. Show that the apportionment produced by Jefferson's method violates the quota rule for a 200-member house and 6 states, with populations given by the following table.

State	A	B	C	D	E	F
Population	28,562	19,218	15,216	4,307	3,912	1,234

39. At a national level, Slovenia elects a head of state (a president) and a legislature. The president is elected for a five year term by the people using the runoff system. The National Assembly, Slovenia's parliament, has 90 members each elected for 4-year terms. All but two of these are elected using the d'Hondt method of list proportional representation. The remaining two members are elected by the Italian and Hungarian ethnic minorities using the Borda count. The results of the 2008 parliamentary elections are presented here. Apportion 90 seats between these 7 parties using Webster's method.

Party	Number of Votes
Social Democrats	320,248
Slovenian Democratic Party	307,735
Zares – new politics	98,526
Democratic Party of Pensioners of Slovenia	78,353
Slovenian National Party	56,832
Slovenian People's Party and Youth Party of Slovenia	54,809
Liberal Democracy of Slovenia	54,771

40. The police department in a large city has 76 new police officers to be apportioned among 6 high-crime precincts. Crimes by precinct are shown in the following table. Use
    (a) Jefferson's method,   (b) Adams' method.   (b) Webster's method;
    to apportion the new officers among the precincts.

Precinct	A	B	C	D	E	F
Crimes	481	567	330	272	371	793

41. Show that the apportionment produced by Webster's method violates the quota rule for a 200-member house and 6 states, with populations given by the following table.

State	A	B	C	D	E	F
Population	37,745	964	527	312	162	123

42. Twenty-two sections are to be offered in algebra, financial mathematics, calculus, differential equations, combinatorics, and numerical analysis. The preregistration figures for the number of students planning to enroll in these sections are given in the following table. Apportion these sections to these six math courses using
    (a) Jefferson's method,   (b) Adams' method.   (b) Webster's method.

Discipline	Algebra	Financial Math	Calculus	DE	Combinatorics	NA
Enrollment	185	227	302	126	34	21

43. Every four years, Turkey elects on the proportional basis a legislature—the Grand National Assembly—that has 550 members. To participate in the distribution of seats, a party must obtain at least 10% of the votes cast at the national level as well as a percentage of votes in the contested district according to a complex formula. The president was elected for a seven-year

252   *Review of Chapter 2*

term by the parliament prior to the 2007 constitutional changes, and will be elected for at most two five-year terms by the people in the future.

Summary of the 22 July 2007 Grand National Assembly of Turkey election is as follows.

Justice and Development Party (AKP)	6,340,534
Republican People's Party (CHP)	7,300,234
Nationalist Movement Party (MHP)	5,004,003
Democratic Party (DP)	1,895,807
Independents	1,822,253
Youth Party	1,062,352

Allocate 550 seats between these six parties using
(a) Jefferson's method,   (b) Adams' method.   (b) Webster's method.

44. A poll was taken by a group of university students about which language they would like to take the most during their spring semester. Use the Webster method to apportion 15 professors based on the data collected to find out which classes they need to teach. Then compare your allocation with distribution obtained using Hamilton's, Lowndes', and Jefferson's methods.

Language	Population	Natural Quota	I.A. $10	F.A. $10	F.A. Lowndes	Threshold Divisor	F.A. $2
Spanish	306						
Italian	204						
French	232						
Chinese	183						
Japanese	97						
Total:	1022	15		15	15		15

45. Each winter, all schools in the Monmouth area of New Jersey choose new members for their education committee for the next four academic years. The committee consists of 18 members that are chosen from the schools based on the schools enrollment. To apportion these seats, use Webster's method based on the data in the following table.

School	Enrollment	Natural Quota	I.A.	Threshold Divisor	Rank	F.A. Webster
Marlboro	16					
Freehold	1326					
Colts Neck	1112					
Manalapan	650					
Howell	471					
Township	195					
Total:	3760	18				18

46. A city has 23 fire trucks that are to be assigned to 1 of 5 fire stations. The new fire chief decides to apportion the trucks to stations in proportion to the number of fires reported in each station's district over the last year, given below. Allocate the fire trucks using the Hill-Huntington method.

Station	A	B	C	D	E
Number of Fires	94	83	77	46	21

47. Apportion a 21-member committee using the Hill-Huntington method in proportion of the races presented in the table.

Race	Hispanic	Asians	African-American	Caucasian	Indians
Population	176	84	31	49	7

48. A sorority is choosing 25 girls for a new pledge class. Potential pledges live in five dorms. Each girl is given a cumulative score out of 100 based on personality, presentation ability, and her response to questions asked. The sorority wants to use the Hill-Huntington method to pick girls from five dorms. Summary scores are presented here.

Dorm	Barlow	Browning	Gorham	Merrow	Weldin
Amount	830	800	790	950	850

49. A store manager wants to represent candy sales in a percentage way based on the data presented. Use the Hill-Huntington method.

Candy	Twix	Skittles	Hershey	Twizzlers	M&M
Sales	903	332	874	536	947

50. A 88-member board of trustees represents the major cities of New Jersey. How many representatives will come from each city based on their populations in 2006 using
    (a) Hamilton's method. **(b)** Lowndes' method. **(c)** Jefferson's method.
    (d) Adams' method. **(e)** Webster's method. **(f)** The Hill-Huntington method.

City	Newark	Jersey City	Peterson	Elizabeth	Edison	Woodbridge
Population	273,546	240,055	148,708	126,179	99,523	99,208

51. The state of New Jersey wants to represent the wealth of municipalities and communities by per capita income in a percentage way. Use the Hill-Huntington method to represent 100% of income based on the following information.

County	Mantoloking	Saddle River	Far Hills	Essex Fells	Alpine	Camden
Income	$114,017	$85,934	$81,535	$77,434	$76,995	$9,815

52. A zoo is getting 15 more animals from Africa. Allocate these animals in proportion to the species already living in the zoo using
    (a) Hamilton's method. **(b)** Lowndes' method. **(c)** Jefferson's method.
    (d) Adams' method. **(e)** Webster's method. **(f)** The Hill-Huntington method.

Animals	Baboons	Lions	Tigers	Giraffes	Elephants	Hippopotamuses
Population	12	8	6	11	5	9

53. A zoo asked 140 people for their favorite reptile out of frogs, turtles, lizards, snakes, crocodiles, and uroplatus.

Reptiles	Frogs	Turtles	Lizards	Snakes	Crocodiles	Uroplatus
Number of Answers	25	35	40	18	19	3

Represent peoples' answers as a percentage using
    (a) Hamilton's method. **(b)** Lowndes' method. **(c)** Jefferson's method.
    (d) Adams' method. **(e)** Webster's method. **(f)** The Hill-Huntington method.

54. A park's manager wants to represent different types of trees as a percentage. The data show that there are six types of trees:

Types	Oak	Willow	Elm	Maple	Pine	Hickory
Number of Trees	1,409	648	973	530	1,218	222

Represent these data as a percentage using
    (a) Hamilton's method. **(b)** Lowndes' method. **(c)** Jefferson's method.
    (d) Adams' method. **(e)** Webster's method. **(f)** The Hill-Huntington method.

55. Interstate 95 is one of the busiest highways in the United States. It stems as far north as the Canadian border, and as far south as Miami. Running along the east coast, it is the most heavily traveled road in the nation. With this high volume of travel comes a high volume of accidents and deaths/injuries. A high percentage of accidents are caused by careless driving and excessive speeding. For example, in Connecticut, the nine mile stretch between West Haven and New Haven (exits 39-50) is the most accident area on I-95. As a general idea, most of 95 is listed as 65 miles an hour; however, there are certain areas that need to be evaluated for lower speed limits. The federal government allocates only a certain amount of miles for each state to have 65-mile an hour speed limits. In order for each state to determine where they need to reduce their speed limits to lower than 65 they need to evaluate their highest accident areas and decide if those accidents are caused by speeding. The committee wants to allocate 25 officers to patrol 6 locations on I-95 based on the number of accidents using the Hill-Huntington method.

Location	A	B	C	D	E	F
Number of Accidents	1107	835	791	215	163	89

56. A flower shop wants to represent their sales in a percentage way using Hill-Huntington's method. The data show that the most popular flowers sold during last year were as follows.

Types	Roses	Lilies	Daisies	Jasmines	Orchids	Dandelions
Sales	5,796	637	179	563	381	264

57. Six Central American countries decide to convene 200 business leaders for a conference to address economic issues. Allocate 200 seats between these 7 countries in proportion to their populations (in 2009) using
    (a)   Hamilton's method.   (b)   Lowndes' method.   (c)   Jefferson's method.
    (d)   Adams' method.   (e)   Webster's method.   (f)   The Hill-Huntington method.

Country	Costa Rica	Salvador	Guatemala	Honduras	Nicaragua	Panama
Population	4,579,000	6,163,000	14,027,000	7,466,000	5,743,000	3,454,000

58. The school system of Long Island (New York) has a committee consisting of 35 people that represent 5 school districts. Each school district is given seats in proportion to their enrollment. Allocate 35 seats between these districts using
    (a)   Hamilton's method.   (b)   Lowndes' method.   (c)   Jefferson's method.
    (d)   Adams' method.   (e)   Webster's method.   (f)   The Hill-Huntington method.

Town	Syosset	Hebrew	Plainview	Huntington	Suffolk	Nassau
Population	7,253	798	1,600	1,200	264,322	211,771

59. The local ice cream shop has a huge selection of ice cream flavors. The number of gallons sold last year for the most popular flavors are presented here. Represent these data as a percentage using
    (a)   Hamilton's method.   (b)   Lowndes' method.   (c)   Jefferson's method.
    (d)   Adams' method.   (e)   Webster's method.   (f)   The Hill-Huntington method.

Flavor	Vanilla	Strawberry	Cookie Dough	Oreo	Chocolate	Candy
Amount	1,759	938	719	556	3,297	217

60. The following five countries are members in the European Union. There are 55 seats that should be distributed between these countries. Use
    (a)   Hamilton's method;   (b)   Lowndes' method;   (c)   Jefferson's method;
    (d)   Adams' method;   (e)   Webster's method;   (f)   The Hill-Huntington method;

to allocate 55 seats between these countries based on the 2008 populations presented.

Country	Belgium	United Kingdom	France	Germany	Portugal
Population	10,708,433	61,414,062	62,277,432	82,110,097	10,622,413

61. There is $500 in prize money to be distributed among the top 6 participants in a dance competition. The girls will receive a total score out of 1000 points. The prize money will be distributed in apportionment according the score using
  (a) Hamilton's method.   (b) Lowndes' method.   (c) Jefferson's method.
  (d) Adams' method.   (e) Webster's method.   (f) The Hill-Huntington method.

Competitor	Anna	Chelsey	Jenna	Michelle	Rachel	Shannon
Score	972	834	491	767	822	753

62. The Nordic countries make up a region in Northern Europe and the North Atlantic that consists of Denmark, Finland, Iceland, Norway, and Sweden (all of which use a Nordic Cross flag) and their associated territories which include the Faroe Islands, Greenland, Svalbard, and Åland. The region's five nation-states and three autonomous regions share much common history as well as common traits in their respective societies, such as political systems and the Nordic model. Politically, Nordic countries do not form a separate entity, but they co-operate in the Nordic Council. The 2009 populations are presented in the table here.

Country	Åland	Denmark	Faroe	Finland
Population	27,456	5,519,287	49,006	5,349,829

Country	Greenland	Iceland	Norway	Sweden
Population	57,600	319,756	4,836,183	9,336,487

Apportion 87 seats in the Nordic Council using
  (a) Hamilton's method.   (b) Lowndes' method.   (c) Jefferson's method.
  (d) Adams' method.   (e) Webster's method.   (f) The Hill-Huntington method.

63. In the tough economy that we live in today, many state agencies are being forced to cut back their services because of their struggles. Police departments in many states are doing that same thing. The Newark Police Department in New Jersey has been forced to cut back its total employment hours from 4200 hours a week in 2007 to almost half that in 2010. This means that the city of Newark will cut back on the number of officers (who are paid hourly) working daily—there will be fewer officers on duty at a single time than there has before. Allocate available 2200 patrol hours between 7 districts based on the number of incidents that have occurred over the past several months using
  (a) Hamilton's method.   (b) Lowndes' method.   (c) Jefferson's method.
  (d) Adams' method.   (e) Webster's method.   (f) The Hill-Huntington method.

District	North	South	East	West	Central	Vailsburg	Dayton
Incidents	249	381	177	530	218	294	122

64. A small country is comprised of four counties with populations given in the table. The number of seats to be apportioned is $h = 33$. Show that there is no suitable divisor for the Hill-Huntington method; therefore, the method does not work. Similarly, show that other populations are not suitable for the Webster method with the house size $h = 36$.

House size	County	A	B	C	D
$h = 33$	Population	879	5274	4201	9834
$h = 36$	Population	12475	42415	52478	71802

References and Further Reading

[1] Balinski, M. L. and Young, H. P., The Quota Method of Apportionment, *American Mathematical Monthly*, **82**, 701 – 730, 1975.

[2] Balinski, M. L. and Young, H. P., The Webster Method of Apportionment, *Proceedings of the National Academy of Sciences, U.S.A.*, **77**, No. 1, 1 – 4, 1980.

[3] Brams, Steven J. and Taylor, Alan D., *Fair Division: from Cake-cutting to Dispute Resolution*, Cambridge University Press, Cambridge; New York, 1996.

[4] Huntington, Edward, The Mathematical Theory of the Apportionment of Representatives, *Proceedings of the National Academy of Sciences, U.S.A.*, **7**, 123 – 127, 1921.

[5] Huntington, Edward, The Apportionment of Representatives in Congress, *Transactions of the American Mathematical Society*, **30**, 85 – 110, 1928.

[6] Meder, Albert E., *Legislative Apportionment*, Houghton Mifflin, Boston, 1966.

[7] Neubauer, Michael and Zeitlin, Joel, Apportionment and the 2000 Election, *The College Mathematical Journal*, **34**, No. 1, 2 – 9, 2003.

[8] Saari, D.G., Apportionment Methods and the House of Representatives, *American Mathematical Monthly*, **85**, 792 – 802, 1978.

[9] Steinhaus, Hugo, *Mathematical Snapshots*, Dover Publications; 3rd edition, New York, 1999.

[10] Tannenbaum, Peter, *Excursions in Modern Mathematics with Mini-Excursions*, Pearson, Upper Saddle River, 2007.

[11] Willcox, F. W., The Apportionment of Representatives, *American Economic Review*, **6**, Part 2, 3 – 16, 1916.

[12] Woodall, D. R., How Proportional is Proportional Representation?, *The Mathematical Intelligencer*, **8**, No. 4, 36 – 46, 1986.

The Internet Sites

http://www.apportionment.us/
* The official apportionment site of the United States.

http://www.census.gov
* The official Census Bureau Web site.

http://www.cut-the-knot.org/Curriculum/SocialScience/HH.shtml
* Hill-Huntington method

Chapter 3

Everyday Finance

Knowledge of financial operations is crucial for everyone throughout the course of their lives. Money is the medium of exchange in our society and financial independence is a common goal that most people share. You may have heard in a variety of circumstances that one must be smart when dealing with money, either spending or working for it. Freedom from financial reliance on creditors is a key goal for many people, where aside from taxes, you are free to do what you want with all your income. Understanding how money works is important in achieving financial independence and in navigating obligations like student debt, car payments, as well as educational and housing loans.

This chapter is devoted to the mathematics of finance, including simple and compound interest, annuities and amortization loans. The financial operations have been utilized for thousands of years in our every day lives. Despite the topics to be covered being centuries old, the modern information technologies have completely changed traditional financial operations. Today, we trade stocks and bonds, buy or sell goods, pay bills, lend and borrow money—all electronically via the Internet. However, as this chapter demonstrates, the principles have remained the same throughout centuries.

3.1 Review of Calculations

The next sections in this chapter present an introduction to financial operations that have existed for more than three thousand years. To facilitate their understanding, we start with a review of needed mathematical operations. Since most of the practical calculations involve multi-step algorithms, we need a special tool to perform all the intermediate operations—a calculator with a memory.

We strongly recommend using the TI-83 – 86 calculator (TI stands for Texas Instruments), if you do not currently have a calculator. These calculators have a large display with the ability to show many operations typed. There are plenty of other calculators with similar properties.

This section recalls the basic arithmetic operations that you can perform on your calculator. The right column of buttons contains five main arithmetic operations ordered from top to bottom: $\wedge$ (raising to a power), $\div$ (division), $\times$ (multiplication), $-$ (subtraction), and $+$ (addition). The last button (in the right column) is marked with RETURN that you have to press in order to see the results of calculations typed before.

The five arithmetic operations ($\wedge$, $\div$, $\times$, $-$, $+$) should be familiar to anyone who has ever attended school. Nevertheless, we recall that these five arithmetic operations are performed in a special order—raising into a power ($\wedge$) is executed first, then $\div$ and $\times$ (division and multiplication have the same priority), and only after that the calculator executes "$+$" or "$-$." Sometimes multiplication is denoted either by $*$ or by a dot, which is usually omitted. A slanted bar for division is also common. For instance, when you type

$$2 \times 3 \div 4 + 5 \times 6 - 7 \qquad \text{then press RETURN,}$$

the calculator will execute these operations in the following order:

$$(2 \times 3 \div 4) + (5 \times 6) - 7 \qquad \text{giving the answer} \quad 24.6.$$

However, if you want to divide 2×3 by $4 + 5$ and then multiply by 6, and subtract 7, you must utilize the parentheses as follows:

$$2 \times 3 \div (4 + 5) \times 6 - 7 \qquad \text{to obtain} \quad -3.$$

You do not need to embrace 2×3 into parentheses because multiplication and division have the same priority of execution (of course, you may type $(2 \times 3) \div (4 + 5) \times 6 - 7$, but the result will be the same).

Calculation of fractions requires some care when the number of terms in the denominator is more than one. For example, suppose you need to evaluate the ratio

$$\frac{8 \times 9}{4 \times 6}.$$

To find its value, you may want to type either

$$8 \times 9 \div 4 \div 6 \qquad \text{or} \qquad 8 \times 9 \div (4 \times 6)$$

before pressing ENTER.

Our next operation to consider is the power operation. For any positive integer n, the product of n copies of a real number x can be written as

$$x^n = \underbrace{x \cdot x \cdots x}_{n \text{ times}}.$$

The reciprocal of x^n is denoted as

$$x^{-n} = \frac{1}{x^n}.$$

For instance,

$$3^4 = 3 \cdot 3 \cdot 3 \cdot 3 = 81 \qquad \text{and} \qquad 3^{-4} = \frac{1}{3^4} = \frac{1}{81} \approx 0.012345679 \,,$$

where we have rounded the value to nine decimal places. Note that the sign $\approx$ denotes an approximate value of. These values can be computed using the exponential key ($\wedge$) on your calculator:

$$3 \wedge 4 \qquad \text{and} \qquad 3 \wedge ((-)4) \quad \text{or} \quad 1 \div 3 \wedge 4 \qquad \text{press ENTER}\,.$$

Raising to the negative power requires the usage of a special button (usually in the right bottom corner) in your calculator, denoted by $\boxed{(-)}$. If you type instead the "subtract" operation, a calculator will return an error message on its display. Note that some types of calculators may have another exponential key such as $\boxed{x^y}$ or $\boxed{y^x}$.

The power function, a^n, exists not only for integer values of n. In many cases we need to find a square root, $\sqrt{a}$, of a positive real number, a, or a cubic root, $\sqrt[3]{a}$, or an arbitrary root $\sqrt[r]{a}$ (where r is a real positive number). These expressions can be written as

$$\sqrt{a} = a^{1/2}, \qquad \sqrt[3]{a} = a^{1/3} \qquad \sqrt[r]{a} = a^{1/r},$$

respectively. The roots are used to express solutions of a power equation:

$$a = x^r \qquad (a \text{ and } r \text{ are given positive real numbers}) \qquad (3.1.1)$$

Indeed, raising both sides of Eq. (3.1.1) to the power $1/r$, we get

$$a^{1/r} = (x^r)^{1/r} = x^{r \cdot (1/r)} = x.$$

Here we used the exponential properties: $(x^p)^q = x^{pq}$ and $x^1 = x$.

Unfortunately, Eq. (3.1.1) may have many solutions. For instance, the equation $1 = x^2$ has two solutions $x = 1$ and $x = -1$. To avoid ambiguity, we consider only *positive real* numbers, then a positive solution of (3.1.1) is unique, which we denote by $a^{1/r}$ or $\sqrt[r]{a}$.

Example 3.1.1. Consider the equation $x^{12} = 1.25 = 5/4$. When both sides are raised to the power $1/12$, we will get

$$\left(x^{12}\right)^{1/12} = 1.25^{1/12} = \left(\frac{5}{4}\right)^{1/12} = \frac{5^{1/12}}{4^{1/12}}$$

$$x = \frac{5^{1/12}}{4^{1/12}} = \frac{5^{1/12}}{2^{1/6}} = 5^{1/12} \cdot 2^{-1/6} \approx 1.018769265\,.$$

Here we used our knowledge that $2^2 = 4$ to obtain

$$4^{1/12} = \left(2^2\right)^{1/12} = 2^{2 \cdot (1/12)} = 2^{1/6}.$$

Example 3.1.2. Solve for x in the equation $(1+x)^{12} = 1.25$, where $1 + x > 0$.

Following Example 3.1.1, we raise both sides to the power $1/12$, which yields

$$\left[(1+x)^{12}\right]^{1/12} = 1.25^{1/12} = 5^{1/12} \times 2^{-1/6}$$
$$1 + x = 1.25^{1/12} = 5^{1/12} \times 2^{-1/6}.$$

After subtracting 1 from both sides, we get the answer

$$x = 5^{1/12} \times 2^{-1/6} - 1 \approx .018769265\,.$$

To solve this equation numerically, we apply the following operations on the calculator:

$$1.25 \wedge (1 \div 12) - 1 \quad \text{or} \quad 5 \wedge (1 \div 12) \times 2 \wedge ((-)1 \div 6) - 1 \qquad (\text{press ENTER}). \qquad \blacksquare$$

A calculator allows you to perform the operations "raising to" and finding a "root" simultaneously. For example, raising the root $\sqrt{2} = 2^{1/2}$ into the third power can be performed as follows

$$2 \wedge (3 \div 2) \qquad \text{or} \qquad 2 \wedge 1.5$$

because $\left(\sqrt{2}\right)^3 = \left(2^{1/2}\right)^3 = 2^{3/2} = 2^{1.5}$.

A calculator is so smart that it could raise any real number into any power, not necessarily to be a positive number. For instance, $2^{-\sqrt{2}}$ can be evaluated as follows

$$2 \wedge ((-)2 \wedge (1 \div 2)) \qquad \text{press RETURN} \quad .375214227246\,.$$

Many financial problems encounter equations in which we want to solve for a variable that appears in the exponent. For instance,

$$2^x = 3$$

is an equation of this type. In mathematics, there is a special operation to solve this kind of equation—logarithm. So the solution of the equation $b^x = a$ can be written as $x = \log_b a$, where b is the base of the logarithm (which is always assumed to be a positive real number). These two notations are equivalent:

$$b^x = a \qquad \Longleftrightarrow \qquad x = \log_b a.$$

Therefore, the logarithm of a base b is the power to which you need to raise b in order to obtain a. From this definition, we immediately obtain two basic properties

$$b^{\log_b a} = a \qquad \text{and} \qquad \log_b b^x = x. \tag{3.1.2}$$

In what follows, we will use the following properties of a logarithm:

$$\log_b a^p = p \log_b a \qquad (\text{power rule}), \tag{3.1.3}$$

$$\log_b(a_1 \times a_2) = \log_b a_1 + \log_b a_2 \qquad \text{(product rule)}, \qquad (3.1.4)$$

$$\log_b\left(\frac{a_1}{a_2}\right) = \log_b a_1 - \log_b a_2 \qquad \text{(quotient rule)}. \qquad (3.1.5)$$

The next property allows us to evaluate a logarithm with an arbitrary base through the logarithm with a specific base:

$$\log_b a = \frac{\log_c a}{\log_c b} = \frac{1}{\log_a b}. \qquad (3.1.6)$$

A typical calculator provides two options to compute logarithms using keys that are labeled with $\boxed{\log}$ and $\boxed{\ln}$. The former has base 10, that is, $\log \equiv \log_{10}$, and the latter has base e. This irrational number was introduced by Leonhard Euler (pronounced oiler) almost 300 years ago; one of its definitions is given by the limit

$$e = \lim_{x \to \infty}\left(1 + \frac{1}{x}\right)^x = 2.7182818284590452354\ldots. \qquad (3.1.7)$$

A logarithm with base e is called the **natural logarithm**, and denoted by $\ln \equiv \log_e$.

Example 3.1.3. Let us verify the rules (3.1.2) – (3.1.6) on some examples. Choosing $b = 2$, $a = 8$, and $x = 3$ in Eq. (3.1.2), we get $\log_2 8 = 3$ because $2^3 = 8$.

Considering the power rule Eq. (3.1.3) with $b = 2$, $a = 2$, $p = 3$. Then

$$\log_2 8 = \log_2\left(2^3\right) = 3\log_2 2 = 3$$

because $\log_2 2 = 1$.

Choosing $b = 2$, $a_1 = 2$, and $a_2 = 8$ in the product rule (3.1.4), we see that

$$\log_2(2 \times 8) = \log_2 16 = \log_2 2^4 = 4 = \log_2 2 + \log_2 8 = 1 + 3.$$

For the quotient rule, we have

$$\log_2\left(\frac{2}{8}\right) = \log_2 2 - \log_2 8 = 1 - 3 = -2 = \log_2\left(\frac{1}{4}\right) = \log_2 2^{-2}.$$

Eq. (3.1.6) yields

$$3 = \log_2 8 = \frac{\ln 8}{\ln 2} \approx \frac{2.0794415416798359283}{.69314718055994530942} = 3.000000000.$$

Example 3.1.4. Suppose we want to solve for x in the equation $(1.25)^x = 3$. Applying the logarithm to both sides of the given equation, we obtain

$$
\begin{aligned}
\ln(1.25)^x &= \ln 3 \qquad \text{use the power rule}\\
x\ln 1.25 &= \ln 3\\
x &= \frac{\ln 3}{\ln 1.25} \approx \frac{1.0986122886681096914}{.22314355131420975577} \approx 4.923343214.
\end{aligned}
$$

Perform these calculations on a TI-83: $\ln(3) \div \ln(1.25)$.

Example 3.1.5. Solve for x in the equation $(1.25)^{4x} = 3$. Again, the power rule (3.1.3) leads to

$$4x \ln 1.25 = \ln 3 \qquad \Longrightarrow \qquad x = \frac{\ln 3}{4 \ln 1.25} \approx 1.230835804 \,.$$

This result can be obtained with your lovely calculator as

$$\ln(3) \div \ln(1.25) \div 4 \qquad \text{or} \qquad \ln(3) \div (\ln(1.25) \times 4).$$

3.1.1   Taxes and Percentage

The word **percent** means "per hundred," which goes back to the Latin term "per centum." Therefore, to convert a percentage into a decimal, you must divide it by 100. For example, 6% means "six per hundred" that can be written either as $\frac{6}{100}$ or in decimal form as 0.06. In the metric system, the word "centi" is used to express 0.01 of a unit (as in centimeter). So 1 percent of a dollar is 1 centidollar or cent (abbreviated as ¢).

It is convenient to remember that when converting a percent to a decimal, you move the decimal point two places to the left and that in converting a decimal to a percent, you move the decimal point two places to the right. For example, 0.16 is expressed as 16%.

The media often uses percentages to explain the change in some quantity. For instance, on a particular day in April 2010, you may hear that the unemployment rate in the state of Rhode Island reached 12.4% (according to the U.S. Bureau of Labor Statistics).

The percent of change is always in relationship to a previous, or **base amount**. The percent of change is computed in regard to a **new amount** as

$$\text{percent of change} = \frac{\text{new amount} - \text{base amount}}{\text{base amount}}.$$

Example 3.1.6.   Change in U.S. Debt

When the government spends more than it receives in tax revenue, it borrows the rest by issuing U.S. Treasury Securities. The United States public debt, or the national debt, is the sum of all these outstanding securities. In 2000, the national U.S. debt was about \$5,628.7 billion, or about 58% of the gross domestic product; while in 2010 it became \$14.078 trillion, or about 90.5% of the gross domestic product. During these past 10 years, we see the relative increase in debt as the fraction $\frac{14,078}{5,628.7} \approx 2.501110381$ or more than 250%. The federal budget was \$385 billion in 2000 while it became \$3.55 trillion in 2010. During 10 years, the annual government spending has grown by $\frac{3550}{383} \times 100\% \approx 926.89\%$ or more than nine times.

Example 3.1.7.   Tuition Increase

Moses Brown/Wheeler schools (in Providence, RI) had tuition of \$10,000 per year in 2000; ten years later, it became \$12,000. To find the percent of tuition increase, we calculate

$$\text{percent of change} = \frac{\text{new tuition} - \text{last tuition}}{\text{last tuition}} = \frac{\text{new amount} - \text{base amount}}{\text{base amount}}$$

$$= \frac{12000 - 10000}{10000} = \frac{2000}{10000} = 0.2 \qquad \text{or} \qquad 20\%$$

Example 3.1.8. Finding Percent of Decrease

Businesses reduce prices, or give **discounts**, to attract customers or reduce inventory. The discount rate or clearance rate is a percent of the original price.

Suppose you found a laser printer on sale for \$350, while its regular price is \$430. Its discount rate is

$$\text{discount rate} = \frac{\text{original price} - \text{reduced price}}{\text{original price}} = \frac{430 - 350}{430} \approx 0.186 \quad \text{or} \quad 18.6\%.$$

When buying the printer, you gave a coupon to the cashier to reduce this price additional 10%. So actually you paid (excluding state's tax) \$315. Therefore, your discount becomes $\frac{430-315}{430} = 0.26744$ or 26.744%.

Example 3.1.9. Washington Capitals Statistics

In the 2009–2010 regular season, the Washington Capitals, of the National Hockey League, had a record of 54 wins, 15 losses, and 13 overtime losses (worth one point). Since they played 82 games, the percent of their games won is

$$\text{percent of victories} = \frac{54}{82} \times 100\% \approx 65.\overline{8536}\%.$$

Example 3.1.10. Percent and Sales Tax

One of the important applications involving percents is determination of the sales tax collected by states, countries, cities, and airports on sales of items to customers:

$$\text{sales tax amount} = \text{tax rate} \times \text{item's cost}.$$

The sales tax in the state of New York is 7.5%. If somebody purchases a TV set for \$800, the sales tax would be

$$7.5\% \times \$800 = 0.075 \times \$800 = \$60.00.$$

The total cost of the TV set is the purchase price, \$800, plus the sales tax, \$60, which adds to \$860.

Example 3.1.11. Back-Load Spending

Due to inflation (see §3.3), the average spending on food and lodging in medical schools may be increased by 0%, 2%, and 4% over 3 years. Sometimes, students are offered the chance "back-load" when this amount increases by 4%, 2%, and 0%. Which offer, back-load or front-load, is better? Let us look at an example, with say, \$100.

	Front-load	Back-load
Original amount	\$100	\$100
Amount in first year	\$100 + 0% = \$100	\$100 + 4% (100) = \$104
Amount of second year	\$100 + 2%(100) = \$102	\$104 + 2% (104) = \$106.08
Amount in third year	\$102 + 4% (102) = \$106.08	\$106.08 + 0 % = \$106.08
	In 3 years the gain is 2 + 6.08 = \$8.08	In 3 years the gain is 4 + 6.08 + 6.08 = \$16.16

So back-load requires a $16.16 increase in spending whereas front-load requires only $8.08, half as much.   □

Calculation of taxes paid by a citizen relies heavily on properties of percentages. Let us consider a particular example of filling out Form 1040 to compute the federal income tax for a non-married person. It is based on your tax bracket and depends on the year you fill it in. The example of 2010 federal tax brackets is presented in the following table.

If your taxable income is **over**	But not **over**	The tax is	Of the amount **over**
$0	$8,375	10%	$0
$8,375	$34,000	$837.50 + 15%	$8,375
$34,000	$82,400	$4,681.25 + 25%	$34,000
$82,400	$171,850	$16,781.25 + 28%	$82,400
$171,850	$373,650	$41,827.25 + 33%	$171,850
$373,650		$108,421.25 + 35%	$373,650

The first column in the table contains the value of your taxable income, which is the difference:

$$\text{taxable income} = \text{adjusted gross income} - (\text{exemptions} + \text{deductions}).$$

On the other hand, adjusted gross income is also the difference:

$$\text{adjusted gross income} = \text{gross income} - \text{adjustments}.$$

Let us explain what each term includes:
– *Gross income* is the sum of all your earnings for the year including wages, tips, consultant's compensations, earnings from investments, and unemployment compensation.
– *Adjustments* include payments to tax-deferred savings plans.
– *Exceptions* include a fixed amount for yourself and the same amount for each dependent.
– *Deductions* include interest on home mortgages, state's income taxes, property taxes, charitable contributions, and medical expenses exceeding 7.5% of adjusted gross income.

Example 3.1.12. Calculating Your Income Tax
Suppose that Chris is unmarried with no dependents. His gross income in 2010 was $76,750.00. The standard deduction for a single person in 2010 was $5,700, with personal exemption of $3,650. He paid $2400 to a tax-deferred 401(k) plan, $6854 as mortgage interest, $2350 as property taxes, and donated charities for $1200. Therefore, his adjusted gross income becomes

$$\text{Adjusted gross income} = \text{Gross income} - \text{Adjustments}$$
$$= \$76,750.00 - \$2400 = \$74,350.00\,.$$

The next step is to determine his taxable income:

$$\text{Taxable income} = \text{Adjusted gross income} - (\text{Exemptions} + \text{Deductions})$$
$$= \$74,350.00 - \$3,650 - \text{Deductions}$$
$$= \$70,700 - (\$6854 + \$2350 + \$1200)$$
$$= \$70,700 - \$10,404 = \$60,296.00.$$

Now we are ready to calculate Chris' income tax, based on his taxable income of \$60,296. Using the third line in the table, we see that Chris must pay $\$4,681.25 + 25\%$ of the amount of taxable income over 34,000. This yields

$$4,681.25 + 0.25 \times (60,296 - 34,000) = 4,681.25 + 0.25 \times 26,296 = 4,681.25 + 6574 = \$11,255.25.$$

So Chris must pay \$11,255.25 to the IRS (Internal Revenue Service).

Let us explain where the amount \$4,681.25 comes from. The second line in the table tells us that Chris' income can be divided into two parts. His \$60,296 is taxed according to the instructions on line 2 of the table: $\$750 + 15\%$, which is

$$837.50 + 0.15 \times (34000 - 8375) = 837.50 + 0.15 \times 25625 = \$4,681.25.$$

This is in accordance with the third line of the table.

Exercises for Section 3.1.

1. Using a calculator, evaluate the following expressions within eight decimal places.

   (a) $\left(\dfrac{7}{8}\right)^{12}$;   (b) $\left(\dfrac{8}{9}\right)^{16}$;   (c) $\left(\dfrac{8}{7}\right)^{1.2}$;   (d) $\left(\dfrac{9}{8}\right)^{0.16}$;

   (e) $(1.875)^{12}$;   (f) $(0.78865)^{16}$;   (g) $(1.875)^{0.12}$;   (h) $(0.78865)^{0.16}$;

   (i) $\dfrac{7 \times 8}{9 \times 11}$;   (j) $\dfrac{18 \times 19}{12 \times 17}$;   (k) $\dfrac{1.3278 \times 0.9845}{3.4557 \times 0.7846}$;   (l) $\dfrac{9.7785 \times 0.6758}{8.7876 \times 0.1214}$;

   (m) $\dfrac{\ln 3.4519}{12 \ln 0.14}$;   (n) $\dfrac{\ln 2.147}{4 \ln 0.23}$;   (o) $\dfrac{\ln 0.1365}{12 \ln 3.45}$;   (p) $\dfrac{\ln 0.9856}{4 \ln 0.56}$.

2. Solve for the unknown variable x and round your answers to eight decimal places.

   (a) $x^{12} = 12$;   (b) $x^{16} = 16$;   (c) $x^{1.2} = 12$;   (d) $x^{0.16} = 16$;

   (e) $x^{-12} = 12$;   (f) $x^{-16} = 16$;   (g) $x^{0.12} = 1.375$;   (h) $x^{0.16} = 9.978$;

   (i) $\dfrac{1}{x^{0.12}} = 1.5$;   (j) $\dfrac{1}{x^{0.16}} = 9.9$;   (k) $(1+x)^{15} = 1.3$;   (l) $(1+x)^{16} = 9.9$;

   (m) $4^x = 56$;   (n) $5^x = 65$;   (o) $6^{3x} = \dfrac{1}{3}$;   (p) $7^{4x} = \dfrac{1}{4}$;

   (q) $(1.3)^{-12x} = 36$;   (r) $(1.4)^{-4x} = 63$;   (s) $6^{3x-1} = 3$;   (t) $4^{7x-2} = 6$.

3. Suppose that the local tax rate is 6.5% and you purchase a new Saturn SKY car for $31,900. How much tax is paid? What is the car's total cost?

4. Suppose that the local tax rate is 7.5% and you purchase a Texas Instruments TI-83 Plus graphing calculator for $99.54. How much tax is paid? What is the calculator's total cost?

5. On clearance, a lady's summer dress was sold for $23.99. By what percent of the original price of $39.49 was it reduced?

6. On sale, a men's long sleeve shirt was sold for $33.99. By what percent of the original price of $49.99 was it reduced?

7. From 2000 to 2008, the population of Ghana grew from 18,412,247 to 23,382,848. What was the percent of increase?

8. From 2000 to 2008, the population of Uruguay grew from 3,334,074 to 3,477,778. What was the percent of increase?

9. In 2000, the population of Russia was 146,001,176 people. Eight years later, its population became 140,702,094. What was the percent of decrease?

10. A Panasonic KX FP215 FAX machine is regularly sold for $88. If the sale price is $79, find the percent decrease of the sale price from the regular price.

11. The price for a Honda Accord LX V-6 sedan car increased from $21,000 in 2001 to $25,000 in 2008. What is the percent of price increase during seven years?

12. In January 2010, there were 247,791 midsize cars sold in the United States, which was a 21.5% increase compared with January sales of the previous year. How many midsize cars were sold in January 2009?

13. If a dealer buys a Civic-LX 2010 sedan car from the manufacturer for $16,932 and then sells it for $18,315, what is his markup (i.e. percentage increase from the original value)?

14. A new 2010 Chevrolet Silverado 2500 pickup truck costs $27,465. If the same pickup truck is estimated to cost $10,162 in 2017, by what percent would it depreciate during 7 years?

15. The price for a Ford Focus drops from $17,170 to $8,150 in four years. What is the percentage drop in its price?

16. If you are given a raise of 3% this year and a raise of 2% the following year, what single raise would give you the same yearly salary in the second year?

17. If a merchant reduces the price of some item by $x\%$ and then later increases the price by $x\%$, is the price of the item the same as the original price? Now suppose the opposite, the price is increased by $x\%$ and then later decreases by $x\%$. Is the price of the item the same as the original price?

18. You left a $50 note on the table to pay your restaurant expenses of $38.45. What percent of the bill did you pay in tips?

19. Founded in 1979, SymphonyIRI Group (www.symphonyiri.com) is a market research company which provides clients with consumer, shopper, and retail market intelligence and analysis focused on the consumer packaged goods industry. SymphonyIRI's clients include 95 percent of the Fortune Global 500 CPG, retail, and healthcare companies. The firm operates in 58 countries through stand-alone operations, wholly owned subsidiaries, partnerships, and alliances. Its 2008 revenue was $700,000,000. Out of 1200 employees, 54% are males and 46% are females. How many males and females are working in the company?

20. How much income tax will a single male pay in 2010 if his gross income is \$86,000? He contributed \$3,000 to a tax-deferred IRA (Individual Retirement Account), paid \$21,000 in mortgage interest, \$3,800 in property taxes, and donated \$1,200 to charitable institutions.

21. How much income tax will a single female pay in 2010 if her gross income is \$77,000? She contributed \$2,500 to a tax-deferred 401(k) plan, paid \$12,000 in mortgage interest, \$3,300 in property taxes, and gave \$2,200 in charitable contributions.

22. Suppose that you invested \$30,000 into an account that gave you a 23% increase on its original value. A year after, it drops by 20%. What is your actual percent gain or loss of your original \$30,000 investment?

23. In 1985, the U. S. footwear production was 336,400 thousand of pairs, while the country imported 957,100 thousand of pairs. Twenty-two years later (in 2007), the footwear production fall to 30,660 thousand of pairs, while imports grew to 2,362,347 thousand of pairs. What is the percent of decrease in footwear production in the USA during twenty-two years? By what percent did the import of footwear grow during the same period of time?

Payroll taxes for Social Security benefits are collected under the authority of the Federal Insurance Contributions Act or FICA. Most people simply refer to these taxes as FICA tax. This tax had its origins back in 1935 when the tax was part of the Social Security program. Generally, FICA taxes are collected at a rate of 7.65% on gross earnings—earnings before any deductions. The breakdown of FICA is 6.2% for Social Security (Old-Age, Survivors, and Disability Insurance or OASDI) and 1.45% for Medicare. The following table shows the FICA limits for 2010:

- FICA Tax Rate = 7.65%.

- Social Security Limit = \$106,800.

- Maximum Social Security Contribution = \$6,621.60.

The employee share of FICA (Social Security) is exactly matched by the employer. This means, for instance, that people who are self-employed pay double the tax. Wages over \$106,800 are applied to 1.45% pay the Medicare (Hospital Insurance) tax alone without any wage base limitation. An employee making \$106,800 or less in any given year can expect to pay the full 7.65% on every paycheck of the year, and his or her employer would contribute the same amount. However, let us say you make \$126,800 in a particular year. You will actually see an artificial "raise" of several hundred dollars in your last couple paychecks because you will pay only 1.45% FICA tax on the \$20,000 over the limit instead of 7.65%. In this case, you will see the additional cash $20,000 \times .062 = \$1240$ in hand in the final couple paychecks in aggregate.

24. If you are not self-employed and earn \$70,000 in 2010, what are your FICA taxes?

25. If you are not self-employed and earn \$90,000 in 2010 and have a \$25,000 bonus coming to you, what are your FICA taxes?

26. If you are self-employed and earn \$190,000 in 2010, what are your FICA taxes?

3.2   Simple Interest

If a person has some spare money, he or she may give this money to another person or organization for a certain period of time. The amount of money involved in this transaction is called the **principal**. When the specific period of time expires, the borrower returns the principal and pays a fee to the owner for using this money. The amount of such payment (or fee) is called the **interest**. Therefore, the owner of the money gets back the initial investment together with the interest. This total amount of money the owner will have at the end of a specified period of time is called the **future value**, which can be expressed by the formula

$$F = P + I, \tag{3.2.1}$$

where P is the principal (initial investment) and I is the interest paid by the borrower for using the money.

Paying interest for using somebody's item is typical in human relations. For example, when a person rents a car or apartment, he or she pays interest for using the property. In this chapter, we are only concerned with financial relations, and do not consider rental relations (however, any borrowed item has a monetary value). So we see that the operation of borrowing money (as well as any property) involves two parties—one party that has the money, called the **owner** or the **lender**, and another party that needs the money, called the **borrower**. The owner could be a person, a family, a college, a company, a party, or any organization. Usually the borrower is a trusted person or company to whom the lender gives the money. It could be a friend, a relative, or any company. A well-known organization that is specialized in handling money is called a bank.

The amount of interest to be paid is determined by two parties (the owner and the borrower) before a money transaction. Historically, one of the first and easiest ways to calculate the interest is the **simple interest**. This amount is paid only once at the end of a specified period of time according to the following formula:

simple interest = principal × rate × time.

The **rate** is a proportion of a payment during one year. Then the simple interest is calculated as

$$I = P \times r \times t \qquad \text{or} \qquad I = Prt, \tag{3.2.2}$$

where I is the interest, P is the principal, t is lifespan of the loan, measured in years, and r is the rate measured in decimals. However, in all bank documents, the annual rate is written as a percentage. From Eq. (3.2.1), we derive

$$F = P + I = Prt = P(1 + rt). \tag{3.2.3}$$

The initial investment, P, is also called the **present value** corresponding to the future value, F. From Eq. (3.2.3), we find the principal to be borrowed in order to have the future value:

$$P = \frac{F}{1 + rt}. \tag{3.2.4}$$

We can also determine the rate

$$r = \frac{1}{t}\left(\frac{F}{P} - 1\right),$$ (3.2.5)

and the time needed:

$$t = \frac{1}{r}\left(\frac{F}{P} - 1\right) = \frac{F - P}{r\,P}.$$ (3.2.6)

Par value, in finance and accounting, means stated value or face value on a bill, note, stamp, bond, or any security.

Example 3.2.1. Calculating Future Value

Suppose that \$16,000 is invested at the simple interest rate of 5% per year for 3 months, how much will the borrower pay at the end of the deal?

Here the principal is $P = \$16000$, the annual interest rate is $r = 0.05$, and the time is $t = 3/12 = 1/4$ year. Using the simple interest formula (3.2.2), we get the interest earned to be

$$I = \$16000 \times 0.05 \times 1/4 = \$200.00,$$

and the future value becomes $F = P + I = \$16,200.00$.

Example 3.2.2. Determining the Future Value

A man borrows \$750 from his friend to be repaid in 45 days at a simple interest rate of 10% annually. Hence his friend will get

$$F = P(1 + rt) = \$750(1 + 0.1 \times 9/73) \approx \$759.25.$$

at the end of 45th day because $t = 45/365 = 9/73$, $r = 0.1$, and $P = \$750$.

Example 3.2.3. Finding a Present Value

Suppose a student wants \$20,000 in 4 years; how much should he deposit in a simple interest account paying 3.6% annually?

Using Eq. (3.2.4) with $F = \$20,000$, $r = 0.036$ and $t = 4$, we get

$$P = \frac{\$20000}{1 + 0.036 \times 4} = \frac{\$20000}{1.144} \approx \$17482.52.$$

Example 3.2.4. Determining a Simple Interest Rate

A man borrows from a pawnshop a \$550.00 loan to be repaid after one month in the amount of \$600.00. What is the simple interest rate of this loan?

To answer the question, we use Eq. (3.2.5) to obtain

$$r = \frac{1}{t}\left(\frac{F}{P} - 1\right) = \frac{1}{1/12}\left(\frac{600}{550} - 1\right) = 12\,\frac{5}{55} = \frac{12}{11} = 1.09\overline{09} \quad \text{or} \quad 109.09\%$$

Example 3.2.5. Calculating a Lifespan of the Deal

If \$1,000 is deposited into an account earning 3.5% annually, how long will it take for the account with the simple interest to be worth \$1,500?

Using Eq. (3.2.6), we get

$$t = \frac{1}{r}\left(\frac{F}{P} - 1\right) = \frac{1}{0.035}\left(\frac{1500}{1000} - 1\right) = \frac{5}{0.35} \approx 14.28571429 \text{ years.}$$

Example 3.2.6. Treasury Bills

A United States Treasury security is a government debt issued by the United States Department of the Treasury through the Bureau of the Public Debt. Treasury securities are the debt financing instruments of the United States federal government, and they are often referred to simply as Treasuries. These securities can be sold to anybody in the world, including foreign countries. Today, China and Japan are the primary holders of U.S. securities. There are four types of marketable treasury securities: Treasury bills, Treasury notes, Treasury bonds, and Treasury Inflation Protected Securities (TIPS).

Treasury bills, or **T-bills**, are short-term debt obligations backed by the U.S. government with a maturity of less than one year: they are sold in terms ranging from a few days to 52 weeks. Treasury bills are sold by single price auctions held weekly in increments of \$100. The minimum purchase—effective April 7, 2008—is \$100. (This amount formerly had been \$1,000.) Bills do not pay interest prior to maturity; instead they are sold at a discount of the par value to create a positive yield to maturity. So the interest one earns on a T-bill is the face value minus the purchase price.

One of the biggest reasons why T-bills are so popular is that they are one of the few money market instruments that are affordable to individual investors. T-bills are usually issued in denominations of \$1,000, \$5,000, \$10,000, \$25,000, \$50,000, \$100,000, and \$1 million, and commonly have maturities of one month (four weeks), three months (13 weeks), or six months (26 weeks). Other positives are that T-bills (and all Treasuries) are considered to be the safest investments in the world because the U.S. government backs them. In fact, they are considered risk-free. Furthermore, they are exempt from state and local taxes. The only downside to T-bills is that you would not get a great return because Treasuries are exceptionally safe.

The popularity of T-bills is also due to their simplicity. Essentially, T-bills are a way for the U.S. government to raise money from the public. Treasury bills (as well as notes and bonds) are issued through a competitive bidding process at auctions. If you want to buy a T-bill, you submit a bid that is prepared in one of two ways: either **non-competitively** or **competitively**.

- With a non-competitive bid, you agree to accept the discount rate determined at auction. With this bid, you are guaranteed to receive the bill you want, and in the full amount you want. To place a non-competitive bid, you may use **TreasuryDirect**[1] or **Legacy Treasury Direct**[2], or a bank, broker, or dealer.

- With a competitive bid, you specify the discount rate you are willing to accept. Your bid may be: 1) accepted in the full amount you want if the rate you specify is less than the discount rate set by the auction, 2) accepted in less than the full amount you want if your bid is equal to the high discount rate, or 3) rejected if the rate you specify is higher than the discount rate set at the auction. To place a competitive bid, you must use a bank, broker, or dealer.

However, a bill auction may result in a price equal to par, which means that the Treasury will issue and redeem the securities at par value.

For instance, you might buy a 13-week $1,000 T-bill for $990. When the bill matures, you would be paid $1,000. The difference between the purchase price and face value is your interest, which means that you earn $10 during 13 weeks. To find the annual rate of your investment, we use the formula (3.2.5) with $F = \$1000$, $P = \$990$, and $t = 13/52 = 1/4$ to obtain

$$r = \frac{1}{1/4}\left(\frac{1000}{990} - 1\right) = 4\frac{10}{990} \approx 0.0404 \quad \text{or} \quad 4.04\%.$$

T-bills could be sold and bought before a mature day on auctions that are typically held every Thursday. Bills are issued in electronic form. You can hold a bill until it matures or sell it before it matures. In a single auction, an investor can buy up to $5 million in bills by non-competitive bidding or up to 35% of the initial offering amount by competitive bidding.

For instance, suppose you bought a 13-week $1,000 T-bill for $990 and sell it for $993 in 2 weeks. The annual interest rate of such investment becomes

$$\frac{1}{2/52}\left(\frac{993}{990} - 1\right) = 26 \cdot \frac{3}{990} = 0.07878(78) \quad \text{or} \quad \approx 7.879\%.$$

Example 3.2.7. Treasury Notes

Treasury notes, or T-notes, are issued in terms of 2, 3, 5, 7, and 10 years, and pay interest every 6 months until they mature. The price of a note may be greater than, less than, or equal to the face value of the note. When a note matures, you are paid its face value.

Notes are sold in TreasuryDirect and Legacy Treasury Direct, and by banks, brokers, and dealers. Effective April 2009, TreasuryDirect permits accounts for both individuals and

---

[1]TreasuryDirect is the first and only financial services website that lets you buy and redeem securities directly from the U.S. Department of the Treasury in paperless electronic form. Web site: http://www.treasurydirect.gov/

[2]Legacy Treasury Direct is a program in which investors buy Treasury bills, Treasury notes, and Treasury Inflation-Protected Securities (TIPS) directly from the U.S. Treasury, without a broker. Legacy Treasury Direct charges fees in only two circumstances. If your account holds more than $100,000, you are charged an annual fee of $100. If you want to sell your security before it matures, you are charged $45.

various types of entities including trusts, estates, corporations, partnerships, etc. The yield on a note is determined at auction. Notes are sold in increments of $100, and the minimum purchase is $100. Notes are issued in electronic form. You can bid for a note in either of two ways, similar to T-bills.

Example 3.2.8. Treasury Bonds

Treasury bonds are issued in terms of 30 years and pay interest every 6 months until they mature. When a Treasury bond matures, you are paid its face value.

The price and yield of a Treasury bond are determined at auction. The price may be greater than, less than, or equal to the face value of the bond. Treasury bonds are sold in TreasuryDirect (but not in Legacy Treasury Direct) and by banks, brokers, and dealers. Effective April 2009, TreasuryDirect permits accounts for both individuals and various types of entities including trusts, estates, corporations, partnerships, etc. There are known two types of bids—non-competitive and competitive—similar to T-bills.

Bonds exist in either of two formats: as paper certificates (these are older bonds) or as electronic entries in accounts. Today, Treasury bonds are issued only in electronic form, not paper. Paper bonds can be converted to electronic form. Bonds are sold in increments of $100. The minimum purchase is $100. You can hold a Treasury bond until it matures or sell it before it matures. In a single auction, an investor can buy up to $5 million in bonds by non-competitive bidding or up to 35% of the initial offering amount by competitive bidding.

Example 3.2.9. Interest Rate Earned on an Investment

Suppose that after buying a new car you decide to sell your old car. After advertising about your car, you get a call from a potential buyer who offers you a 180-day note for $5500 at 6% simple interest as payment. (Both principal and interest will be paid at the end of 26 weeks.) Thirteen weeks later, you find that you need the money and sell the note to a third party for $5,555. What annual interest rate will the third party receive for the investment?

Solution. According to the formula (3.2.3), the future value of the note is

$$F = \$5500 \left(1 + 0.06 \times \frac{1}{2}\right) = \$5500 \times 1.03 = \$5665.$$

The interest that the third party will make becomes the difference:

$$\$5665 - \$5555 = \$110.$$

The lifespan of the note is 26 weeks, which is half of the year, so the third party will have the interest for only 13 weeks, or one quarter of the year. Using Eq. (3.2.5) with $F = \$5665$, $P = \$5555$, $t = 1/4$, we get the rate to be

$$r = \frac{1}{1/4} \frac{5665 - 5555}{5555} = \frac{4 \times 110}{5555} \approx 0.0792 \quad \text{or} \quad 7.92\%.$$

3.2.1 Consumer Loans

Since ancient times, people have borrowed money to make purchases. We start with the following example that describes refund loans.

Example 3.2.10. Refund Anticipation Loans
Refund Anticipation Loans (RALs) are loans arranged by tax preparation agencies like H&R Block, Jackson Hewitt, and Liberty Tax Service. They are also sometimes called "rapid refunds," "fast cash refunds," or "express money." A refund anticipation loan (RAL) is a high interest rate short-term loan secured by a taxpayer's expected tax refund, and designed to offer customers quicker access to funds than waiting for their tax refund. Thus, the RAL fee is equivalent to the interest charge for a loan. □

In the modern world, there exists a universal and convenient method to borrow relatively small amounts of money—credit cards. It becomes very convenient for customers to make purchases, especially on the Internet. Once you use a credit card to make a money transaction or purchase, the credit card company bills you for this amount. The main advantage of credit cards consists of the possibility getting fast access to money and extending your payments by making them in small amounts, called **installments**. Customers who do not make cash advances or pay in full the amount borrowed within a month do not pay interest on their debt.

The interest charged on a loan is often called a **finance charge**. There are two main ways to make finance charges depending on whether the number of installments is known or not; we start with the so-called **add-on interest method**. This method is used usually only for relatively short periods of time and when the number of monthly payments is known. A loan given for a fixed period of time is called a **closed-ended credit** agreement. If P is the amount of the loan, I is the interest due on the loan, and n is the number of monthly payments, then

$$\text{monthly payment} = \frac{P+I}{n}. \tag{3.2.7}$$

Example 3.2.11. Payments for an Add–On Interest Loan
Suppose that you take out an add-on loan of $1000 for 18 months at an annual interest rate of 20%. We first use the simple interest formula (3.2.2) to calculate the interest:

$$I = Prt = \$1,000 \times 0.2 \times \frac{18}{12} = \$300.$$

Next, we add the interest to the loan: $P + I = \$1,300$. Then, according to Eq. (3.2.7), we get the monthly payments:

$$\frac{1300}{18} \approx \$72.22. \qquad \square$$

Most credit card companies use so-called **open-ended credit** when the number of install-ments is not determined a priori. However, calculations of finance charges become more complicated. We present the two most popular ways to determine financial monthly charges for open-ended credits. The first method, called the **unpaid balance method**, uses the simple interest formula $I = Pr/12$ because it calculates the interest only for one month. However, the present value, P, in this formula is

previous month's balance + finance charge + purchases made − returns − payments

Note that credit card companies make finance charges only if a customer does not pay off his or her current monthly balance. This means that a customer can use company's money within a month for free if the monthly balance is paid off. So a customer has an option either to pay off the outstanding balance within a grace period with no finance charge or to make a smaller payment and pay off the debt later, but accept finance charges. This may lead to the debt increasing even though a customer is sending a minimum payment every month.

Example 3.2.12. The Unpaid Balance
Suppose that you used a credit card, charging an annual interest rate of 19%, and your unpaid balance at the beginning of last month was $1,200. Since then, you purchased gasoline for $68.46, spent $234.54 for food, and sent in the minimum payment of $120. So your unpaid balance becomes

$$\$1200 + \$68.46 + \$234.54 - \$120 = \$1,383.00.$$

You need to add the finance charge on the last month (which is 1/12 of a year): $1200 \times 0.19 \times \frac{1}{12} = \$228 \times \frac{1}{12} = \$19$. Therefore, at the end of the month you still owe

$$\$1,383 + \$19 = \$1,402.$$

Thus, your $120 payment does not reduce your debt, but instead you have a bigger balance next month. The finance charge for the next month will be $22.20.   □

The previous example illustrates how difficult it is to pay off a large credit card bill. Be aware that taking a cash advance on your credit card may lead to higher debt because those rates are often bigger than regular interest rates.

There exists a more complicated method, called the **average daily balance method**, which is one of the most common ways used by credit card companies to determine finance charges. The method includes two steps:

1. Add the outstanding balance for your account for each day of the month and divide the total by the number of days—the average daily balance.

2. Use the simple interest formula, $I = Prt$, to determine the finance charge, where P is the average daily balance found in step 1, r is the annual interest rate, and t is the number of days in the month divided by 365.

Example 3.2.13. The Average Daily Balance

Suppose that you begin the month of January (which has 31 days) with a credit card balance of $1,000. Assume that your card has an annual interest rate of 19% and that during January the following adjustments are made on your account:

- January 9: A payment of $500 is credited to your account.

- January 11: You charge $11.67 for gasoline.

- January 18: You charge $78.43 for food.

- January 24: You charge $16.74 for gasoline.

- January 28: You charge $88.21 for food.

First, we find the average daily balance for January. The easiest way to calculate the balance is to keep a day-to-day record of what you owe the credit card company. So we make a table:

Day	Balance	Number of Days × Balance
1, 2, 3, 4, 5, 6, 7, 8	$1,000	$8 \times 1000 = \$8,000$
9, 10	$500	$2 \times 500 = \$1,000$
11, 12, 13, 14, 15, 16, 17	511.67	$7 \times 511.67 = \$3,581.69$
18, 19, 20, 21, 22, 23	$590.1	$6 \times 590.1 = \$3,540.6$
24, 25, 26, 27	$606.84	$4 \times 606.84 = \$2,427.36$
28, 29, 30, 31	$695.05	$4 \times 695.05 = \$2,780.20$

Adding the numbers in the last column and dividing the result by 31, we get the average daily balance to be

$$\frac{8000 + 1000 + 3581.69 + 3540.6 + 2427.36 + 2789.2}{31} = \frac{21329.85}{31} \approx \$688.06.$$

We next apply the simple interest formula:

$$I = Prt = \frac{21329.85}{31} \times 0.19 \times \frac{31}{365} = \frac{21329.85 \times 0.19}{365} \approx \$11.10.$$

Your finance charge on the February statement will be $11.10.

Example 3.2.14. Comparing Two Financial Methods

Consider the previous example with given financial adjustments. Since we know the finance charge from Example 3.2.13 based on the average daily balance, we calculate charges using the unpaid balance method.

First, we calculate the finance charge on the previous balance: $1000 \times 0.19/12 = \$15.83$. Then we calculate P:
$P = 1000 - 500 + 11.67 + 78.43 + 16.74 + 88.21 + 15.83 = \$710.88.$

Then we evaluate the interest: $I = Prt = 710.88 \times 0.19 \frac{1}{12} \approx \$11.26$.
With the unpaid balance method, the finance charge is $11.26, while with the average daily balance method the charge is $11.10.

Example 3.2.15. Discounted Loans

Some lenders collect the interest from the amount of the loan at the time that the loan is made. A loan on which the interest and finance charges are deducted from the face amount (par value) when the loan is issued is called the **discounted loan**. The borrower only receives the principal after the finance charges and interest are taken out but must repay the full amount of the loan. Discount loans are often issued when the borrower desires nothing more than a short-term loan (usually, no longer than one year).

Discount loans are normally taken by borrowers who need resources quickly to cover expenses in the near future. For the lender, the discount loan is also beneficial since this type of loan does not usually allow for breaks on the interest charges applicable to the loan, and it has the assurance that the loan can reasonably be repaid in full in a short period of time.

For example, you choose a discounted loan for the amount of $50,000 for 10 months. If the interest and finance charges were $5,000, you would receive $45,000 from the lender, but still have to pay back the whole $50,000. The loan's simple interest rate can be calculated using Eq. (3.2.5) with $F = 50,000$, $P = 45,000$, and $t = 10/12 = 5/6$:

$$r = \frac{6}{5}\left(\frac{50000}{45000} - 1\right) = \frac{6}{5} \cdot \frac{5000}{45000} = \frac{6}{5} \cdot \frac{1}{9} = \frac{2}{15} = 0.1\overline{3} \quad \text{or} \quad 13.33\%.$$

Exercises for Section 3.2.

1. In the following exercises, find the simple interest, I, earned on the given principal, P, for the time period, t, and interest rate, r, specified.
   (a) $P = \$10,000$, $r = 3.6\%$, $t = 4$ years;   (b) $P = \$2,500$, $r = 2.75\%$, $t = 4$ years;
   (c) $P = \$500$, $r = 6.25\%$, $t = 15$ months;   (d) $P = \$750$, $r = 6.75\%$, $t = 30$ months;
   (e) $P = \$2,500$, $r = 4.75\%$, $t = 45$ days;   (f) $P = \$1,750$, $r = 3.275\%$, $t = 90$ days.

2. In the following exercises, find the future value, F, of the given principal, P, earning simple interest for the time period, t, and interest rate, r, specified.
   (a) $P = \$20,000$, $r = 3.6\%$, $t = 4$ years;   (b) $P = \$5,000$, $r = 2.75\%$, $t = 4$ years;
   (c) $P = \$5,500$, $r = 6.2\%$, $t = 15$ months;   (d) $P = \$850$, $r = 5.7\%$, $t = 32$ months;
   (e) $P = \$7,250$, $r = 4.25\%$, $t = 20$ weeks;   (f) $P = \$1650$, $r = 2.85\%$, $t = 28$ weeks;
   (g) $P = \$7,000$, $r = 8.125\%$, $t = 15$ days;   (h) $P = \$575$, $r = 8.375\%$, $t = 42$ days.

3. In the following exercises, find the present value, P, under simple interest given the future value, F, time period, t, and interest rate, r, specified.
   (a) $F = \$22,000$, $r = 4.6\%$, $t = 5$ years;   (b) $F = \$7,000$, $r = 3.75\%$, $t = 5$ years;
   (c) $F = \$500$, $r = 6.2\%$, $t = 20$ months;   (d) $F = \$750$, $r = 6.7\%$, $t = 10$ months;
   (e) $F = \$5,000$, $r = 5.75\%$, $t = 10$ weeks;   (f) $F = \$750$, $r = 7.75\%$, $t = 30$ weeks;
   (g) $F = \$7,500$, $r = 7.125\%$, $t = 45$ days;   (h) $F = \$7,650$, $r = 8.775\%$, $t = 30$ days.

4. In the following exercises, find the time t, in years, for the given principal, P, to reach the future value, F, under simple interest for the interest rate, r, specified.

    (a) $F = \$33,000$, $P = \$30,000$, $r = 4.6\%$;   (b) $F = \$7,700$, $P = \$7,000$, $r = 3.75\%$;

    (c) $F = \$600$, $P = \$550$, $r = 6.215\%$;   (d) $F = \$750$, $P = \$700$, $r = 6.725\%$;

    (e) $F = \$25,500$, $P = \$25,000$, $r = 7.5\%$;   (f) $F = \$17,500$, $P = \$17,000$, $r = 8.7\%$.

5. In the following exercises, find the annual interest rate, r, for the given principal, P, to reach the future value, F, under simple interest for the time period, t, specified.

    (a) $F = \$33000$, $P = \$30000$, $t = 1$ year;   (b) $F = \$7700$, $P = \$7000$, $t = 1$ year;

    (c) $F = \$5500$, $P = \$5000$, $t = 8$ months;   (d) $F = \$750$, $P = \$700$, $t = 8$ months;

    (e) $F = \$600$, $P = \$550$, $t = 45$ days;   (f) $F = \$50$, $P = \$45$, $t = 40$ days.

6. Suppose a loan of \$9 at the beginning of the week should be repaid with \$10 at the end of the week. What annual rate of simple interest did the loan charge?

7. Suppose a woman borrows \$1,700 from a friend, agreeing to pay it back in 15 months with 3.5% simple interest. How much will the woman owe her friend at the end of 15 months?

8. Five years ago, a woman made an investment paying 8.375% simple interest. If her account now contains \$23,389.44, how much did she originally invest?

9. If \$27,000 is invested at 7.825% simple interest, how long will it take for the investment to be worth \$30,000?

10. Suppose you borrow \$200 from a pawn shop charging 200% simple interest. How much would you owe if you paid back the loan after 40 days?

11. A \$5,000 investment in Toyota Motors Corporation in 2000 rose in value to \$7,389 in 2008. What was the rate of return, figured as an annual interest rate, for this investment?

12. A woman borrowed \$325 from her parents, agreeing to pay them back \$350 in 3 months. What annual simple interest rate did she pay?

13. Congress borrowed \$548 million at 5.25% simple interest. How much would it have to pay in 7 years?

14. A woman received a salary bonus of \$4,500. She needs \$5,000 for the down payment on a car. If she can invest her bonus into an account paying a simple interest of 3.785%, how long must she wait until she has enough money to make the down payment?

15. At the University of Rhode Island in 2009, if a student did not pay tuition and fees on time, the student's account would be assessed a monthly late-payment fee of 12.5% annual simple interest. If a student owed \$6,142 and the payment was 3 months late, what is the total amount the student would have to pay?

16. Suppose some bonds pay 2.37% simple interest. How much should you invest in the bonds if you want them to be worth \$30,000 in 5 years?

17. Use the simple interest formula $I = Prt$ and elementary algebra to find the missing quantities in the table below.

I	P	r	t
	\$15,000	2.5%	7 years
\$960		3.2%	3 years
\$272	\$5,000		4 years
\$80	\$1,000	4%	

18. Use the future value formula $F = P(1 + rt)$ and elementary algebra to find the missing quantities in the table below.

F	P	r	t
	$21,000	2.4%	5 years
$21,736		3.6%	4 years
$19,350	$18,000		3 years
$16,680	$15,000	1.6%	

19. You plan to take a trip to Cancun in three years. You want to save $1,500 for this trip by depositing $1,200 into an account that pays simple interest. What annual interest rate are you looking for?

20. You have borrowed $2,000 from your friend to be paid in 2 years in the amount of $2,500. What is the simple interest rate that corresponds to this deal?

21. Ashley purchased a bond for $4,900 and 8 months later she sold it for $4,970. What annual rate, calculated using simple interest, did she earn on this transaction?

In Problems 22 – 25, use the unpaid balance method to find the finance charge on the credit card account. Last month's balance, the payment, the annual interest rate, and any other transactions are given.

22. Last month's balance, $673.8; payment, $255; interest rate, 18%; bought jacket $112.39; returned shoes, $99.99.

23. Last month's balance, $2180; payment, $525; interest rate, 19.24%; bought sweater, $87.99, and gasoline, $48.39; returned books, $102.

24. Last month's balance, $1207.38; payment, $725; interest rate, 11.24%; bought clothes $48.99; returned sweater, $87.99.

25. Last month's balance, $1672.43; payment, $550; interest rate, 12.8%; bought basketball tickets $80; returned clothes, $48.99.

In Problems 26 – 29, use the average daily balance method to find the finance charge on the credit card account. The starting balance and transactions on the account for the month are given. Assume an annual interest rate of 20% in each case.

26. Month: February (28 days); previous month's balance: $985.

Date	Transaction
February 7	Made payment of $280
February 14	Charged $17.66 for gasoline
February 19	Charged $49.99 for clothes
February 25	Charged $82.34 for food

27. Month: March (31 days); previous month's balance: $627.34.

Date	Transaction
March 6	Charged $50 for theater
March 12	Made payment of $300
March 17	Charged $178.29 for ski vacation
March 23	Charged $58.72 for gasoline

28. Month: April (30 days); previous month's balance: $709.31.

Date	Transaction
April 5	Charged $61.08 for gasoline
April 10	Charged $40 for hockey ticket
April 14	Made payment of $250
April 22	Charged $120 for shoes

29. Month: May (31 days); previous month's balance: $1507.39.

Date	Transaction
May 4	Made payment of $500
May 9	Charged $50 for baseball ticket
May 17	Charged $47.41 for gasoline
May 26	Charged $128.56 for food

30. Alisha purchased a laptop for $899.99 and can pay it off in 10 months with an add-on interest loan at an annual rate of 18%, or she can use her credit card that has an annual rate of 19%. If she uses her credit card, she will pay $90 per month (beginning next month) plus the finance charges for the month. Assume that Alisha's credit card company is using the unpaid balance method to compute her finance charges and that she is making no other transactions on her credit card. Which option will have the smaller total finance charges on her loan?

31. Boris purchased an entertainment center for $8,428.99, the rate for the add-on loan is 3.6%, and he is paying off the loan in 24 months. How much interest will he pay? What are his monthly payments?

32. Jeffrey took out an add-on interest loan for $1,149 to buy a new desktop computer. The loan will be paid back in 2 years, and the annual interest rate is 4.5%. How much interest will he pay? What are his monthly payments?

33. Angela's bank gave her a 3-year add-on interest loan for $7,850 to pay for new heating equipment. The annual interest rate is 3.5%. How much interest will she pay? What are her monthly payments?

34. Anna is buying $25,000 worth of rare coins as an investment. The dealer is charging her an annual interest rate of 3.25% and is using the add-on method to compute her monthly payments. If Anna pays off the coins in 32 months, what are her monthly payments?

35. Ben is buying a new swimming pool for $1,539.99. The dealer is charging him an annual interest rate of 3.15% and is using the add-on method to compute his monthly payments. If Ben pays off the pool in 40 months, what are his monthly payments?

36. Patricia is buying a music synthesizer for her rock band for $1,249.95. The music store is charging her an annual interest rate of 3.35% and is using the add-on method to compute her monthly payments. If Anna pays off the synthesizer in 30 months, what are her monthly payments?

37. Prof. Pakula is buying a new boat for $29,900. The dealer is charging him an annual interest rate of 2.45% and is using the add-on method to compute his monthly payments. If Prof. Pakula pays off the boat in 48 months, what are his monthly payments?

38. An investor purchases 50 shares at $18.38 a share, holds the stocks for a year, and then sells the stocks for $19.08 a share. What is investor's interest earned on this transaction? What is the rate of such investment?

39. An investor purchases 77 shares at $43.57 a share, holds the stocks for a year, and then sells the stocks for $45.32 a share. What is investor's interest earned on this transaction? What is the rate of such investment?

40. A client receives a $875 RAL, which is paid back in 35 days. What is the annual interest rate for this loan if he paid a $39.00 fee?

41. A client receives a $1855 RAL, which is paid back in 25 days. What is the annual interest rate for this loan if she paid a $69.00 fee?

42. You borrow $10,000 on an 11% discounted loan for a period of 1 year. Determine the loan's discount and the actual interest rate.

43. In order to start a small business, a person takes out a discounted loan for $5,000 for 10 months at a rate of 9.25%. How much interest must the person pay? Find the actual interest rate.

44. In order to pay for dental work, you borrow $2,200 on a 9.25% discounted loan for 3 months. What is the loan's discount? What is the loan's actual interest rate?

45. You buy a T-bill for $990.85 that pays $1,000 in 13 weeks. What simple interest rate, to the nearest tenth of a percent, does this T-bill earn?

46. To borrow money, you pawn your camera. Based on the value of the camera, the pawnbroker loans you $450. One month later, you get the camera back by paying the pawnbroker $550. What annual interest rate did you pay?

47. Rachel finds a company that charges 46¢ per day for each $1,000 borrowed. If she borrows $5,000 for 3 months, what amount will she repay, and what annual interest rate will she be paying the company?

48. For services rendered, an attorney accepts a 60-day note for $3,550 at 5.75% simple interest from a client. (The future value, including the principal and the interest will be repaid at the end of 60 days.) After 30 days, the attorney sells the note to a third party for $3,555. What annual interest rate will the third party receive for the investment?

49. Is it more profitable to double your money in 12 years or put it into an account paying 6.25% simple interest?

50. Suppose some municipal bonds pay 3.25% simple interest. How much should you invest in the bonds if you want them to be worth $10,000 in 10 years?

51. If you deposit some money in a bank account paying 6.5% annual (simple) interest, how long should you keep the account in order to double your investment?

52. To help your friend, you accept his note promising to pay you back the borrowed amount of $3,000 plus a simple interest of 9% in 3 months. After a month, you sell the note to a third party for $3,020. What annual interest rate will the third party receive for the investment?

53. You bought a T-bill of $5,000 for $4,975. What simple interest did you earn in 26 weeks when the bill matured?

54. A T-bill of $5,000 that you purchased for $4,970 was sold twice. The first time you sold it to your friend for $4,979 in a week, and then your friend sold it to a third party for $4,989 in 2 weeks. What simple interest will this third party earn when the bill matures in two weeks?

3.3 Compound Interest

A simple interest investment has one drawback—it pays back only once when the loan is terminated. People do not want to keep such investments for a long time because they could withdraw their money and deposit it again by earning interest on a previous investment.

For example, a man has $P = \$50,000$ to invest into an account under simple interest earning 6% annually for 3 years. The future value becomes

$$F = P(1 + rt) = 50000(1 + 0.06 \times 3) = 50000 \times 1.18 = \$59,000.$$

On the other hand, if after one year the money will be withdrawn, its value would be

$$F_1 = P(1 + r) = 50000(1 + 0.6) = 50000 \times 1.06 = \$53,000.$$

Redepositing this money for one year again yields

$$F_2 = F_1(1 + 0.06) = 53000 \times 1.06 = \$56,180.00.$$

Putting this amount into the account for the next year, the final investment becomes

$$F_3 = F_2 \times 1.06 = \$59,550.80.$$

So we see that such 3 investments followed by 3 withdrawals increase the final outcome by $550.80.

It is no surprise that people do not usually use a simple interest investment unless it is made for a short period of time. Now all banks offer savings accounts for which interest is paid at regular intervals. This means that interest is paid not only on the initial principal, but also on any previous interest earned. This kind of interest is called **compound interest**.

Let us derive the formula to calculate F_3, the future value of the initial investment of $P = \$50,000$. After the first year we got $F_1 = P(1 + r) = 50000(1 + 0.6) = \$53,000$, which was reinvested again under the same conditions (simple interest). This yields $F_2 = F_1(1 + r) = p(1 + r)(1 + r) = P(1 + r)^2 = 50000(1 + 0.6)^2 = \$56,180$. After the third year, we get $F_3 = F_2(1 + r) = P(1 + r)^3 = 50000(1.06)^3 = \$59,550.80$.

In a similar way, we derive the general formula: $F = P(1 + r)^t$, where P is the principal initially invested, r is the annual rate of interest compounded annually, and F is the future value after t years.

In the general situation, interest may be credited more often than once per year. Suppose that in our example, the initial investment of $P = \$50,000$ was made for half a year, so $t = 1/2$. The future value of this principal is $F_{1/2} = P(1 + r/2) = 50000(1 + 0.3) = \$51,500.00$. If the amount of $F_{1/2}$ will be deposited into the same account, it would be worth

$$F_1 = F_{1/2}(1 + r/2) = P(1 + r/2)^2 = 50000(1 + 0.3)^2 = 50000 \times 1.0609 = \$53045.00$$

after one year. If the interest is compounded monthly, then after the first month the investment of $P = \$50,000$ will earn the interest of $I_{1/12} = P r \frac{1}{12}$, which in our case will be $I_{1/12} = 50000 \times 0.06/12 = \$250$. So the account will be worth $F_{1/12} = P + I_{1/12} = P\left(1 + \frac{r}{12}\right) = \$50,250.00$. Depositing $F_{1/12}$ for the next month (with the same annual rate) will generate the interest of $I_{2/12} = F_{1/12} \times r/12 = 50250 \times 0.06/12 = \$251.25$, which yields the future value $F_{2/12} = F_{1/12} + I_{2/12} = F_{1/12}\left(1 + \frac{r}{12}\right) = P\left(1 + \frac{r}{12}\right)^2$. In our case, it becomes $F_{2/12} = \$50,501.25$. Similarly, after three months, the future value will be $F_{3/12} = P\left(1 + \frac{r}{12}\right)^3$, and after m months we get

$$F_{m/12} = P\left(1 + \frac{r}{12}\right)^m.$$

If the annual interest rate, r, is compounded n times per year, then the **periodic interest rate** is given by

$$\text{periodic interest rate} = \frac{r}{n}.$$

For instance, if interest is compounded quarterly, $n = 4$, if it is compounded monthly, $n = 12$, if weekly—$n = 52$, and if daily—$n = 365$.

The amount in the account compounded n times per year that is held during t years is given by

$$F = P\left(1 + \frac{r}{n}\right)^{nt}, \qquad \text{where} \quad \begin{cases} F = \text{future value after } t \text{ years,} \\ P = \text{principal or present value,} \\ r = \text{annual interest rate,} \\ n = \text{number of periods per year,} \\ t = \text{time (in years).} \end{cases} \tag{3.3.1}$$

From Eq. (3.3.1), we find the principal

$$P = F\left(1 + \frac{r}{n}\right)^{-nt}, \tag{3.3.2}$$

the annual interest rate

$$r = n\left[\left(\frac{F}{P}\right)^{1/nt} - 1\right], \tag{3.3.3}$$

and the time (in years)

$$t = \frac{\ln \frac{F}{P}}{n \ln\left(1 + \frac{r}{n}\right)} = \frac{\ln F - \ln P}{n \ln\left(1 + \frac{r}{n}\right)} = \frac{1}{n}\log_{\left(1 + \frac{r}{n}\right)}\left(\frac{F}{P}\right). \tag{3.3.4}$$

Note that the product, nt, is the total number of periods (= number of times the interest is paid).

Example 3.3.1. Comparing Interest for Various Compounding Periods
Suppose $50,000 is deposited in an account earning 6% interest compounded quarterly. After 6 years, there will be

$$F = P\left(1 + \frac{r}{n}\right)^{nt} = 50000\left(1 + \frac{0.06}{4}\right)^{4\times 6} = \$71,475.14$$

in the account because $n = 4$, $r = 0.06$, and $t = 6$ in Eq. (3.3.1). This value can be obtained using the calculator:

$$50000 \times (1 + 0.06 \div 4) \wedge (4 \times 6)$$

Suppose that the interest is paid either monthly ($n = 12$) or weekly ($n = 52$). The future value for $n = 12$ becomes $F = 50000\left(1 + \dfrac{0.06}{12}\right)^{12\times 6} = \$71,602.21$, and for $n = 52$ it is

$$F = 50000\left(1 + \frac{0.06}{52}\right)^{52\times 6} = \$71651.59.$$ Therefore, we see that the difference in earning between weekly payments and monthly payments is $49.38, and the account where interest is compounded monthly has $127.07 more than the account with quarterly payments. □

What would happen to the amount if interest were compounded more frequently, say daily, or every minute, or every second ($n > 1000$)? As we see in Example 3.3.1, the differences in amounts suggest that as the number n of compounding periods per year increases without bound, the amount will approach some limiting value. To see that this is indeed the case, we rewrite the future value as follows:

$$F = P\left(1 + \frac{r}{n}\right)^{nt} = P\left(1 + \frac{r}{n}\right)^{(n/r)\,rt} = P\left(1 + \frac{1}{x}\right)^{xrt}$$

$$= P\left[\left(1 + \frac{1}{x}\right)^{x}\right]^{rt}, \qquad \text{where } x = \frac{n}{r}.$$

When the number n of compounding periods increases without bounds, so does x. As Leonhard Euler showed in 1748, the expression in square brackets gets close to the irrational number, denoted by e. This allows us to write the limiting expression as

$$F = P\,e^{rt}, \qquad e = \lim_{x\to\infty}\left(1 + \frac{1}{x}\right)^{x} = 2.718281828\ldots, \tag{3.3.5}$$

where F is the future value after t years when the principal, P, was initially deposited, and r is the annual interest rate. In this case, we say that the interest is **compounded continuously**.

By comparing the simple interest formula, $F = P\left(1 + rt\right)$ with the compounded interest formula, $F = P\left(1 + \dfrac{r}{n}\right)^{nt}$, we see that the former is a linear function in t, and the latter is a power function. Since a power function grows faster than a linear function, the compound interest curve always eventually outpaces simple interest, no matter what the simple interest rate is. Similarly, since the exponential function in Eq. (3.3.5) always majorates the polynomial function, the interest compounded continuously would be larger than any interest no matter how often it is compounded.

Example 3.3.2. Compounding Daily and Continuously

Following Example 3.3.1, we consider an investment of \$50,000 in two accounts for 5 years: one pays interest daily, and the other one pays interest continuously. Using the compounded interest formula (3.3.1) with $P = 50000$, $r = 0.06$, $n = 365$, and $t = 5$, we get

$$F_{365} = \$50000 \left(1 + \frac{0.06}{365}\right)^{365 \times 5} = \$50000 \left(1 + \frac{0.06}{365}\right)^{1825} \approx \$67,491.27 \,.$$

The future value, F_c, of the account that pays the same interest of 6% continuously is

$$F_c = 50000\, e^{0.06 \times 5} = 50000\, e^{0.3} \approx \$67,492.94 \,.$$

Example 3.3.3. Long-term Investment

The following story is said to be true. George Washington, the first president of the United States, borrowed \$24 from a man to be repaid with the interest of 8% compounded annually. Now the descendants of the man try to collect the money. What is the amount of their claim, if we can assume that it has happened 250 years ago?

To answer this question, we use Eq. (3.3.1):

$$F = 24\,(1 + 0.08)^{250} = \$5,446,908,922$$

or around five and a half billion dollars.

Example 3.3.4. Inflation

Our next example discusses a very important topic—**inflation**, which is an increase in the prices and services as time elapses. Economists use the term "inflation" to denote an ongoing rise in the general level of prices quoted in units of money during a certain period of time. The magnitude of inflation—the inflation rate—is usually reported as the annualized percentage growth of some broad index of money prices. With U.S. dollar prices rising, a one-dollar bill buys less each year. Inflation thus means an ongoing fall in the overall purchasing power of the monetary unit.

Inflation rates vary from year to year and from currency to currency. In recent years, Japan has experienced negative inflation, or "deflation," of around 1 percent per year, as measured by the Japanese Consumer Price Index (CPI). Central banks in most countries today profess concern with keeping inflation low but positive. Some specify a target range for the inflation rate, typically 1 – 3 percent.

In the United States, the inflation rate is most commonly measured by the percentage rise in the Consumer Price Index (CPI), which is reported monthly by the Bureau of Labor

Statistics (BLS). Since 1950, the U.S. dollar inflation rate, as measured by the December-to-December change in the U.S. CPI, has ranged from a low of -0.7 percent—deflation in 1954—to a high of 13.3 percent (1979). Starting in 1991, the rate has stayed between 1.6 percent and 3.3 percent per year. A CPI of 120 in the current period means that it now takes $120 to purchase a representative basket of goods that $100 once purchased. Because the CPI basket is not identical with the specific basket of goods and services that you consume, the percentage rise in the CPI is, at best, only a rough approximation of the percentage rise in your cost of living. The history of the United States has seen a series of fluctuations in the inflation rate, especially around 1915 when the inflation rate was about 17%, followed by deflation in 1920s.

Inflation works just as compound interest because prices on goods and services increase with time. If the annual rate of inflation is r, then, on average, we expect that next year a consumer will pay more for the same product by $P(1 + r)$, where P is the present value.

Example 3.3.5. Car's Price Growth

Now we consider a particular example involving inflation. Suppose the rate of inflation for the next 5 years is projected to be 3.5% per year. If a new car price is $19,320, to purchase it today will cost $19320\,(1 + 0.07) = \$20,672.40$ assuming the local tax to be 7%. Next year, the same car will cost about $19320\,(1 + 0.035) = \$19996.20$, so adding 7% tax, a man should pay $19320\,(1 + 0.035)(1 + 0.07) = \$21395.93$ or $723.53 more than this year.

In 5 years, the price of this car is projected to be $\$19,320\,(1 + 1.035)^5 = \$22,946.10$. However, 5 years ago, a consumer paid about $\$19,320\,(1 + 1.035)^{-5} = \$16,266.92$ for the same car.

Example 3.3.6. Hyperinflation

Hyperinflation is an inflation that is very high or "out of control," a condition in which prices increase rapidly as a currency loses its value. There is no precise numerical definition to hyperinflation. The main cause of hyperinflation is a massive and rapid increase in the amount of money that is not supported by a corresponding growth in the output of goods and services. This results in an imbalance between the supply and demand for the money, accompanied by a complete loss of confidence in the money.

Hyperinflation is generally associated with paper money because this can easily be used to increase the money supply: add more zeros to the plates and print, or even stamp old notes with new numbers. Historically there have been numerous episodes of hyperinflation in various countries that suffered from runaway inflation. In 1923, German banknotes had lost so much value that they were used as wallpaper. An inflation rate averaging over 10% per day in Hungary in 1945 caused the largest denomination of currency ever circulated. Since 1950 at least 18 countries have experienced episodes of hyperinflation, in which the CPI inflation rate has soared above 50 percent per month. More recent examples of hyperinflation give the dollarization in Ecuador (now the country circulates its own coins, but the currency in use is U.S. dollar), initiated in September 2000 in response to a massive

75% loss of value of the Sucre currency, and the Malagasy franc of Madagascar had a turbulent time in 2004, losing nearly half its value. Early in the 21st century Zimbabwe started to experience chronic inflation. Inflation reached 624% in 2004, then fell back to low triple digits before surging to a new high of 1,730% in 2006. During that time, the Reserve Bank of Zimbabwe revalued its currency on 1 August 2006 at a rate of 1,000 old Zimbabwean dollars to 1 revalued Zimbabwean dollar. In June 2007 inflation in Zimbabwe had risen to 11,000% year-to-year.

The nature of the hyperinflation in Russia in the 1990's was due to the expansion of the money supply by the Central Bank of Russia. During 1993, Russia had inflation of 874.62% in consumer prices. Such hyperinflation was caused by offering bad loans and printing money by the Central Bank, which led to removal of the market equilibrium prices.

When the money supply expands in excess of the real growth of the economy there is an increase in the price level—inflation. In the case of Russia at the time there was no real growth in the economy, it was in fact contracting. If the money supply expands 5 percent and production is contracting 20 percent this is an expansion of the money supply by about 25 percent per unit of output and there would be correspondingly high impact on the price level. When inflation develops the velocity of money increases and thus the impact on the rate of inflation is due to the sum of the rate of growth of the money supply and the rate of increase of the velocity of money. This is how inflation rates can be driven to astronomical levels leading to the Russian default on its debt in 1998. The consumer prices grew from 100% in 1993 to almost 20000% in 1998.

The hyperinflation in Russia in the 1990's followed the standard scenario. The blame for the social devastation which resulted from the hyperinflation such as the impoverishment of the pensioners and the destruction of the value of lifetime savings lies with the monetary policies of the head of the Central Bank, who was replaced in 1998. After that, the hyperinflation quickly disappeared although significant inflation continued.

Example 3.3.7. Saving for College

A couple wants to invest money for their child's college education. If they want to have $120,000 in 14 years and the investment pays about 5.5% compounded monthly, how much should they deposit?

Using Eq. (3.3.2), we obtain

$$P = F\left(1 + \frac{r}{n}\right)^{-nt} = 120000\left(1 + \frac{0.055}{12}\right)^{-12 \times 14} = \$55,659.40.$$

You get the answer by typing in your calculator:

$120000 \times (1 + 0.055 \div 12) \wedge ((-)12 \times 14)$ or $120000 \div (1 + 0.055 \div 12) \wedge (12 \times 14)$

Example 3.3.8. Finding Time

In some situations, the amount you have to deposit is fixed as well as the interest rate, but you have a goal in mind—a particular amount as the result of your investment. How long should you keep up this investment rate in order to achieve your goal?

Suppose $50,000 is invested with interest compounded monthly with an annual rate of 3.75%, how long will it take for the account to be worth $55,000? Using Eq. (3.3.4), we get

$$t = \frac{\ln \frac{F}{P}}{n \ln \left(1 + \frac{r}{n}\right)} = \frac{\ln \frac{55}{50}}{12 \ln \left(1 + \frac{0.0375}{12}\right)} \approx 2.545573987 \text{ years or } 30.5 \text{ months.}$$

Using a calculator, you can type:

$$\ln(55 \div 50) \div 12 \div \ln(1 + 0.0375 \div 12) \quad \text{ENTER.}$$

Example 3.3.9. Finding the Interest Rate

Suppose you have $50,000 to be invested for 4 years, and you are shopping around for the best investment. Different banks offer different options, so you try to find the best rate that earns at least $5,000 in 4 years. Assuming that banks offer interests compounded monthly, what annual rate satisfies your constraint?

Solution: Using Eq. (3.3.3) with $n = 12$, $F = 55,000$, $P = 50,000$, and $t = 4$, we get

$$r = 12 \left[\left(\frac{54000}{50000}\right)^{1/48} - 1 \right] \approx .02385122 \quad \text{or} \quad 2.385\%.$$

Hence, you may want to choose a rate of at least 2.385%. This result can be obtained by typing in your calculator:

$$12 \times ((55000 \div 50000) \wedge (1 \div 12 \div 4) - 1) \quad \text{ENTER.}$$

Example 3.3.10. Reduction of Player's Taxes

Suppose that a football player expects to retire after playing one more year. His agent is negotiating a contract, and he agrees to defer a bonus of $2.5 million in order to reduce the player's taxes. The agent, who is paid 20% of the contract amount, suggests to accept the bonus of $2.65 million in 2 years. At what annual interest rate that is compounded monthly should the amount of $2.5 be invested in order to generate $2.65 million in two years?

To answer this question, we use Eq. (3.3.4) to obtain

$$r = 12 \left[\left(\frac{2.65}{2.5}\right)^{1/24} - 1 \right] \approx 0.02916985$$

or 2.917%.

Example 3.3.11. Double the Initial Investment

When comparing offers for possible investment offered by different banks or unions, people sometimes are confused with financial terminology, corresponding rates, and conditions on their investments. There exists another way to compare such offers understandable to almost everyone: doubling time. This means that the profitability of your investment can be compared by the time needed to double your initial investment.

Suppose that you put P amount of money into an account paying interest at a rate r compounded n times per year. To find the time period that will be enough to double the initial investment, we use Eq. (3.3.1) with $F = 2P$ to obtain

$$2P = P\left(1 + \frac{r}{n}\right)^{nt} \quad \text{or} \quad 2 = \left(1 + \frac{r}{n}\right)^{nt}.$$

Applying logarithms to both sides, we have

$$t_{\text{double}} = \frac{\ln 2}{n \ln\left(1 + \frac{r}{n}\right)}. \tag{3.3.6}$$

If, instead, you would like to determine the rate that allows you to double your investment, we raise both sides of the equation $2 = \left(1 + \frac{r}{n}\right)^{nt}$ to the power $1/(nt)$ to obtain

$$2^{1/(nt)} = 1 + \frac{r}{n} \quad \Longrightarrow \quad r_{\text{double}} = n\left[2^{1/(nt)} - 1\right]. \tag{3.3.7}$$

For instance, if you deposit into an account that pays 6% interest compounded monthly, you should wait about 11 and a half years to double your initial investment because

$$t = \frac{\ln 2}{12\left(1 + \frac{0.06}{12}\right)} = \frac{\ln 2}{12 \ln 1.005} \approx 11.58.$$

If you want to get twice more money in 10 years, you need to find an account with a rate at least

$$r = 12\left[2^{1/(12\times 10)} - 1\right] = 12\left[2^{1/(120)} - 1\right] \approx 0.0695 \quad \text{or} \quad r = 6.95\%. \qquad \square$$

For an account paying interest compounded continuously, the doubling time is

$$t_{\text{double}} = \frac{1}{r} \ln 2. \tag{3.3.8}$$

For an account with an annual interest rate r compounded n times per year there exists a simple interest rate that would make the investment under this simple interest worth the same amount as the compound interest deposit (see Figure 3.1 on page 285). Such simple interest is called the **annual percentage yield** (APY), given by

$$\text{APY} = \left(1 + \frac{r}{n}\right)^n - 1 \quad \Longrightarrow \quad r = n\left[(1 + \text{APY})^{1/n} - 1\right]. \tag{3.3.9}$$

For an account paying interest compounded continuously the annual percentage yield is evaluated as

$$\text{APY} = e^r - 1. \tag{3.3.10}$$

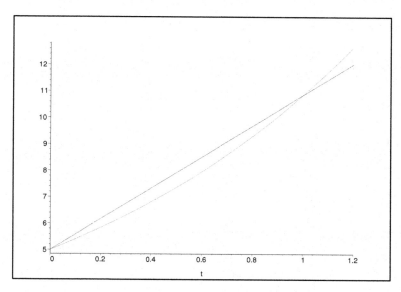

Figure 3.1: Annual percentage yield versus compound interest.

The APY is sometimes referred to as the annual effective yield or annual equivalent yield. Some banks offer distinct rates during years for a deposit to make promotions for customers. The accumulative APY is called the **blended APY**, and it is explained in detail in the following example.

Commercial banks and credit unions commonly offer to consumers a savings certificate entitling the bearer to receive interest. This promissory note is called a **certificate of deposit** or CD. It is a time deposit that restricts holders from withdrawing funds on demand. Although it is still possible to withdraw the money, this action will often incur a penalty.

A CD bears a maturity date and a specified fixed interest rate and can be issued in any denomination. CDs are insured by the FDIC (Federal Deposit Insurance Corporation) for banks or by the NCUA (National Credit Union Administration) for credit unions. Thus, they are virtually risk-free because of the insurance, which guarantees the safety of deposits in member banks, currently up to $250,000 per depositor per bank.

Example 3.3.12. Blended APY
In 2009, Citizens bank offered a 3-year certificate of deposit (CD) with distinct rates for each year—the first year it offered 1.50%, the second year—1.75%, and the third year—2.0%, each compounded daily.

Suppose that you deposit $50,000 in such a 3-year CD account. After the first year, your account would be worth

$$F_1 = 50000 \left(1 + \frac{0.015}{365}\right)^{365} \approx \$50,755.63 \,.$$

The next year you earn more:

$$F_2 = F_1 \left(1 + \frac{0.0175}{365}\right)^{365} = 50755.63 \left(1 + \frac{0.0175}{365}\right)^{365} \approx \$51,651.65 \,.$$

Finally, after three years, you will have

$$F_3 = F_2 \left(1 + \frac{0.02}{365}\right)^{365} = 51651.65 \left(1 + \frac{0.02}{365}\right)^{365} \approx \$52,695.06 \,.$$

To find the simple interest (APY) that gives the same future value after three years, we need to solve the equation:

$$52695.06 = 50000 \left(1 + \text{APY} \times 3\right) \qquad \Longrightarrow \qquad \text{APY} = \frac{2695.06}{3 \times 50000} \approx 0.017967.$$

Therefore, the annual percentage yields of such a 3-year blended investment is 1.7967% .

Example 3.3.13. Using APY and Doubling Time to Compare Investments
Consider advertisements of four banks offering different conditions for certificates of deposits listed in the following table.

Bank	Abbreviation	Rate	Compounded
Alpha	α	2.24%	continuously
Beta	β	2.25%	daily
Gamma	γ	2.26%	monthly
Delta	δ	2.27%	quarterly

We consider two approaches to determining which of these four CDs gives the greatest return. We start with calculating the annual percentage yield for each bank. Using Eqs. (3.3.9) and (3.3.10), we get

$$\text{APY}_\alpha = e^{0.0224} - 1 \approx 0.02265276, \qquad \text{APY}_\beta = \left(1 + \frac{.0225}{365}\right)^{365} - 1 \approx 0.022754,$$

$$\text{APY}_\gamma = \left(1 + \frac{.0226}{12}\right)^{12} - 1 \approx 0.022733, \qquad \text{APY}_\delta = \left(1 + \frac{.0227}{4}\right)^{4} - 1 \approx 0.02289.$$

Comparing these APYs, we conclude that bank Delta will have the largest return and bank Alpha will have the smallest.

Now we compare these certificates using doubling time formulas (3.3.6) and (3.3.8):

$$t_\alpha = \tfrac{1}{.0224} \ln 2 \approx 30.944, \qquad t_\beta = \frac{\ln 2}{365 \ln\left(1 + \frac{.0225}{365}\right)} \approx 30.807,$$

$$t_\gamma = \frac{\ln 2}{12 \ln\left(1 + \frac{.0226}{12}\right)} \approx 30.699, \qquad t_\delta = \frac{\ln 2}{4 \ln\left(1 + \frac{.0227}{4}\right)} \approx 30.62168 \,.$$

Since bank Delta has the lowest doubling time, it gives the best return for an investment. The conclusion is the same as we found previously using APYs; however, the best return on investments gives the bank with the highest APY, while its doubling time is the smallest one.

Exercises for Section 3.3.

1. In the following exercises, find the future value, F, of the given principal, P, earning compound interest for the time period, t, and interest rate, r, specified.

   (a) $P = \$26,000$, $t = 8$ years, $r = 4.25\%$ compounded annually.

   (b) $P = \$34,000$, $t = 5$ years, $r = 3.75\%$ compounded annually.

   (c) $P = \$48,000$, $t = 4$ years, $r = 6.275\%$ compounded quarterly.

   (d) $P = \$8,500$, $t = 12$ years, $r = 5.725\%$ compounded quarterly.

   (e) $P = \$9,550$, $t = 7$ years, $r = 7.25\%$ compounded monthly.

   (f) $P = \$11,500$, $t = 6$ years, $r = 8.5\%$ compounded monthly.

   (g) $P = \$17,000$, $t = 9$ months, $r = 6.55\%$ compounded monthly.

   (h) $P = \$45,000$, $t = 6$ months, $r = 7.35\%$ compounded monthly.

   (i) $P = \$22,000$, $t = 4$ years, $r = 4.75\%$ compounded daily.

   (j) $P = \$33,500$, $t = 6$ years, $r = 5.35\%$ compounded daily.

2. In the following exercises, find the present value, P, under compound interest given the future value, F, time period, t, and interest rate, r, specified.

   (a) $F = \$6,500$, $t = 8$ years, $r = 5.25\%$ compounded annually.

   (b) $F = \$4,500$, $t = 5$ years, $r = 4.75\%$ compounded annually.

   (c) $F = \$8,000$, $t = 4$ years, $r = 7.275\%$ compounded quarterly.

   (d) $F = \$7,500$, $t = 12$ years, $r = 6.725\%$ compounded quarterly.

   (e) $F = \$550$, $t = 7$ years, $r = 8.25\%$ compounded monthly.

   (f) $F = \$300$, $t = 6$ years, $r = 7.5\%$ compounded monthly.

   (g) $F = \$7,000$, $t = 9$ months, $r = 4.55\%$ compounded monthly.

   (h) $F = \$5,000$, $t = 6$ months, $r = 5.35\%$ compounded monthly.

   (i) $F = \$2,000$, $t = 4$ years, $r = 3.75\%$ compounded daily.

   (j) $F = \$3,500$, $t = 6$ years, $r = 4.35\%$ compounded daily.

3. In the following exercises, find the time t, in years, for the given principal, P, to reach the future value, F, at the interest rate, r, specified.

   (a) $F = \$7,500$, $P = \$7,000$, $r = 6.55\%$ compounded annually.

   (b) $F = \$5,500$, $P = \$5,000$, $r = 5.88\%$ compounded annually.

   (c) $F = \$28,000$, $P = \$22,000$, $r = 7.875\%$ compounded quarterly.

   (d) $F = \$37,500$, $P = \$33,000$, $r = 5.25\%$ compounded quarterly.

   (e) $F = \$750$, $P = \$600$, $r = 9.75\%$ compounded monthly.

   (f) $F = \$450$, $P = \$400$, $r = 9.5\%$ compounded monthly.

   (g) $F = \$19,000$, $P = \$16,000$, $r = 3.65\%$ compounded monthly.

   (h) $F = \$16,000$, $P = \$15,000$, $r = 6.45\%$ compounded monthly.

   (i) $F = \$22,000$, $P = \$16,000$, $r = 2.25\%$ compounded daily.

(j) $F = \$37,500$, $P = \$31,000$, $r = 3.45\%$ compounded daily.

4. In the following exercises, find the annual interest rate, r, that will yield the given future value, F, starting from the given principal, P, over the specified time period, t.

    (a) $F = \$27,500$, $P = \$25,000$, $t = 8$ years compounded annually.

    (b) $F = \$15,500$, $P = \$12,000$, $t = 7$ years compounded annually.

    (c) $F = \$58,000$, $P = \$52,000$, $t = 12$ years compounded quarterly.

    (d) $F = \$47,500$, $P = \$43,000$, $t = 11$ years compounded quarterly.

    (e) $F = \$850$, $P = \$800$, $t = 8$ months compounded monthly.

    (f) $F = \$250$, $P = \$200$, $t = 9$ months compounded monthly.

    (g) $F = \$39,000$, $P = \$36,000$, $t = 12$ years compounded monthly.

    (h) $F = \$56,000$, $P = \$55,000$, $t = 10$ years compounded monthly.

    (i) $F = \$72,000$, $P = \$66,000$, $t = 9$ years compounded daily.

    (j) $F = \$50,500$, $P = \$44,000$, $t = 9$ years compounded daily.

5. In the following exercises, find the annual percentage yield (APY) for the given interest rate, r, compounded as specified.

    (a) $r = 7.33\%$ compounded quarterly;   (b) $r = 4.73\%$ compounded quarterly;

    (c) $r = 4.63\%$ compounded monthly;   (d) $r = 5.44\%$ compounded monthly;

    (e) $r = 2.25\%$ compounded daily;   (f) $r = 3.55\%$ compounded daily.

6. For an account with an annual interest rate of 3.275%, find the annual percentage yield (APY) when it is compounded
    (a) quarterly,   (b) monthly,   (c) daily.

7. For an account with an annual interest rate of 5.785%, find the annual percentage yield (APY) when it is compounded
    (a) quarterly,   (b) monthly,   (c) daily.

8. Find the future value of an account after 8 years if $850 is deposited initially and the account earns 3.66% interest compounded
    (a) annually,   (b) quarterly,   (c) monthly,   (d) daily.

9. Find the future value of an account after 14 years if $75,000 is deposited initially and the account earns 4.27% interest compounded
    (a) annually,   (b) quarterly,   (c) monthly,   (d) daily.

10. Find the future value of an account after 25 months if $755 is deposited initially and the account earns 5.65% interest compounded
    (a) quarterly,   (b) monthly,   (c) weekly,   (d) daily.

11. Find the future value of an account after 50 months if $45,000 is deposited initially and the account earns 3.75% interest compounded
    (a) quarterly,   (b) monthly,   (c) weekly,   (d) daily.

12. Bank A offers an interest rate of 2.76% compounded quarterly, whereas bank B offers 2.75% compounded weekly. If you want to invest in a 1-year certificate of deposit, which bank should you choose?

13. Tuition at a state university grew by about 3.4% per year between 1999 and 2009. In 2009, the tuition was $7,500 for residents. Assuming that it will continue to grow at a rate of about 3.4% per year, what will the tuition be at the university in 2019?

14. Car prices grew on average by about 1.7% per year between 2005 and 2009. In 2009, the price for a new Honda Civic-LX car was $18,255. Assuming that it will continue to grow at a rate of about 1.7% per year, what will be the cost of this car in 2011?

15. One-hundred-year railroad bonds were issued in 1909 when interest rates on railroad bonds averaged 6.12%. How much would a bond purchased for $1,000 in 1909 be worth in 2009 assuming

    (a) simple interest of 6.12% is paid?

    (b) the interest is compounded monthly with an annual rate of 6.12%?

16. Twenty-year recreation bonds were issued in 1989 when interest rates on recreation bonds averaged 6.67%. How much would a bond purchased for $1,000 in 1989 be worth in 2009 assuming

    (a) simple interest of 6.67% is paid?

    (b) the interest is compounded monthly with an annual rate of 6.67%?

17. Suppose you lend $2,500 to a friend, and your friend agrees to pay the loan back in 2 years at 5% interest compounded quarterly. How much will the friend owe you at the end of 2 years?

18. Suppose you borrow $955 from your friend, and you agree to pay back the loan in 9 months at 4.5% interest compounded monthly. How much will you owe to your friend at the end of 9 months?

19. In 1959, the rate for a first-class letter mailed within the United States was 4 cents. In 2009, it was 42¢. What is the rate of inflation based on stamp price?

20. A house was bought in 1979 for $40,000; in 2009, the house was sold for $254,000. What is the annual rate of growth (which is the inflation rate) in the cost of the house from 1979 and 2009?

21. Based on the Consumer Price Index (CPI) and Producer Price Index (PPI), the rate of inflation in the United States during 1989 and 1999 was about 5% per year. What salary in 1989 would be equivalent to a $50,000 salary in 1999?

22. Thousand dollars in 2008 has the same buying power as $990.42 in 2009. What lump sum of money in 2008 is equivalent to $1,000 in 2009? What is the inflation rate for this year?

23. Suppose that the rate of inflation for the next 7 years is predicted to be 2.5% per year. How much would a house that costs $275,000 today be expected to cost in 7 years?

24. If you increase the compound interest rate for an investment by 10%, will the future value increase by 10%?

25. During 1996 the Bulgarian economy collapsed due ton slow production and mismanaged economic reforms, its disastrous agricultural policy, and an unstable and decentralized banking system, which led to an inflation rate of 311% and the collapse of the lev. If this rate had continued compounding annually for two more years, how much would an item that cost 10 levs (the Bulgarian currency) in 1996 be expected to cost in the year 1998?

26. A couple wants to set up a college savings account for their daughter. If the account earns 3.275% compounded monthly, how much should they invest today so that the account will be worth $80,000 in 16 years?

27. A person wants to get 1.75 million dollars in 8 years. How much should he invest today into a mutual fund that earns 7.25% annually?

28. A man has $26,000 to invest today. What interest rate, compounded weekly, would he have to earn on his investment so that it would grow to $30,000 in 4 years?

29. A woman invests $75,000 in a retirement account. What interest rate, compounded daily, would she have to earn in order for the account to be worth $500,000 in 35 years?

30. A bank offers a money market account with an annual percentage yield (APY) of 3.75%. The interest is compounded monthly. Find the annual interest rate for this account.

31. A bank offers a money market account with an annual percentage yield (APY) of 4.25%. The interest is compounded daily. Find the annual interest rate for this account.

32. A bank paid 6.25% interest, compounded weekly. How long would it have taken a $10,000 deposit to grow to $20,000?

33. A family has set a goal of saving $18,255 to be used toward the purchase of a new car. If they put $15,500 into an account earning 4.75% interest compounded weekly, how long will they have to wait until the account has the required amount?

34. If a deposit is made into an account earning 3.375% interest compounded weekly, how long will it take the investment to double in value?

35. If a deposit is made into an account earning 4.75% interest compounded monthly, how long will it take the investment to triple in value?

36. Suppose $1,000 was deposited into a savings account in a bank. For the first 2 years the bank paid interest at a rate of 2.9% compounded daily. For the next 3 years the bank paid interest at a rate of 3.75% compounded weekly. How much will the investment be worth after the entire 5 years?

37. Suppose $2,000 is deposited into a savings account earning 5.75% interest compounded weekly. After 3 years, all of the money is withdrawn from the account and reinvested in an account earning 6.25% interest compounded monthly for 4 more years. How much was in the account at the end of the entire 6 years?

38. A man deposits $6,000 into an account earning 3.66% interest compounded weekly. Two years later he deposits an additional $1,000 into the same account. What will the total investment be worth after the entire 5 years?

39. A woman deposits $12,000 into an account earning 4.33% interest compounded daily. Three years later she deposits an additional $8,000 into the same account. What will be the total amount in the account 6 years after the first deposit was made?

40. Government Medicare insurance premiums paid every month by seniors ware $5.30 in 1970; as of 2008, they were $96.40. What is the rate of inflation based on Medicare insurance?

41. Boris purchased a bond valued at $10,000 for highway construction for $5,950. If the bond pays 4.25% annual interest compounded every half of a year, how long must he hold it until it reaches its full face value?

3.4 Systematic Savings/Annuities

In this section, we discuss how to accumulate money by making a series of regular payments over many years. This is what, for instance, people do to secure their retirement. This type of investment is called an **annuity** or **systematic savings plan**. An annuity is an interest-bearing account into which a person makes a series of fixed payments at the end (or beginning) of every compounding period. The value of the annuity can become enormous due to the compounding of interest.

From a theoretical point of view, an annuity is a sequence of equal payments made at equal periods of time. If the payments are made at the end of the time period, and if the frequency of compounding is the same as the frequency of payments, the sequence of such payments is called an **ordinary annuity**. In this book, we will use and discuss only ordinary annuities, omitting, for simplicity, the word "ordinary." The time between payments is the **payment period**, and the time from the beginning of the first payment period to the end of the last period is called the **term** of the annuity. The **future value of an annuity** or the **value of an annuity** is the total lump sum of money including all deposits and interest earned. An annuity plan in which payments are made at the beginning of each period is called an **annuity due**.

There are two common uses of annuities:

- Saving money for the future. Since the amount of money in the savings account is accumulated due to a sequence of deposits, we refer to this type of annuity as an **increasing annuity**.

- Withdraw funds from the account. Since the amount of money in the savings account is decreasing each time a payment is made, we call this type of annuity a **decreasing annuity**.

You may want to save regularly to have a fixed amount available in the future. It could be a down payment for a new car, a fixed amount for vacation, renovation of your home, and many others. The account that you establish for your deposits is called a **sinking fund**.

Suppose $100 is deposited at the end of each quarter for 1 year into a savings account earning 4% interest compounded quarterly. So the total investment made is $4 \times \$100 = \$400$. However, each deposit, except the last one, earns some interest. The first investment of $100 will earn interest for 3 quarters, or 9 months. Using the compound interest formula (3.3.1), page 282, with $P = \$100$, $r = 0.04$, $t = 9/12$ years, and $n = 4$, we find that the first deposit will be worth

$$100 \left(1 + \frac{0.04}{4}\right)^{4 \times 9/12} = 100\,(1 + 0.01)^3 = 100\,1.01^3 \approx \$103.01.$$

The second deposit will only be in the account for 2 quarters, or half a year, so $t = 6/12 =$

1/2 in the compound interest formula. Hence the second investment will be worth

$$100 \left(1 + \frac{0.04}{4}\right)^{4\times 6/12} = 100\,(1+0.01)^2 = 100 \times 1.01^2 = \$102.01\,,$$

and the third deposit will be worth $100\,(10.01) = \$101$. The last deposit is made at the end of the year, so it will have earned no interest and is worth \$100. Summing all the future values of each of the separate investments, we get

$$F = 100\,(1+0.01)^3 + 100\,(1+0.01)^2 + 100\,(1+0.01) + 100\,.$$

To calculate the exact value of this future value, we use the formula for the sum of a geometric series:

$$a^{m-1} + a^{m-2} + \cdots + a + 1 = \frac{a^m - 1}{a - 1} \qquad (a \neq 1). \tag{3.4.1}$$

Setting $a = 1.01$ in the above identity, we find the future value to be

$$F = 100\left[1.01^3 + 1.01^2 + 1.01 + 1\right] = 100\,\frac{1.01^4 - 1}{1.01 - 1} = 10000\left(1.01^4 - 1\right) \approx \$406.04\,.$$

This formula can be expanded for an arbitrary number of payments:

$$F = D\,\frac{n}{r}\left[\left(1 + \frac{r}{n}\right)^{nt} - 1\right], \qquad \text{where} \quad \begin{cases} F = \text{future value after } t \text{ years,} \\ D = \text{amount of each deposit,} \\ r = \text{annual interest rate,} \\ n = \text{number of periods per year,} \\ t = \text{time (in years),} \end{cases} \tag{3.4.2}$$

and deposits are made at the end of each period. Note that the intervals at which the deposits are made have to be the same as the intervals at which interest is compounded in order for the formula (3.4.2) to be valid. If it is not the case, we have to modify the formula.

For instance, suppose that in the previous example the interest is compounded monthly. Then the first deposit will worth

$$100 \left(1 + \frac{0.04}{12}\right)^{12\times 9/12} = 100\,(1+0.01/3)^9 \approx \$103.04.$$

The second investment will be worth

$$100 \left(1 + \frac{0.04}{12}\right)^{12\times 6/12} = 100\,(1+0.01/3)^6 =\approx \$102.01,$$

and the third one—

$$100 \left(1 + \frac{0.04}{12}\right)^{12\times 3/12} = 100\,(1+0.01/3)^3 \approx \$101.00.$$

Adding up the future values of each of the separate deposits, we obtain

$$
\begin{aligned}
F_{12} &= 100\left(1+\frac{0.04}{12}\right)^9 + 100\left(1+\frac{0.04}{12}\right)^6 + 100\left(1+\frac{0.04}{12}\right)^3 + 100 \\
&= 100\left[\left(1+\frac{0.04}{12}\right)^9 + \left(1+\frac{0.04}{12}\right)^6 + \left(1+\frac{0.04}{12}\right)^3 + 1\right] \\
&= 100\,\frac{\left(1+\frac{0.04}{12}\right)^{12}-1}{\left(1+\frac{0.04}{12}\right)^3-1} \approx \$406.06.
\end{aligned}
$$

In general, if the interest is compounded k times per period, such systematic savings plan will be worth

$$
F = D\,\frac{\left(1+\frac{r}{n}\right)^{nt}-1}{\left(1+\frac{r}{n}\right)^{k}-1}. \tag{3.4.3}
$$

From Eq. (3.4.2), we find the value of deposit in an ordinary annuity:

$$
D = \frac{Fr/n}{\left(1+\frac{r}{n}\right)^{nt}-1}, \tag{3.4.4}
$$

and the time of the term:

$$
t = \frac{\ln\left(1+\frac{Fr}{Dn}\right)}{n\ln\left(1+\frac{r}{n}\right)}. \tag{3.4.5}
$$

There is no exact formula to determine the rate, r, when F, D, t, and n are specified. However, we present in Example 3.4.4 the algorithm of how to find its value approximately. When payments are made at the beginning of each period, the value of an annuity due is

$$
F = D\,\frac{n}{r}\left[\left(1+\frac{r}{n}\right)^{nt+1}-1\right] - D. \tag{3.4.6}
$$

Example 3.4.1. Future Value of an Ordinary Annuity

Suppose you deposit $D = \$120$ at the end of each month into an account earning $r = 3.6\%$ interest compounded monthly. How much will be in the account, how much of the future value will be from deposits, and how much of the future value will be from interest after making deposits for

> (a) 15 years?   (b) 30 years?

Solution: Using Eq. (3.4.2), we find the future value after 15 years to be

$$
F_{15} = D\,\frac{n}{r}\left[\left(1+\frac{r}{n}\right)^{nt}-1\right] = 120\,\frac{12}{0.036}\left[\left(1+\frac{0.036}{12}\right)^{12\times15}-1\right] \approx \$28,584.80\,.
$$

Similarly, after 30 years of deposits, the future value of systematic savings would be

$$
F_{30} = 120\,\frac{12}{0.036}\left[\left(1+\frac{0.036}{12}\right)^{12\times30}-1\right] \approx \$77,596.90\,.
$$

The total amount of deposits after 15 years is $120 \times 12 \times 15 = \$21,600$, so the interest earned during this time period is $\$28,584.80 - \$21,600 = \$6,984.80$. After 30 years of investments, the total deposit is $120 \times 12 \times 30 = \$43,200$, so the interest earned during this time period is $\$77,596.90 - \$43,200 = \$34,396.90$.

Example 3.4.2. Using Calculators
We show two ways to find the future value of an annuity from the previous example. Using the formula for F_{30}, the future value of 30 years of deposits, we type on the calculator:

$$120 \times 12 \div 0.036 \times ((1 + 0.036 \div 12) \wedge (12 \times 30) - 1) \qquad \text{ENTER}.$$

Another way is to use the TVM Solver (TVM stands for "Time Value of Money") available on your TI-83/84 Plus graphical calculator. It uses five numbers: duration (denoted by N), interest rate ($I\%$), present value (PV), periodic payment (PMT), and future value (FV). Once four of the five numbers are known, TVM Solver can be used to calculate the fifth number. This involves the following steps.

- Enter all information except for the one value to be determined.

- Move the cursor to the line containing the value to be calculated.

- Press ALPHA [SOLVE]. Note: the green word SOLVE is located above the ENTER key.

To open TVM Solver, press the button APPS , choose from the scroll menu the first option—Finance. You should see the options, including the first one—TVM Solver. Then you enter in the screen the following numbers:

$N = 360$	because there are 360 payments during 30 years.
$I\% = 3.6$	the interest rate in a percent.
$PV = 0$	you start from scratch.
$PMT = -120$	is your monthly deposit—it is a negative number because you pay.
$FV = 0$	this is the unknown value to be determined, so put the cursor into this position.
$P/Y = 12$	number of payments per year.
$C/Y = 12$	number of times the interest is compounded.

By pressing ENTER, you will get the required future value: $FV = 77596.90144$.

Example 3.4.3. Sinking Funds
Parents want to save $200,000 to pay for the estimated cost of their child's education. They plan to invest their savings in a mutual fund, and based on the past performance of the fund and its projected future performance, they estimate that it will give returns that are equivalent to earning about 7.25% interest compounded quarterly. How much should the parents deposit at the end of each quarter into this fund so that they can achieve their goal in 15 years?

Solution: In this case, we have $F = 200000$, $r = 0.0725$, $t = 15$, and $n = 4$; then we use the formula (3.4.4):

$$D = \frac{200000 * 0.0725/4}{\left(1 + \frac{0.0725}{4}\right)^{60} - 1} \approx \$1870.39 \,.$$

This answer is achieved on the calculator:

$$200000 \times 0.0725 \div 4 \div ((1 + 0.0725 \div 4) \wedge (4 \times 15) - 1) \qquad \text{ENTER}.$$

Example 3.4.4. Finding Time

Suppose a man wants \$5,000 available for the down payment on a car. If he deposits \$145 every month, at the end of the month, into an account earning 3.6% compounded monthly, how long will it take until he has the down payment?

Solution: In this case, we have $F = 5000$, $r = 0.036$, $D = 145$, and $n = 12$; then we use the formula (3.4.5):

$$t = \frac{\ln\left(1 + \frac{Fr}{Dn}\right)}{n \ln\left(1 + \frac{r}{n}\right)} = \frac{\ln\left(1 + \frac{5000 \times 0.036}{145 \times 12}\right)}{12 \ln\left(1 + \frac{0.036}{12}\right)} \approx 2.74 \text{ years} \qquad \text{or} \qquad 33 \text{ months.}$$

We obtain the answer by typing on the calculator

$$\ln(1 + 5000 \times 0.036 \div 12 \div 145) \div 12 \div \ln(1 + 0.036 \div 12) \qquad \text{ENTER}.$$

Example 3.4.5. Finding the Interest Rate

Suppose you want to have \$20,000 available to purchase a car in 6 years. Your estimation shows that you can afford to make monthly payments of \$150. What annual interest rate are you looking for such investment when the interest is compounded monthly?

Solution: Since there is no formula to calculate the rate, we use a calculator to estimate its value. Setting $r = 0.01, 0.02, \ldots, 0.9$ in the formula (3.4.2), we obtain the following future values with $D = 150$, $n = 12$, and $t = 6$:

r	1%	2%	3%	4%	5%	6%	7%	8%
F	11125.89	11464.58	11816.91	12183.38	12564.64	12961.33	13374.14	13803.8

r	9%	10%	11%	12%	13%	14%	15%	16%
F	14251.05	14716.7	15201.55	15706.49	16232.41	16780.26	17351.04	17945.79

r	17%	18%	19%	20%	21%	22%	23%	24%
F	18565.59	19211.58	19884.96	20586.98	21318.94	22082.22	22878.26	23708.55

So we see that you are looking for the rate between 19% and 20%. Then we calculate similar values:

r	19.11%	19.12%	19.13%	19.14%	19.15%	19.16%	19.17%	19.18%
F	19960.76	19967.66	19974.58	19981.49	19988.4	19995.33	20002.25	20009.17

Therefore we conclude that the rate should be at least $r = 19.167\%$. *Note* that you cannot use TVM Solver in this case because it would provide you with the annual percent rate (APR) instead of the annual rate, r. For instance, in this problem, TVM Solver will give you the answer APR $= 18.431354\%$. The APR is explained in §3.6.

Exercises for Section 3.4.

1. In the following exercises, find the future value, F, of a systematic savings plan with the given deposits, D, made at the end of the compounding periods for the time period, t, and interest rate, r, specified.

   (a) $D = \$1000$, $t = 17$ years, $r = 5.22\%$ compounded annually.

   (b) $D = \$3400$, $t = 18$ years, $r = 3.25\%$ compounded annually.

   (c) $D = \$500$, $t = 22$ years, $r = 4.66\%$ compounded quarterly.

   (d) $D = \$600$, $t = 23$ years, $r = 5.75\%$ compounded quarterly.

   (e) $D = \$20$, $t = 8$ years, $r = 4.25\%$ compounded monthly.

   (f) $D = \$75$, $t = 12$ years, $r = 5.75\%$ compounded monthly.

   (g) $D = \$30$, $t = 25$ years, $r = 6.25\%$ compounded biweekly.

   (h) $D = \$55$, $t = 30$ years, $r = 4.75\%$ compounded biweekly.

2. In the following exercises, find the amount of each deposit, D, for a systematic savings plan with the deposits made at the end of the compounding periods given the future value, F, time period, t, and interest rate, r, specified.

   (a) $F = \$100,000$, $t = 20$ years, $r = 4.22\%$ compounded annually.

   (b) $F = \$200,000$, $t = 25$ years, $r = 5.11\%$ compounded annually.

   (c) $F = \$500,000$, $t = 22$ years, $r = 4.275\%$ compounded quarterly.

   (d) $F = \$60,000$, $t = 23$ years, $r = 5.275\%$ compounded quarterly.

   (e) $F = \$2,250$, $t = 18$ months, $r = 4.225\%$ compounded monthly.

   (f) $F = \$7,500$, $t = 12$ months, $r = 5.65\%$ compounded monthly.

   (g) $F = \$3,000,000$, $t = 25$ years, $r = 6.235\%$ compounded biweekly.

   (h) $F = \$4,000,000$, $t = 30$ years, $r = 4.275\%$ compounded biweekly.

3. In the following exercises, find the time, t, in years, for a systematic savings plan with the given deposits, D, made at the end of the compounding periods to reach the given future value, F, under the specified interest rate, r.

   (a) $F = \$100,000$, $D = \$1,000$, $r = 3.25\%$ compounded quarterly.

   (b) $F = \$100,000$, $D = \$2,500$, $r = 6.66\%$ compounded quarterly.

   (c) $F = \$8,750$, $D = \$78$, $r = 5.25\%$ compounded monthly.

   (d) $F = \$4,250$, $D = \$90$, $r = 5.44\%$ compounded monthly.

   (e) $F = \$37,000$, $D = \$225$, $r = 6.35\%$ compounded biweekly.

(f) $F = \$46,000$, $D = \$175$, $r = 5.25\%$ compounded biweekly.

4. In the following exercises, use a calculator to find a minimum interest rate that guarantees achieving the given future value, F, for a systematic savings plan with the given deposits, D, made at the end of the compounding periods for the specified time period, t.

   (a) $F = \$114,301.86$, $D = \$1,500$, $t = 15$ years compounded quarterly.
   (b) $F = \$682,965.19$, $D = \$3,500$, $t = 24$ years compounded quarterly.
   (c) $F = \$1098.44$, $D = \$48$, $t = 22$ months compounded monthly.
   (d) $F = \$3070.31$, $D = \$60$, $t = 45$ months compounded monthly.
   (e) $F = \$104296.44$, $D = \$125$, $t = 16$ years compounded biweekly.
   (f) $F = \$468,568.05$, $D = \$275$, $t = 22$ years compounded biweekly.

5. Suppose a 25-year-old woman bought an annuity (a systematic savings plan) that calls for deposits of $120 at the end of each month and pays an interest rate of 5.75% compounded monthly.

   (a) How much will the annuity be worth when the woman is 60 years old?
   (b) How much of the future value will be from deposits?
   (c) How much of the future value will be from interest?

6. A man has $55 deducted from his paycheck every month and the company matches this amount in a 401(k) plan. Suppose he continues to work for this company and contributes such payments for 15 years. The pension plan earns 6.23% interest compounded monthly.

   (a) How much will the plan be worth after 15 years?
   (b) How much of the future value will be from deposits?
   (c) How much of the future value will be from interest?

7. A couple predicts that they will need to save $90,000 for their child's university education. Suppose their investment will earn 4.37% interest, compounded monthly.

   (a) If they begin the deposits at the end of each month for 18 years, how large must each deposit be?
   (b) If they deposit at the end of each month for 10 consecutive years, how large must each deposit be?

8. A man needs $7,000 for a down payment to purchase a car. Suppose that he can deposit $25 every month into a savings account that pays 3.75% interest compounded monthly.

   (a) How long should he wait until the investment will be worth $7,000?
   (b) How much of the future value will be from deposits?
   (c) How much of the future value will be from interest?

9. A fire company wants to take advantage of a state program in order to purchase a new fire truck. The truck will cost $500,000, and members of the finance committee estimate that with community and state contributions, they can save $14,750 per month in an account paying 3.25% interest compounded monthly. How long will it take to save for the truck?

10. Keith is saving to buy a new scooter. If he deposits $125 at the end of each month in an account that pays an interest rate of 2.725% compounded monthly, how much will he have saved in 48 months?

11. A city must replace its water treatment equipment within 3 years. The new equipment will cost $124,000, and the city will transfer $4,550 per month into a special account that pays 4.5% interest compounded monthly. How long will it take to save for this new equipment?

12. Gary has set up an ordinary annuity to save for his retirement in 25 years. If his monthly payments are $225 and the annuity has an interest rate of 3.5% compounded monthly, what will be the value of the annuity when he retires?

13. Mary is making monthly payments into a sinking fund. She wants to have $2,500 in 14 months for her vacation. Her account pays 2.75% interest compounded monthly. What are her monthly payments to the account?

14. Ashley wants to save for an African safari. She is putting $125 each month in a sinking fund that pays an interest rate of 3.75% compounded monthly. If she makes payments for 3 years, how much will she have for her trip?

15. Beatrice wants to save $40,000 for a down payment on a condominium at a ski resort. She feels that she can save $475 per month in a sinking fund that has a 4.5% interest rate compounded monthly. How long will it take to acquire her down payment?

16. Mark wants to save $26,500 in 10 years by making monthly payments into an ordinary annuity for a down payment on a condominium at the shore. If the annuity pays 0.01% monthly interest, what will his monthly payment be?

17. Ilya deposits $50 each month in an ordinary annuity to save for a new car. If the annuity pays a monthly interest rate of 0.005%, how much will he be able to save in 4 years?

18. Olga is making monthly payments into an annuity that pays 0.004% monthly interest to save enough for a down payment to start her own business. If she wants to save $25,000 in 4 years, what should her monthly payments be?

19. Wendy has set up an ordinary annuity to save for a retirement home in Florida in 20 years. If her monthly payments are $375 and the annuity has an annual interest rate of 4.5% compounded monthly, what will be the value of the annuity when she retires?

20. Leo needs to save $20,000 for a down payment to start a restaurant business. He intends to save $250 per month in an account that pays an annual interest rate of 3.5% compounded monthly. How long will it take for him to save for the down payment?

21. Rebbecca is saving for her retirement in 15 years by putting $350 each month into an ordinary annuity. If the annuity has an annual interest rate of 3.6% compounded monthly, how much will she have when she retires?

22. Joseph began to save for his retirement at age 28, and for 15 years he put $250 per month into an ordinary annuity at an annual interest rate 3% compounded monthly. After the 17 years, he could no longer make payments, so he placed the value of the annuity into another account that paid 4.5% annual interest compounded monthly. He left the money in this account for 25 years until he was ready to retire. How much did he have for retirement?

23. A typical smoker spends about $120 per month on cigarettes. Suppose the smoker invests that amount each month in a savings account at 3.9% interest compounded monthly. What would the account be worth after 30 years?

24. Melissa made payments of $320 at the end of each month to purchase a piece of property. At the end of 35 years, she completely owned a piece of property, which she sold for $375,000. What annual interest rate compounded monthly would she need to earn on an annuity for a comparable rate of return?

25. The owner of a small business knows that he must buy a new machine in 5 years. The machine costs $32,000. In order to accumulate enough money to pay for the machine, the owner decides to deposit a sum of money at the end of each 3 months in an account paying 3.24% compounded quarterly. How much should each payment be?

26. A corporation sets up a sinking fund to replace an aging warehouse. The cost of the warehouse today would be $12 million. However, the corporation plans to replace the warehouse in 4 years. It estimates that the cost of the warehouse will increase by 3.5% annually. The sinking fund will earn 7.5% interest compounded monthly. What should be the monthly payment to the fund?

27. A corporation sets up a sinking fund to replace some aging machinery. It deposits $11,000 into the fund at the end of each month for 3 years. The annuity earns 4.8% interest compounded monthly. The equipment originally cost $412,000. However, the cost of the equipment is rising 2.3% each year. Will the annuity be adequate to replace the equipment? If not, how much additional money is needed?

28. To cover the costs of in-state tuition at a 4-year university for their newborn baby, the parents decide to put the money into a special program. The plan has two options: either pay $5214 immediately, and make payments of $200 at each month's end for 5 years beginning immediately, or payments of $62.11 at each month's end for 18 years. Show that the future values after 13 and 18 years of each of these deposits into an account with an interest rate of 4% compounded monthly are fairly close in value.

29. A family currently has $15,000 of the $35,000 they want to have available for a down payment on a house. They invest this $15,000 in a savings account paying 3.6% interest compounded monthly, and at the end of each month they deposit another $100 into the account. How long will it take them to save enough for the down payment?

30. A grateful alumnus decides to donate a permanent scholarship of $800 per year. How much money should be deposited in the bank at 4.25% interest compounded quarterly in order to be able to supply the money for the scholarship at the end of each quarter for 10 years?

31. Is it more profitable to receive a lump sum of $32,200 at the end of 5 years or to receive $500 every month at the end of each month for 5 years? Assume that money can earn 2.9% interest compounded monthly.

32. Is it more profitable to receive $500 at the end of each quarter for 5 years or to receive $32,200 at the end of 5 years? Assume that money can earn 2.9% interest compounded monthly.

33. When Jennifer takes a new job, she is offered a $3,500 bonus now or the option of an extra $300 each month for the next year. If interest rates are 3.2% compounded monthly, which choice is better and by how much?

34. Suppose $500 is deposited at the end of each month into an account earning 5.76% interest compounded monthly. After 10 years the deposits are discontinued, but the money is left in the account and continues to earn the same interest. How much will be in the account 20 years after it was first opened?

3.5   Amortized Loans/Mortgages

Every time a person uses a credit card, he or she signs an obligation to pay back the amount which was borrowed. On average, credit card debt per U.S. household was \$15,788 at the end of 2009 (at the same time, 76% of undergraduates have credit cards, and the average undergrad has \$2,200 in credit card debt). Credit card companies require minimum monthly payments toward the card holder debt. However, you can repay your loan whenever you want. A loan that you pay with some sequence of payments (monthly, weekly, quarterly, or in some other time period) is called an **installment loan**. A credit card loan is an example of an open-end installment loan because there is no deadline for paying the debt off.

There are two kinds of loans—direct and reverse. Direct installment loans are used to repay a certain lump sum of money that was borrowed previously. One may need to buy a real estate property or land, pay for tuition and lodging, or use some money for a vacation, house renovation, wedding, and many other purposes. There is one particular and important example of direct loans that deserves its own name—a **mortgage**, which is a long-term installment loan given against a real estate property (a house, a condominium, or apartment), which is used as collateral to guarantee the loan. Every mortgage requires monthly payments in equal installments. If payments are not made on the loan, the borrower may lose possession of the property. The process of paying off a loan together with interest by making a series of regular, equal payments is called **amortization**, and such a loan is called an **amortization loan**. If a loan is paid off through a sequence of payments, we say that the loan is **amortized**, which literally means "killed off."

In a reverse loan, you are paid in some installments from previously accumulated funds. It could be your savings account, a lottery prize, inheritance money, or reverse mortgage when the price of the property is repaid in certain installments. It is assumed that the funds were deposited into accounts that pay some interest while keeping the money. In the previous section, we considered annuities that are used either for accumulating funds or for spending money. Decreasing annuities can be considered as an example of a reverse loan when a person is paid regularly with equal installments.

The **present value of a decreasing annuity** (or principal) is the amount of money necessary to finance the sequence of installments. In other words, the present value of an annuity is the amount one needs to deposit in order to secure the desired sequence of payments, called the **rent**, and leave a balance of zero at the end of the term (amortize it). Let us find a formula for the present value of a decreasing annuity by looking at it from two different prospectives.

Let P be the present value (or principal) of a decreasing annuity that pays the rent, denoted by R, n times per year. Assuming that the account of the annuity pays interest (a fee) for keeping the money at a rate r compounded n times per year, let i denote the rate per period, that is, $i = r/n$.

Suppose that P dollars were put into an account that pays interest compounded n times per year. Then the future value, F, of such investment after t years is given by the compound interest formula (3.3.1):

$$F = P(1+i)^{nt}, \qquad i = r/n.$$

Suppose that the same future value was accumulated in a savings account (annuity) by making regular payments, R, during t years. Using the systematic savings plan formula (3.4.2), we get

$$F = R\frac{1}{i}\left[(1+i)^{nt} - 1\right], \qquad i = r/n.$$

Setting both formulas equal to each other, we have

$$P(1+i)^{nt} = R\frac{1}{i}\left[(1+i)^{nt} - 1\right].$$

After multiplication by $(1+i)^{-nt}$, we obtain the following formula for a decreasing annuity:

$$P = R\frac{1}{i}\left[1 - (1+i)^{-nt}\right], \qquad i = r/n. \tag{3.5.1}$$

Solving this equation for R, we get the rent for a decreasing annuity to be

$$R = \frac{iP}{1 - (1+i)^{-nt}}, \qquad i = r/n. \tag{3.5.2}$$

It is interesting to compare this formula with the deposit formula for a systematic savings plan:

$$D = \frac{iF}{(1+i)^{nt} - 1}, \qquad i = r/n. \tag{3.5.3}$$

Example 3.5.1. Determining the Principal for a Decreasing Annuity
How much money must you deposit now at 2.46% compounded quarterly in order to be able to withdraw \$2,000 at the end of each quarter for 5 years?

Solution: Using the present value formula (3.5.1) with $R = 2000$, $t = 5$, $i = 0.0236/4 = 0.00615$, we obtain

$$P = \frac{2000}{0.00615}\left[1 - 1.00615^{-20}\right] \approx \$37,529.49.$$

Using a calculator, we type

$$2000 \times (1 - 1.00615 \wedge ((-)20)) \div 0.00615 \qquad \text{ENTER.}$$

Example 3.5.2. Determining the Rent for a Decreasing Annuity
If you deposit \$150,000 into a fund that pays 3.24% interest compounded monthly, how much can you withdraw at the end of each month for 10 years?

Solution: For this decreasing annuity, $P = 150000$, $n = 12$, $t = 10$, $r = 0.0324$, so $i = 0.0324/12 = 0.0027$. Using Eq. (3.5.2), we get the value of rent to be

$$R = \frac{0.0027 \times 150000}{1 - 1.0027^{-120}} = \frac{405}{1 - 1.0027^{-120}} \approx \$1,465.08 \,.$$

Using a calculator, we type

$$0.027 \times 150000 \div (1 - 1.0027 \wedge ((-)120)) \qquad \text{ENTER}. \qquad\qquad \square$$

Now we turn our attention to amortized loans. Since the sequence of payments in an amortization loan constitutes a decreasing annuity for the bank, with the borrower taking out the loan paying the interest, we can use Eq. (3.5.1) as the **amortization loan formula**:

$$P = R\frac{n}{r}\left[1 - \left(1 + \frac{r}{n}\right)^{-nt}\right], \qquad \text{where} \quad \begin{cases} P = \text{principal or present value,} \\ R = \text{rent,} \\ r = \text{annual interest rate,} \\ n = \text{number of periods per year,} \\ t = \text{time (in years).} \end{cases} \qquad (3.5.4)$$

Solving for R, the amount of each payment, or using Eq. (3.5.2), we get

$$R = \frac{Pr/n}{1 - \left(1 + \frac{r}{n}\right)^{-nt}}. \qquad (3.5.5)$$

We need one more formula for determination of the time (lifespan of the loan):

$$t = -\frac{\ln\left(1 - \frac{Pr}{Rn}\right)}{n \ln\left(1 + \frac{r}{n}\right)}. \qquad (3.5.6)$$

Note that the value of the logarithm in the numerator is negative because the expression $1 - \frac{Pr}{Rn}$ is less than 1.

Example 3.5.3.   Cost of Buying a Car

Suppose you want to purchase a new car for \$21,374. Since you do not need your old car anymore, you decide to trade it in and the dealership gives you \$7,500 for this car. So you have to pay the difference between the price of the new car and the old car, which is \$13,874. Adding 7% for taxes, you take a car loan of \$14,845.18 at an interest rate of 3.6% compounded monthly. You are to repay the loan with monthly payments for five years.

1. What are the monthly payments?

2. What is the total amount you make in payments over the lifespan of the loan?

3. How much interest will you pay over the life of the loan?

Solution: We use the amortization loan formula with $P = 14845.18$, $r = 0.036$, and $t = 5$.

1. Using Eq. (3.5.5), we calculate the monthly payment (rent) to be

$$R = \frac{14,845.18 \times 0.036/12}{1 - \left(1 + \frac{0,036}{12}\right)^{-12\times5}} = \frac{44.53554}{1 - 1.003^{-60}} \approx \$270.72.$$

Note that the last fraction could be rounded also to \$270.73 because its exact value is $270.7250551\ldots$.

2. Since you make 60 payments, the total amount you make in payments over the lifespan of the loan will be

$$270.72 \times 60 = \$16,243.20 \qquad (\text{or } \$16,243.80 \text{ if your rent is } 270.73)$$

3. The interest you will pay over the loan's life is the difference between the total amount you pay by making 60 installments of \$270.72 and the loan amount:

$$\$16,243.20 - \$14,845.18 = \$1,398.02.$$

Example 3.5.4. Determining How Much You Can Afford to Borrow

The loan payment formula (3.5.4) can be used to determine how much your regular (usually monthly) payments will be on fixed amortization loans other than the mortgage. One such loan could be a student loan to cover expenses for education. A student with such a loan starts paying only half a year after graduation. The government provides federal student loans that cover school expenses, including tuition and fees, room and board, books and school supplies, as well as any transportation. Loans can also help pay for technology needs (i.e., a computer) and for necessary dependent care. For instance, nearly all students are eligible to receive a Stafford student loan regardless of credit score or other financial issues. The rate of such a loan can be as low as 4.5%, but there is an upper limit for borrowed funds.

Suppose that a college senior does not have enough money to cover expenses and plans to get extra money from a private student loan. She estimates that after graduation she can afford to make monthly payments of \$200 on a loan for five years. If the loan is made at an interest rate of 7.9% compounded monthly, how much can she afford to borrow? The student learned from her friends that a private lending agency may impose a monthly fee until the debt is amortized if she makes one late payment.

Solution. In this situation, we use the loan formula (3.5.4) with $R = 200$, $r = 0.079$, $n = 12$, and $t = 5$ to obtain

$$F = 200\,\frac{12}{0.079}\left[1 - \left(1 + \frac{0.079}{12}\right)^{-12\times5}\right] = \frac{2400}{0.079}\left[1 - \left(1 + \frac{0.079}{12}\right)^{-60}\right] \approx \$9887.$$

Using a calculator, we type:

$$200 \times 12 \div (1 - (1 + 0.079 \div 12) \wedge ((-)60)) \qquad \text{ENTER.}$$

Example 3.5.5. Saving by Paying Off a Loan Quickly
A couple purchased a condo by taking out a \$175,000 mortgage with monthly payments for 30 years. The interest rate on the loan was 6.25% compounded monthly.

They calculated their monthly payment by typing on the calculator

$$175000 \times 0.0625 \div 12 \div (1 - (1 + 0.0625 \div 12) \wedge ((-)360)) \qquad \text{ENTER},$$

according to the formula (3.5.5):

$$R = \frac{175000 * 0.0625/12}{1 - \left(1 + \frac{0.0625}{12}\right)^{-12 \times 30}} \approx \$1077.51.$$

So during a course of the mortgage's life, the couple will pay

$$\$1077.51 \times 360 = \$387,903.60,$$

which is more than a double the original borrowed amount. Therefore, the couple decided to pay an extra \$122.49 each month that makes their monthly rent be \$1200. Using Eq. (3.5.6), they calculated the new lifespan of the loan:

$$t = -\frac{\ln\left(1 - \frac{175000 * 0.0625}{1200 * 12}\right)}{12 \ln\left(1 + \frac{0.0625}{12}\right)} \approx 22.86 \quad \text{years or about 275 months}$$

by typing on the calculator

$$(-)\ln(1 + 175000 \times 0.0625 \div 1200 \div 12) \div 12 \div \ln(1 + 0.0625 \div 12) \qquad \text{ENTER}.$$

By paying an extra \$122.49 each month, the loan is paid off in 275 months rather than 360. With such new payment, the couple will pay \$1200 \times 275 = \$330,000. So the couple will save \$57,903.60 over the course of the mortgage. $\square$

What happens if you are amortizing a debt with equal periodic installments and at some point decide to pay off the remainder of the loan in one lump-sum payment? This occurs each time a home with an outstanding mortgage is sold or refinanced. The procedure of paying off the remaining debt on the loan by taking another loan at a lower interest rate is called **refinancing** the original loan.

The amount that you owe on the loan at a given point in time is called the **loan balance**. The loan balance is the present value of the series of payments that are left to make. Therefore, the loan balance is the principal that would have to be borrowed in order to pay off the loan. (It is your sinking fund.) If you borrow this amount at the same conditions that were made for the original loan, you can apply the loan formula (3.5.4) for the remaining time. Let t be the loan term of the loan and t^* ($t^* < t$) be a point of time at which you would like to know the balance. Then the balance at time t^* is

$$B_{t^*} = R \frac{n}{r}\left[1 - \left(1 + \frac{r}{n}\right)^{-n(t-t^*)}\right], \qquad (3.5.7)$$

where R is rent, the amount of each payment, r is the annual interest rate, n is the number of periods per year. The formula can be simplified a little bit after introducing the rate per period: $i = r/n$. Let $N = nt$ be the number of payments needed to amortize the loan, and $m = nt^*$ be the number of payments made. Then the balance of the loan after m payments, B_m, is

$$B_m = R \frac{1}{i} \left[1 - (1+i)^{-N+m} \right] = R \frac{n}{r} \left[1 - \left(1 + \frac{r}{n}\right)^{-N+m} \right]. \qquad (3.5.8)$$

Example 3.5.6. A Loan Balance

Consider the situation from Example 3.5.5 in which $175,000 is borrowed at 6.25% compounded monthly and repaid with 360 monthly payments of $1077.51. After 8 years of payments, the couple decides to refinance the mortgage with another amortization loan having a lower rate of 5.5%.

First, the couple calculated the amount of money that remains to be paid on an amortizing loan after eight years using Eq. (3.5.8):

$$B_8 = 1077.51 \frac{12}{0.0625} \left[1 - \left(1 + \frac{0.0625}{12}\right)^{-360+8\times12} \right] = \$154,386.99.$$

They were shocked: after making 96 payments of total $103,440.96 they actually paid only $20,613.01 toward their loan (about 20% of the money they spent). The couple has to make another 22-year mortgage on the current outstanding balance. Calculating the new rent, they get

$$R_{22} = \frac{154386.99 * 0.055/12}{1 - \left(1 + \frac{0.055}{12}\right)^{-12\times22}} \approx \$1009.46.$$

On this mortgage, they will make 264 ($= 12 \times 22$) payments of $1009.46 of total $266,497.44. The couple's saving by switching to the new mortgage will be $17,965.20.

Example 3.5.7. Comparing Two Mortgages

Suppose a couple has to decide which one of the following two mortgages to choose in order to finance their real estate purchase. One offer from bank A gives them a 15-year mortgage which requires regular monthly payments at 6%. Another bank B offers them a 30-year mortgage at an annual interest rate of 6.5%. The couple was confused about which of the two lenders to choose. Is a lower rate a better offer? Sure, but only if both offers require the same loan term.

The couple starts with calculating rent for their mortgage of $175,000. The monthly payment for the mortgage offered by bank A would be

$$R_A = \frac{175000 * 0.06/12}{1 - \left(1 + \frac{0.06}{12}\right)^{-12\times15}} \approx \$1476.75.$$

Similar calculations for the offer from bank B yield

$$R_B = \frac{175000 * 0.065/12}{1 - \left(1 + \frac{0.065}{12}\right)^{-12\times30}} \approx \$1106.12.$$

During the lifespan of a 15-year mortgage, the couple would pay $\$1476.75 \times 12 \times 15 = \$265,815$, while they would spend $\$1106.12 \times 12 \times 30 = \$398,203.20$ on a 30-year mortgage. So they would save $\$132,388.20$ if they could afford the higher monthly rent.

3.5.1 Amortization Schedules

In order to understand the amortizing process better, let us consider an amortization schedule that is important for both the borrower and the lender. An **amortization/mortgage schedule** is a list of equal payments to be made on a loan/mortgage that breaks down each payment into principal and interest. In particular, the schedule identifies the balance of the loan—how much is owed after each payment. The portions of each payment going toward the principal and toward the interest vary with each payment: the former increases and the later decreases. Sometimes amortizing loans stipulate a **balloon payment** at the end of the term, which will lead to a larger last payment.

As we saw in the previous examples, the amount of each payment is always rounded to the nearest cent. The exact value of the rent in Eq. (3.5.5) has usually (unless you use unrealistic numbers for its entries) an infinite number of digits in its decimal representation. Any rounding of the rent value leads to an error that may affect all further calculations. In particular, the last payment may either not pay the loan in full or may overpay the balance. In the latter case, the lender has to reimburse the borrower with this amount overpaid. If the final balance exceeds the regular payment, the borrower has to cover the difference to amortize the loan.

Let B_m be the loan balance after m payments. We assume that a loan of P dollars has been taken at the annual interest rate r compounded n times per year (for a mortgage, $n = 12$). To work out the formula for B_m, we calculate the balance at the end of the first period, second period, and maybe a couple more, and look for a pattern.

When the loan is first taken out, no payments have been made, so the loan balance is the same as the loan amount: $B_0 = P$, the principal of the loan. What happens at the end of the first period? Interest has accrued. The interest rate per period is $i = r/n$, and the balance is B_0, so the accrued interest is i times B_0, which equals $iP = rP/n$; this gets added to the loan balance. On the other hand, the payment R is subtracted. Therefore,

$$B_1 = P + iP - R = P(1 + i) - R.$$

What happens to the loan balance at the end of the second period? We add in the interest on the previous balance, which is i times B_1, and subtract the next payment R:

$$B_2 = B_1 + iB_1 - R = (1+i)B_1 - R = (1+i)\left[P(1+i) - R\right] - R = P(1+i)^2 - R(1+i) - R.$$

The next step is to calculate B_3:

$$B_3 = (1+i)B_2 - R = P(1+i)^3 - R\left[1 + (1+i) + (1+i)^2\right].$$

Using our observation between the new balance and the previous balance

$$B_{\text{new}} = (1+i)\, B_{\text{previous}} - R = \left(1 + \frac{r}{n}\right) B_{\text{previous}} - R, \tag{3.5.9}$$

the process repeats at the end of each period: add in the accrued interest and subtract the payment. There is no need to show all the steps for B_4 that lead to

$$\begin{aligned}
B_4 &= P(1+i)^4 - R(1+i)^3 - R(1+i)^2 - R(1+i) - R \\
&= P(1+i)^4 - R\left[1 + (1+i) + (1+i)^2 + (1+i)^3\right].
\end{aligned}$$

Now we have enough information to write a general expression for B_m, the loan balance after the mth payment. Look at the first term in B_0, B_1, B_2, B_3, B_4: it is always the principal, P, times the mth power of $(1+i)$. (Remember that $(1+i)^0 = 1$, so $B_0 = P(1+i)^0 = P$.) The expression in brackets, which is a multiple of R, is the sum of the geometric progression (3.4.1) on page 296:

$$1 + (1+i) + (1+i)^2 + (1+i)^3 + \cdots + (1+i)^{m-1} = \frac{(1+i)^m - 1}{(1+i) - 1} = \frac{(1+i)^m - 1}{i}.$$

This yields the loan balance formula after m payments have been made:

$$B_m = P\left(1 + \frac{r}{n}\right)^m - \frac{Rn}{r}\left[\left(1 + \frac{r}{n}\right)^m - 1\right]. \tag{3.5.10}$$

(For a savings account or other investment, just change the first minus sign to a plus.)

Example 3.5.8. Energy Bank
A man borrows from the Energy Bank \$9,000 to finance a new more efficient heating system, including an air conditioner. The bank gives the borrower a one-year break to pay the loan without any charge. After one year, the borrower has to amortize the loan including 4.8% interest. It is assumed that the borrower pays monthly a fixed amount during three years to pay off the loan completely.

The first step in preparing an amortization schedule consists in calculating the regular payment according to Eq. (3.5.5):

$$R = \frac{9000 \times 0.048/12}{1 - \left(1 + \frac{0.048}{12}\right)^{-12\times 3}} = \frac{9000 \times 0.004}{1 - 1.004^{-36}} \approx \$268.93,$$

where we round the rent to the nearest cent by disregarding digits in the fraction $268.930\ldots$. Than we arrange all information about the loan in a table, called the loan amortization schedule. Each row contains the payment number, the amount of each payment, the amount of interest and principal in each payment, and the loan balance. In our case, the first row will look as follows.

Payment Number	Payment	Interest Paid	Principal Paid	Balance
				$9000
1				
2				
⋮				

To fill out the first row in the table, we enter $268.93 in the payment column. To find what portion of this rent is used to pay interest, we multiply the balance appearing in the previous row (9000) by the interest rate for one period: $i = r/n = 0.048/12 = 0.004$. We get

$$\text{interest paid} = \$9000 \times 0.004 = \$36.$$

The remainder of the first payment goes toward paying the principal, so we have

$$\text{principal paid} = \$268.93 - \$36 = \$232.93.$$

We compute the new balance by subtracting the principal paid from the previous balance:

$$\text{balance} = \$9000 - \$232.93 = \$8767.07.$$

Entering this information into the amortization table, we get

Payment Number	Payment	Interest Paid	Principal Paid	Balance
				$9000.00
1	$268.93	$36	$232.93	$8767.07
2				
⋮				

To fill out the second row, we first enter the payment (which is the same) and then compute the interest paid. It is given by the new balance of $8767.07 multiplied by the monthly interest rate, $i = 0.004$, as follows

$$\text{interest paid} = \$8767.07 \times 0.004 = \$35.07.$$

The principal paid is then given by the difference:

$$\$268.93 - \$35.07 = \$233.86.$$

The new balance becomes $8767.07 - \$233.86 = \$8533.21$. Entering this information into the amortization schedule, we have

Payment Number	Payment	Interest Paid	Principal Paid	Balance
				$9000.00
1	$268.93	$36	$232.93	$8767.07
2	$268.93	$35.07	$233.86	$8533.21

Continuing in this way, we complete the next two rows of the amortization table:

Payment Number	Payment	Interest Paid	Principal Paid	Balance
				$9000.00
1	$268.93	$36	$232.93	$8767.07
2	$268.93	$35.07	$233.86	$8533.21
3	$268.93	$34.13	$234.80	$8298.41
4	$268.93	$33.19	$235.74	$8062.67

We can check our calculations, at least for the balance, if we apply Eq. (3.5.10):

$$B_4 = 9000\,1.004^4 - \frac{268.03}{0.004}\left[1.004^4 - 1\right] = 8062.674758\ldots,$$

which is exactly the same (after rounding) as our last balance in the fourth row. Now we check the last few payments:

Payment Number	Payment	Interest Paid	Principal Paid	Balance
33	$268.93	$4.26	$264.67	$800.40
34	$268.93	$3.20	$265.73	$534.67
35	$268.93	$2.14	$266.79	$267.88
36	$268.93	$1.07	$267.86	$0.02

As we see, the final payment does not match the balance by 2 cents—this is due to rounding errors that occurred during all calculations. During the loan term, the borrower paid $9638.48, where the interest $638.48 is about 7% of the principal. □

Notice that with each successive payment, the amount of interest to be paid goes down, while the amount to be paid toward principal goes up. This phenomenon is typical with amortization loans. The amount of interest paid during the course of the loan is relatively low—this is also typical for short-term installation loans. Mortgages and other long-term loans require that much higher portions of total payments go toward interest.

Example 3.5.9. Interest Portion Equals Principal Portion

In the early years of a mortgage, most of each monthly payment is applied to interest. For a 30-year mortgage with fixed rate r, when will the repayment of the principal portion of the monthly payment bypass the interest portion? The time t^*, for which the principal portion of the monthly payment equals the interest portion, is called the *equilibrium point*. So at time t^* half of the balance is used to pay the interest accrued.

Let R be the monthly payment on a 30-year mortgage with the principal P and the annual interest rate r. The balance after t years of payments is, according to Eq. (3.5.7),

$$B = R\frac{12}{r}\left[1 - \left(1 + \frac{r}{12}\right)^{-360+nt}\right].$$

The interest on the balance is $B \times r/12$, which we equate to $R/2$ to obtain

$$R \frac{12}{r} \left[1 - \left(1 + \frac{r}{12} \right)^{-360+12t} \right] \frac{r}{12} = \frac{R}{2} \quad \text{or} \quad 1 - \left(1 + \frac{r}{12} \right)^{-360+12t} = \frac{1}{2}.$$

After isolating the exponential part, we get

$$\left(1 + \frac{r}{12} \right)^{-360+12t} = \frac{1}{2}.$$

Raising to the exponent, we obtain

$$(-360 + 12t) \ln \left(1 + \frac{r}{12} \right) = \ln \frac{1}{2} = -\ln 2.$$

This yields the value of the equilibrium point:

$$t^* = \frac{360 \ln \left(1 + \frac{r}{12} \right) - \ln 2}{12 \ln \left(1 + \frac{r}{12} \right)} = 30 - \frac{\ln 2}{12 \ln \left(1 + \frac{r}{12} \right)}. \qquad (3.5.11)$$

The equilibrium point does not depend on the initial amount borrowed, but on the interest rate and the number of payments $(n = 12)$ per year over a 30-year mortgage. This point may be used for a comparison of two 30-year mortgages. Of course, if the term of the mortgage is not 30 years, the formula should be modified.

For instance, consider three 30-year mortgages (the amount of the loan does not matter) with different interest rates: 5.5%, 6.5%, and 7.5%. According to Eq. (3.5.8), the equilibrium points are

$$t^*_{5.5\%} = 30 - \frac{\ln 2}{12 \ln \left(1 + \frac{0.055}{12} \right)} \approx 17.36,$$

$$t^*_{6.5\%} = 30 - \frac{\ln 2}{12 \ln \left(1 + \frac{0.065}{12} \right)} \approx 19.307,$$

$$t^*_{7.5\%} = 30 - \frac{\ln 2}{12 \ln \left(1 + \frac{0.075}{12} \right)} \approx 20.729.$$

Therefore, about two-thirds of the time, the borrower pays more for interest than toward the principal.   □

As we see from the previous example, the borrower pays tons of interest during most of the lifespan of the mortgage. However, to make the cost of owning real estate more affordable, the tax code permits a deduction of all mortgage interest (but not the principal) that the borrower pays during a fiscal year.

Exercises for Section 3.5.

1. In the following exercises, find the principal, P, of a loan with the given payment, R, time period, t, and interest rate, r.

   (a) $R = \$1,500$, $t = 7$ years, $r = 6.12\%$ compounded quarterly.
   (b) $R = \$1,388.39$, $t = 16$ years, $r = 5.87\%$ compounded quarterly.
   (c) $R = \$500$, $t = 15$ years, $r = 7.25\%$ compounded monthly.
   (d) $R = \$799.99$, $t = 20$ years, $r = 8.725\%$ compounded monthly.
   (e) $R = \$80$, $t = 30$ years, $r = 6.325\%$ compounded biweekly.
   (f) $R = \$88.49$, $t = 17$ years, $r = 5.75\%$ compounded biweekly.

2. In the following exercises, find the amount of each payment, R, for a loan with the given principal, P, time period, t, and interest rate, r.

   (a) $P = \$1,500,000$, $t = 45$ years, $r = 7.62\%$ compounded quarterly.
   (b) $P = \$1,377,390$, $t = 30$ years, $r = 8.87\%$ compounded quarterly.
   (c) $P = \$500,000$, $t = 17$ years, $r = 9.75\%$ compounded monthly.
   (d) $P = \$799,900$, $t = 25$ years, $r = 7.725\%$ compounded monthly.
   (e) $P = \$5,000$, $t = 15$ months, $r = 6.25\%$ compounded monthly.
   (f) $P = \$7,999$, $t = 20$ months, $r = 7.725\%$ compounded monthly.
   (g) $P = \$80,400$, $t = 10$ years, $r = 8.325\%$ compounded biweekly.
   (h) $P = \$98,690$, $t = 11$ years, $r = 9.75\%$ compounded biweekly.

3. In the following exercises, find the time, t, in years, required to pay off a loan with the given principal, P, payment, R, and interest rate, r.

   (a) $P = \$2,500,000$, $R = \$127,250$, $r = 5.85\%$ compounded quarterly.
   (b) $P = \$1,800,000$, $R = \$870,000$, $r = 6.47\%$ compounded quarterly.
   (c) $P = \$450,000$, $R = \$3,500$, $r = 8.5\%$ compounded monthly.
   (d) $P = \$799,900$, $R = \$5,700$, $r = 7.725\%$ compounded monthly.
   (e) $P = \$250,500$, $R = \$1,530$, $r = 8.325\%$ compounded biweekly.
   (f) $P = \$875,000$, $R = \$2,750$, $r = 7.55\%$ compounded biweekly.

4. Suppose you can obtain a 12-year loan at an interest rate of 6.75% compounded quarterly. If you can afford to make payments of $900 each quarter, how much can you borrow?

5. How much money can you borrow at 7% interest compounded quarterly if you agree to pay $600 at the end of each quarter-year for 4 years and in addition a balloon payment of $1000 at the end of the fourth year?

6. How much money can you borrow at 10% interest compounded semiannually if the loan is to be repaid at half-year intervals for 5 years and you can afford to pay $1200 per half-year?

7. A university is buying 12 new copy machines for $3,199.98 each. After making a down payment of $4,000, the university takes a loan that will be amortized in 4 annual payments at 4.36% compounded annually. Determine the annual payment.

8. Find the monthly rent on a $220,000, 30-year mortgage at 6.75% interest compounded monthly.

9. A trucking company buys 5 large semitrailer trucks for $105,000 each. It agrees to pay for the trucks by a loan that will be amortized with 8 semiannual payments at 5.24% compounded semiannually. Find the value of each installment.

10. Fill out the amortization schedule for the first 4 months of a 30-year mortgage at 5.74% compounded monthly.

Payment Number	Payment	Interest Paid	Principal Paid	Balance
				$340,900
1				
2				
3				
4				

11. Fill out the amortization schedule for the first 4 months of a 30-year mortgage at 6.25% compounded monthly.

Payment Number	Payment	Interest Paid	Principal Paid	Balance
				$220,500
1				
2				
3				
4				

12. Fill out the amortization schedule for the first 4 months of a 30-year mortgage at 7.25% compounded monthly.

Payment Number	Payment	Interest Paid	Principal Paid	Balance
				$275,500
1				
2				
3				
4				

13. Fill out the amortization schedule for the first 4 months of a 30-year mortgage at 6.35% compounded monthly.

Payment Number	Payment	Interest Paid	Principal Paid	Balance
				$375,500
1				
2				
3				
4				

14. Assume that you have taken out a 30-year mortgage for $145,000 at an annual rate of 6.75%. Assume that you have decided to pay an extra $100 per month to amortize the loan more quickly. How much interest will you save compared with the original mortgage?

15. A person bought a new boat for $87,900. The person paid $10,000 for the down payment and financed the rest for 6 years at an interest rate of 6.8% compounded monthly. What is the monthly payment, and what is the total interest paid?

16. In terms of paying less in interest, which is more economical for a $120,000 mortgage: a 30-year fixed rate at 6.0% or a 20-year fixed rate at 6.75%? How much is saved in interest?

17. In terms of paying less in interest, which is more economical for a $250,000 mortgage: a 30-year fixed rate at 5.75% or a 20-year fixed rate at 6.25%? How much is saved in interest?

18. A couple buys a house for $425,000. They put 20% for the down payment and then finance the rest at 6.25% interest compounded monthly for 30 years.

    (a) Find the monthly rent.

    (b) Find the total amount to be paid for the house.

    (c) Find the total amount to be paid in interest.

    (d) Find the balance after 15 years.

19. A couple buys a house for $375,000. They put 20% for the down payment and then finance the rest at 7.15% interest compounded monthly for 30 years.

    (a) Find the monthly payment.

    (b) Find the total amount to be paid for the house.

    (c) Find the total amount to be paid in interest.

    (d) Find the balance after 16 years.

20. A family takes out a $235,000 mortgage at 5.76% interest compounded monthly with monthly payments for 25 years.

    (a) Find the monthly rent.

    (b) Find the total amount to be paid for the house.

    (c) If they decide to pay an extra $100 each month on the house payments, how long will it take to amortize the loan?

    (d) How much will they save in payments over the life of the mortgage by paying $150 extra each month?

21. A family takes out a $425,000 mortgage at 6.25% interest compounded monthly with monthly payments for 30 years.

    (a) Find the monthly payment.

    (b) Find the total amount to be paid for the house.

    (c) If they decide to pay an extra $250 each month on the house payments, how long will it take to amortize the loan?

    (d) How much will they save in payments over the life of the mortgage by paying $250 extra each month?

22. A mortgage at 7.36% interest compounded monthly with a monthly payment of $1096.24 has an unpaid balance of $17,250 after 340 months. Find the unpaid balance after 341 months.

23. A mortgage at 6.725% interest compounded monthly with a monthly payment of $1238 has an unpaid balance of $9,345 after 350 months. Find the unpaid balance after 351 months.

24. Consider a loan for 9 years at 8.25% interest compounded quarterly and a payment of $1450 per quarter-year.

    (a) What is the total amount of interest paid on the loan?
    (b) Compute the unpaid balance after 5 years.
    (c) How much interest is paid by the end of the fifth year?

25. Consider a loan for 7 years at 7.5% interest compounded quarterly and a payment of $1155.80 per quarter-year.

    (a) What is the total amount of interest paid on the loan?
    (b) Compute the unpaid balance after 5 years.
    (c) How much interest is paid by the end of the fifth year?

26. The price of a home is $366,000. The bank requires a 20% down payment. The buyer is offered two mortgage options: a 20-year fixed rate at 7% or 30-year fixed rate at 6.5%. Calculate the amount of interest paid for each option. How much does the buyer save in interest with a 20-year option?

27. The price of a home is $425,000. The bank requires a 20% down payment. The buyer is offered two mortgage options: a 15-year fixed rate at 7% or 30-year fixed rate at 6.6%. Calculate the amount of interest paid for each option. How much does the buyer save in interest with a 15-year option?

28. Samuel is taking a car loan for $14,566 for a term of 4 years at 3.6% interest compounded monthly.

    (a) Find his monthly payment.
    (b) Find the total amount he pays for the loan.
    (c) Find the total amount of interest he pays.
    (d) What is his unpaid balance after 3 years of payments?

29. Emily is taking a car loan for $9,451 for a term of 4 years at 2.9% interest compounded monthly.

    (a) Find her monthly rent.
    (b) Find the total amount she pays for the loan.
    (c) Find the total amount of interest she pays.
    (d) What is her unpaid balance after 3 years of payments?

30. A company is buying 12 new laptop computers paying $1200 each. They make a down payment of 15% of the total cost and agree to amortize the balance during 3 years with monthly payments at 4% compounded monthly. What is the monthly payment? How much interest would be paid during the lifespan of the loan?

31. A couple purchased a house with a $327,400 loan. The mortgage has an interest rate of 6.74% compounded monthly with monthly payments for 30 years.

    (a) Fill out the amortization schedule for the first 4 months of the mortgage.

Payment Number	Payment	Interest Paid	Principal Paid	Balance
1				
2				
3				
4				

    (b) How much would they save in total payments over the life of the loan by choosing the 15-year mortgage with the same interest rate rather than the 30-year mortgage?

    (c) In the 30-year mortgage with the rate 6.74%, how long will it take to pay off the loan, if the couple pays $100 extra each month?

    (d) After 15 years of payments, the couple decided to refinance the loan since they got an offer to take a new mortgage with the rate of 5.74% compounded monthly. How much will they save with this new 15-year mortgage?

32. A couple purchased a house with a $445,200 loan. The mortgage has an interest rate of 6.125% compounded monthly with monthly payments for 30 years.

    (a) Fill out the amortization schedule for the first 4 months of the mortgage.

Payment Number	Payment	Interest Paid	Principal Paid	Balance
1				
2				
3				
4				

    (b) How much would they save in total payments over the life of the loan by choosing the 20-year mortgage with the same interest rate rather than the 30-year mortgage?

    (c) In the 30-year mortgage with the rate 6.125%, how long will it take to pay off the loan, if the couple pays $150 extra each month?

    (d) After 12 years of payments, the couple decided to refinance the loan since they got an offer to take a new mortgage with the rate of 4.785% compounded monthly. How much will they save with this new 18-year mortgage?

33. Wealthy parents of a graduate student decide to buy a condominium for $185,000 instead of renting an apartment for their child. With a very good credit rating, the parents got a 30-year mortgage without a down payment to purchase the two-bedroom condo with the interest rate of 6.125% compounded monthly. Suppose they rent one room to another student, how much should they charge the tenant to cover the monthly mortgage payment?

34. Derive the formula similar to Eq. (3.5.11) to determine the equilibrium point for a loan with life span t and the number of compounds n.

35. A couple has a 30-year mortgage of \$345,800. After 12 years of paying the monthly rent on the loan, they find that the interest rates drop, and they decide to refinance their home. They found a deal with a 6.12% interest rate instead of the previous one of 7.25%. How much will they save if they accept a 18-year mortgage with this new rate?

36. A couple has a 30-year mortgage of \$275,300. After 14 years of paying the monthly rent on the loan, they find that the interest rates drop, and they decide to refinance their home. They found a deal with 5.75% interest rate instead of previous one of 6.785%. How much will they save if they accept a 16-year mortgage with this new rate?

37. In 1988, Congress renamed the Federal Guaranteed Student Loan program the Robert T. Stafford Student Loan program, in honor of U.S. Senator Robert Stafford, a Republican from Vermont, for his work on higher education. No payments are expected on the loan while the student is enrolled as a full- or half-time student. This is referred to as in-school deferment. Deferment of repayment continues for six months after the student leaves school either by graduating, dropping below half-time enrollment, or withdrawing. This is referred to as the grace period.

    Suppose that your Stafford student loans were of total \$17,000 at the time you start repayment. As a former undergraduate student, your plan calls for a 5.6% interest rate with a fixed amount of payment every month for 10 years. What is your monthly payment? How much will you pay in interest?

38. Suppose that your Stafford student loans were of total \$9,000 at the time you start repayment. As a former graduate student, your plan calls for a 6.8% interest rate with a fixed amount of payment every month for 10 years. What is your monthly payment? How much will you pay in interest during the course of the loan?

39. The Perkins Loan is the best student loan available: it has no origination or default fees, its interest rate is 5%, and it has 9-month grace periods. There is a 10-year repayment period with fixed monthly payments.

    Suppose that your Perkins student loan had accumulated to \$16,500 at the time you start repayment. What is your monthly payment? How much will you pay in interest during the course of the loan?

40. Your parents took out a federal PLUS loan to pay for the total cost of your graduate education. They borrowed \$18,900 at a 7.9% interest rate payable monthly with a fixed amount during 10 years. The origination fee of 3% and 1% federal default fee are included in the loan amount. What is their monthly payment?

41. To promote selling 2010 Ford Focus hatchback, a dealership was offering the choice of 3-year loan at a 3.5% interest rate compounded monthly, or \$1,000 cash back on the purchase of an \$18,330 car.

    (a) If you take the loan offer, how much will your monthly payment be?
    (b) If you take the \$1,000 cash-back offer and can borrow money from your local bank at 6.175% interest compounded monthly for three years, how much will your payment be?
    (c) Which of the two offers is more favorable for you?

42. Redo the previous exercise for the case when the dealership offers \$1,500 cash back on the purchase of a 2010 Ford Ranger compact pickup truck for \$17,820.

3.6 Personal Financial Decisions

Every student faces several financial decisions after graduation from a college or university. When you find a job, you need to find a place to live, and probably buy a car and many other appliances for where you will live. So after graduation, a student usually needs some funds and she or he is going to take out some loans to cover expenses. Also, a person with a steady income usually puts some money into a retirement plan or savings account. This section discusses these issues in detail.

Consumer loans are paid off with a sequence of regular payments very similar to mortgages. However, there are many different ways in which lenders could apply finance charges (it has become a custom to say finance charge instead of financial charge) on consumer loans that make the mathematics of borrowing money complicated. A borrower usually shops around to find the most suitable deal for a loan. How to compare different lenders' offers is a natural question in that matter. Annual interest rates do not disclose true information about the loans simply because other fees and conditions associated with borrowing are not taken into account.

One of the most widely used methods for determining finance charges on consumer loans is the **add-on** method, which was discussed in Example 3.2.11 on page 273. The main disadvantage of the add-on method is that each month the borrower is being charged interest on the entire principal, even though a part of it has been already paid with the regular payment. To understand the rate problem better, let us consider an example.

Suppose that you agree to repay a loan for $2000 plus interest in two yearly payments using an add-on interest rate of 6%. The simple interest formula calls for the value of interest to be $I = Prt = 2000 \times 0.06 \times 2 = \$240$. Thus, the amount to be paid in two installments is the principal ($2000) plus the interest $I = \$240$, which leads to $2,240. Each payment is therefore $\dfrac{\$2240}{2} = \$1120$, of which $1000 is being paid on principal and $120 is interest. After the first year, you pay $120 of interest on the amount of $2000 borrowed. Using the equation (3.2.5) for the rate with $F = 2120$, $P = 2000$, and $t = 1$, we get

$$r = \frac{1}{t}\left(\frac{F}{P} - 1\right) = \frac{2120}{2000} - 1 = 0.06 \quad \text{or} \quad 6\%.$$

The next year you pay $1120, of which $1000 goes to reduce the principal and $120 is interest. Thus, for the second year you have paid $120 interest on a $1000 loan. Using the same equation, we find the rate to be

$$r = \frac{1120}{1000} - 1 = 0.12 \quad \text{or} \quad 12\%.$$

This means that you are charged an additional $60 in interest per year. What is the true interest rate?

In 1968, Congress passed the federal **Truth in Lending Act** (TILA) as a part of the Consumer Protection Act. The sole purpose of TILA is to disclose the costs and charges associated with the loan in a uniform or standardized form so that consumers can shop around. TILA does not regulate the charges that may be imposed on consumer credit. It requires a lender to disclose what is known as the **annual percentage rate** (or **APR**) for any advertised loan. The APR must be disclosed to the borrower within three days of applying for a mortgage.

The definition of APR may vary from country to country. In the United States, the annual percentage rate is expressed as a single percentage number that represents the actual yearly cost of funds over the term of a loan. This includes any fees or additional costs associated with the transaction. Therefore, APR gives a more accurate picture of the cost of borrowing than the interest rate alone. Although it is not perfect, it gives you a suitable standard for comparing the percentage costs on different loans.

Loans or credit agreements can be different in terms of interest-rate structure, transaction fees, late penalties, and other factors. The APR is what credit costs you each year, expressed as a percentage of the loan amount. By law, credit card companies and loan lenders must show customers the APR to facilitate a clear understanding of the actual rates applicable to their agreements. Credit card companies are allowed to advertise interest rates on a monthly basis (e.g., 2% per month), but are also required to clearly state the APR to customers before any agreement is signed.

Let us return to our simple example about the loan of $2000 to be amortized in two years at 6% interest. Denoting by X the annual percentage rate, we use the simple interest formula (3.2.2) on page 268:

$$2000 \times X + 1000 \times X = 240 \qquad \text{(interest to be paid)}$$

Collecting like terms, we get $3000\, X = 240$. Solving this equation gives us $X = 0.08$. Hence, the annual percentage rate is 8%.

The annual percentage rate (APR) differs from annual percentage yield (APY). The APR is the annual rate of interest without taking into account the compounding of interest within that year. Alternatively, APY does take into account the effects of intra-year compounding. This seemingly subtle difference can have important implications for investors and borrowers.

For example, a credit card company might charge 1% interest each month; therefore, the APR would equal 12% (1% × 12 months = 12%). The APY for a 1% rate of interest compounded monthly would be 12.68% $[(1+0.01)^{12} - 1 \approx 12.68\%]$ a year. If you only carry a balance on your credit card for one month's period you will be charged the equivalent yearly rate of 12%. However, if you carry that balance for the year, your effective interest rate becomes 12.68% as a result of compounding each month.

As a borrower, you are always searching for the lowest possible rate. When looking at the difference between APR and APY, you need to be worried about how a loan might be "disguised" as having a lower rate. For example, when looking for a mortgage you are likely to choose a lender that offers the lowest rate. Although the quoted rates appear low, you could end up paying more for a loan than you originally anticipated. This is because banks will often quote you the annual percentage rate (APR). The APR is simply the periodic rate of interest multiplied by the number of periods in the year. For instance, you may come across with the following bank's quotes:

APR	APY or What You are Actually Paying		
	Semi-annual	Quarterly	Monthly
4%	4.04%	4.06%	4.07%
5%	5.06%	5.09%	5.116%
6%	6.09%	6.136%	6.168%
7%	7.12%	7.186%	7.23%
9%	9.20%	9.30%	9.38%

As you can see, even though a bank may have quoted you a rate of 4%, 6%, or 9% depending on the frequency of compounding (this may differ depending on the bank, state, country, etc.), you could actually pay a much higher rate. In the case of a bank quoting an APR of 9%, this rate does not consider the effects of compounding. However, if you were to consider the effects of monthly compounding, as APY does, you will pay 0.38% more on your loan each year—a significant amount when you are amortizing your loan over a 25- or 30-year period. This example should illustrate the importance of asking your potential lender what rate he or she is quoting when seeking a loan.

Example 3.6.1. Determining the Interest Rate on a Consumer Loan

Suppose a mortgage of $200,000 for 20 years has the annual interest of 6% compounded monthly and a monthly payment of $1432.86 (see Eq. (3.5.5) on page 306). What interest rate would be stated with the add-on method?

Solution: The total amount paid by the borrower is $20 \times 12 \times \$1432.86 = \$343,886.40$. Therefore, the interest paid on the loan is $\$343,886.40 - \$200,000 = \$143,886.40$. The interest paid per year is $\$143,886.40/20 = \$7,194.32$, which is about 3.597% of $200,000. Hence, with the add-on method, the interest rate would be given as 3.597%. $\square$

The calculation of APR for your mortgage is slightly complex because it requires solving the decreasing annuity equation (3.5.1) with respect to r. This equation has no analytical solution and different numerical methods are used for calculating the APR by lenders. The computation of APR can be easily done with TVM Solver available on your TI-83/84 Plus graphical calculator. Another way is to use the Internet—there are plenty of free APR calculators.

Example 3.6.2. Computing the Cost of "Leasing to Own"

Kyle uses a car that he leases from a nearby dealership. According to the lease agreement, Kyle drives a $20,000 new car for three years by paying $280 per month. He is allowed to drive at most 12,000 miles per year; after three years, he has an option either to terminate the lease agreement and return the car or to buy the car at 50% discount (that is, he will pay $10,000 for the car).

During three years renting period, Kyle would make 36 payments of $280, so he would pay a total of $36 \times \$280 = \$10,080$. Assuming that the rate of depreciation of the car is about 17%, the cost of the car after three years of use would be

$$\$20,000 \times (0.83)^3 = \$11,435.74.$$

Therefore, buying the 3-year old car for $10,000, Kyle saves $1,435.74. This is actually the markup of the dealership for this 3-year lease deal. The interest per year is $\$1435.74/3 = \$478.58$, which is about 2.39% of $20,000 using the add-on method.   □

When people want to find out how much their mortgages cost, lenders often give them quotes that include both loan rates and mortgage points or **discount points**. Discount points refer to an amount of money paid to a lender in order to obtain a loan at a specific interest rate (usually 0.25% off the original interest rate). By purchasing discount points, the borrower does not reduce the loan amount, but only the interest rate. Each mortgage point is a fee based on one percent of the total amount of the loan. The more points you pay, the lower the interest rate on the loan and vice versa. Borrowers typically can pay anywhere from zero to three or four points, depending on how much they want to lower their rates. An important feature of mortgage points is that they are tax deductible as a home mortgage interest if the deductions are itemized on Form 1040, Schedule A.

Both lenders and borrowers gain benefits from discount points. Borrowers gain the benefit of lowered interest payments down the road, while lenders benefit by receiving cash upfront instead of waiting for money in the form of interest payments over time. The decision to get mortgage points depends on two main factors: the length of time one plans to live in the house and the amount of the upfront payment. If you plan to stay in your home for a while, you would benefit from obtaining discount points because they will lower the interest rates for the long term. But if you need the lowest possible closing costs, choose the zero-point option on your loan program.

You can easily calculate the break-even payment point when refinancing by determining how much lower your new mortgage payment will be each month and dividing your total closing costs including points by the amount of your savings. This will tell you how many months it will take to recoup your expenses from refinancing; if you can live with the length of time it takes to get your money back then refinancing your mortgage probably makes sense in your situation.

Let R be your monthly rent without buying mortgage points and R_{discount} be the monthly payment with discount points. Then every month you save $R_{\text{save}} = R - R_{\text{discount}}$ amount of money. By paying your rent m months on the road of the loan, you reach the break-even point when your total savings exceeds the cost of the discount points:

$$m \, R_{\text{save}} = \text{Cost of Discount Points}. \qquad (3.6.1)$$

The formula (3.6.1) does not take into account a possible loss in an investment of the upfront amount paid to purchase the discount points.

Example 3.6.3. Discount Points and Monthly Payments for Mortgages
Congratulations, you just received a letter from the bank that you are qualified for a mortgage of $175,000—this is the exact amount needed to buy a nice condominium. The result of visiting the bank was quite confusing: you were given a couple of different options for your mortgage. Remember, the lender does not know all your circumstances and it is the bank's obligation to offer you different options. You have to make the right decision; otherwise you will wind up paying more for your loan.

Most lenders will offer you an option to buy some discount points. To be ready to accept their offer, you need to do your homework that consists of two parts. First, you estimate the period of time you plan to live in your property. The second step is to calculate break-even points for the number of months (or years) according to Eq. (3.6.1) by preparing the following table for a 30-year mortgage:

Number of Points	Interest Rate	Monthly Payment	Monthly Savings	Upfront Cost of Points	Break-even After
0	5.5%	$993.63			
1	5.4%	$982.68	$10.95	$1750	13.3 years
1	5.35%	$977.22	$16.41	$1750	8.9 years
1	5.25%	$966.36	$27.27	$1750	5.3 years
2	5.35%	$977.22	$16.41	$3500	17.8 years
2	5.25%	$966.36	$27.27	$3500	10.7 years
2	5.15%	$955.55	$38.08	$3500	7.7 years

The monthly payment (rent) in each row is calculated in accordance with Eq. (3.5.5) on page 306:

$$R = \frac{175000 * r/12}{1 - \left(1 + \frac{r}{12}\right)^{-360}}$$

for different values of rate, r. If a lender offers you a 30-year fixed mortgage of $175,000 at a 6.5 percent interest, you should prepare a similar table:

Number of Points	Interest Rate	Monthly Payment	Monthly Savings	Upfront Cost of Points	Break-even After
0	6.5%	$1106.12			
1	6.4%	$1094.64	$11.48	$1750	12.7 years
1	6.35%	$1088.91	$17.21	$1750	8.5 years
1	6.25%	$1077.51	$28.61	$1750	5.1 years
2	6.35%	$1088.91	$17.21	$3500	17 years
2	6.25%	$1077.51	$28.61	$3500	10.2 years
2	6.15%	$1066.15	$39.97	$3500	7.3 years

With all these data at hand, you make a second visit to the bank being ready to negotiate the lender's offer about discount points. If you are sure you will own the house for more than nine years, you save money by paying one point with a discount of at least 0.15%.

Remember that calculation of break-even points (presented in the last column of the tables) does not take into account a possible investment of the amount spent for purchasing the discount points. For instance, if you put $1750 into a savings account with, say, 2.5% interest compounded monthly, your balance after 6 years will be $1750 $\left(1 + \frac{0.025}{12}\right)^{6 \times 12} = \$2032.89$. To recoup this loss, you should have a saving of at least

$$\frac{\$2032.89}{12 \times 6} = \$28.23 \qquad \text{per month.}$$

Therefore, you are looking for a monthly payment below $1106.12 - \$28.23 = \$1077.89$, which corresponds to 6.25% interest.

Example 3.6.4. Calculating the APR for a Mortgage with Discount Points

Regarding the previous example, suppose you accept an offer for a mortgage of $175,000 with one discount point at 6.25% interest compounded monthly. If you pay $1750 upfront, you actually reduce the amount of the loan to $173,250. Your monthly payment would be $1,077.51 (see Eq. (3.5.5), page 306) since you borrow the original amount of $175,000. To overcome the cost ($1750) of buying one point you have to make more than 61 payments during 5.1 years. On average, the typical mortgage is refinanced or terminated after approximately five to seven years.

To find the APR for your mortgage, you have to solve Eq. (3.5.1) for a decreasing annuity:

$$173250 = \frac{1 - (1 + i)^{-360}}{i} * 1077.51 \qquad \text{where APR is } = 12 * i$$

There is no exact formula for the solution of the above equation; therefore, only a numerical option is available. Using a software package (for instance, Excel, Maple, Mathematica, or Maxima), one can find its solution to be $i = 0.05287886684\ldots$. Multiplication by 12 yields an APR for a 6.25% mortgage to be $\approx 6.345464021\%$. The same result provides the calculator's TVM Solver. The Truth-in-Lending Act requires the loan lender to specify that the mortgage with one discount point has an APR of 6.345%.

Now suppose you decide to sell your condo after six years. During that time, you will make $6 \times 12 = 72$ payments of total $72 \times \$1077.51 = \$77,580.72$. Your balance after 6 years will be $160,540.28 since you would pay only $14,459.72 toward your principal. This amount, $14,459.72, is called the **equity** of the real estate. Suppose that the net market value of your condo (amount received after subtracting all costs involving in selling the property) is $185,000, after 6 years. Then the equity of your condominium becomes $24,459.72.   □

Discount points may be different from an *origination fee* or broker fee. Discount points are always used to buy down the interest rates, while origination fees sometimes are fees the lender charges for the loan or sometimes are just another name for buying down the interest rate. The origination fee is a charge paid to the company originating your loan to cover their costs associated with creating, processing, and closing your mortgage. This payment is a part of the closing costs, usually the biggest part. You can usually have this amount added to your loan amount so you do not have to pay it in cash at closing. Most banks charge you a fee to give you a loan, but some lenders do not.

Investing is a lifelong process, and the sooner you start, the better off you'll likely be in the long run. The first part of that process is developing consistent savings habits. Regardless of whether you are saving for retirement, a new house, or that once-in-a-lifetime vacation, you will need a dedicated focus on saving. Regular contributions to savings or investment accounts are often the most productive.

Every penny that a person earns is subject to state and federal income taxes. However, there are special types of savings accounts that provide a shelter from these taxes. One of them is an **individual retirement account** (IRA), which is known in five options: the **traditional IRA, Education IRA, Roth IRA, SIMPLE IRA**, and **SEP IRA**— Simplified Employee Pension. Typically, money in an IRA cannot be withdrawn without penalty until the person reaches 59.5 years of age.

You can contribute up to $5,000 (if you are under 49 years old) per year into a traditional IRA. Your contributions may range from fully deductible to totally non-deductible depending on your status (single, joint, etc.), and your adjusted gross income. You can put away up to $500 per year into an education IRA, the money grows tax-free and has preferential tax treatment upon distribution to the beneficiary who uses it for authorized education expenses. These plans are not very common in that they are very restrictive on who can make contributions to them, the amount of total contributions allowable each year, and the limitations on what exact education expenses qualify. An employer can establish and fund a Simplified IRA, where the employer can put up to 15% of your compensation into a special IRA account. Contributions to a Roth IRA are not deductible when the funds are contributed, but the Roth IRA earnings accumulate tax-free and remain tax-free upon distribution. To be eligible to contribute, your adjusted gross income must be under $95,000 for singles and $150,000 for married couples, as of December 2000. You cannot withdraw your funds within the first five years after the establishment of the Roth IRA without a penalty. A SIMPLE IRA is a simplified employee pension plan that allows both employer and employee contributions, similar to a 401(k) plan, but with lower contribution limits and simpler (and thus less costly) administration.

The main advantage of a traditional IRA, compared to a Roth IRA, is that contributions are often tax-deductible. All withdrawals from a traditional IRA are included in gross income and subject to federal income tax.

Individual retirement accounts were introduced in 1974 with the enactment of the Employee Retirement Income Security Act (ERISA). The Roth IRA is named after Senator William V. Roth, Jr. The Roth IRA was introduced as part of the Taxpayer Relief Act of 1997. In the United States, a 401(k) retirement savings plan allows a worker to save for retirement and have the savings invested while deferring current income taxes on the saved money and earnings until withdrawal. These plans are mainly employer-sponsored: employees elect to have a portion of their wages paid directly into their individual 401(k) account, which is managed by the employer. Such payments are known as contributions. As a benefit to the employee, the employer can optionally choose to match part or all of the employee's contribution by depositing additional amounts in the employee's 401(k) account or simply offering a profit-sharing contribution to the plan.

Example 3.6.5. Calculating Values Associated with IRAs
Suppose that you are in a 28% marginal federal tax bracket. This means that you have to pay in tax 28% of your income. If you want to invest $5,000 into a savings account, you need to make $6,944.44 and pay $1,944.44 in tax. However, if you open a traditional IRA account, you do not need to pay taxes on it, which is $0.28 \times \$5000 = \$1,400$. Therefore, $1,400 will be saved in this transaction. On the other hand, you have to pay $1,944.44 in tax if you want to deposit this amount into a Roth IRA account.

Assume that you will keep $5,000 in the IRA account for 30 years that earns, on average, 3% per year. The balance after 30 years would be

$$\$5000 \times 1.03^{30} = \$12136.31.$$

If it is a Roth IRA account, this money is yours (you do not pay taxes on it). The percent of increase is $(12136.31 - 6944.44)/6944.44 \times 100\% \approx 74.76\%$. On the other hand, if you earn $5,000 that you would like to invest into a Roth IRA account, you actually deposit only 72% of it: $0.72 \times \$5000 = \$3600$. After 30 years, you earn $\$3600 \times 1.03^{30} = \$8738.14$, this money can be withdrawn without paying tax.

If you have a traditional IRA account, you have to pay 28% on income of $12136.31, and withdraw $8,738.14. This is the same amount you get when you deposit $3600 into a Roth IRA account. The rate of growth would be $(8738.14 - 5000)/5000 \times 100\% \approx 74.76\%$. □

The conclusion of Example 3.6.5 supports the observation that it does not matter in what account you deposit money—traditional or Roth—they both will accumulate the same amount. Deciding whether to open a traditional IRA or Roth IRA is a major choice with potentially large financial consequences. Both forms of the IRA are great ways to save for retirement, although the traditional IRA has a potential for an advantage: when people retire, they usually drop to a lower tax bracket.

Equivalence of traditional and Roth IRAs. Suppose that your tax bracket does not change during the period of all transactions. The net earning upon withdrawal from contributing P dollars into a traditional IRA account is the same amount as would result from paying taxes on the P dollars but then contributing the remaining funds into a Roth IRA account.

Example 3.6.6. Comparing Savings Account with a Traditional IRA
Suppose that you just received a bonus of $5,000, which you would like to invest. If you

prefer to put this money into a savings account, you have to pay 28% tax (assuming that you are in this marginal tax bracket), which means that only $3,600 is available for an investment. Suppose that the account pays 3.5% interest compounded annually. Since the interest earned each year will be taxed during that year, your actual interest earned will be $0.72 \times 3.5\% = 2.52\%$. After 30 years, the balance in the account would be

$$P\left(1 + i\right)^{30} = \$3600 \times 1.0252^{30} = \$7595.57 \qquad (i = r/1 = 0.0252).$$

Individual retirement accounts are investments with tax advantages for the investor. If you put this money into a traditional IRA, with the same interest rate, the balance would be

$$\$3600 \times 1.035^{30} = \$10104.45.$$

This amount is greater than the savings account by $2508.88, which is more than 1.3 times the amount of money in an ordinary savings account.

Example 3.6.7. Advantage of Starting an IRA Early
Suppose that two young people, call them Rachel and Ryan, begin full-time jobs on the same day. They both plan to retire in 40 years. Assume that any money they deposit into IRAs earns 4.2% interest compounded annually.

Suppose Rachel opens a traditional IRA account on her first working day and deposits $5,000 into it at the end of each year for fifteen years. After that she makes no contributions in this account and just lets the money earns the interest.

Since Rachel's account is an increasing annuity, Eq. (3.4.2) on page 296, with $D = 5000$, $n = 1$, $r = 0.042$, and $t = 15$, the future value in the account will be

$$F_{15} = D\frac{n}{r}\left[\left(1 + \frac{r}{n}\right)^{nt} - 1\right] = \frac{5000}{0.042}\left[\left(1 + \frac{0.042}{1}\right)^{1 \times 15} - 1\right] = \$101,618.92.$$

This money then earns interest compounded annually for 25 years. Using the formula $F = F_{15}(1 + r/n)^t$ with $r = 0.042$ and $n = 1$, it grows to

$$F_{\text{she}} = \$101,618.92 \times 1.042^{25} = \$284,228.45.$$

Suppose that Ryan waits 15 years before opening his traditional IRA and only after that starts depositing $5000 into the account at the end of each year until he retires (for 25 years). His balance after 25 years would be

$$F_{\text{he}} = D\frac{n}{r}\left[\left(1 + \frac{r}{n}\right)^{nt} - 1\right] = \frac{5000}{0.042}\left[\left(1 + \frac{0.042}{1}\right)^{1 \times 25} - 1\right] = \$213,928.96.$$

During 25 years, Ryan pays in $25 \times \$5000 = \$125000$ versus the $15 \times \$5000 = \$75000$ paid in by Rachel. Therefore, Ryan paid $50,000 more than Rachel did. Nevertheless, Rachel has $70,299.49 more in the IRA account than Ryan. □

When you deposit money into a bank account, you are making an **investment**. This means that you expect to gain some interest in your account. The income that you earn in an account as a percent of your investment is called its **return**. The bank investment is not risky because the government guarantees accounts up to $250,000. There are different instruments or assets for investments, such as property, commodity, stock, bond, and some others.

Used in the plural, **stocks** is often used as a synonym for shares, which is a unit of ownership. When you buy stocks of a certain company, you purchase a percent of its ownership. Any investor who owns some percent of the company is called a **shareholder**. This also means that you earn this percent of the profit the company makes. A shareholder (or stockholder) is either an individual or a company (including a corporation) that legally owns one or more shares of stocks in a joint stock company. When the company distributes all or part of its profits to shareholders, we say that a stockholder earns **dividends**.

Although ownership of 50% of shares does result in 50% ownership of a company, it does not give the shareholder the right to use a company's building, equipment, materials, or other property. In most countries, boards of directors and company managers have a fiduciary responsibility to run the company in the interests of its stockholders.

The shares of a company may in general be transferred from shareholders to other parties by sale or other mechanisms. The desire of stockholders to trade their shares has led to the establishment of **stock exchanges**. A stock exchange is an organization that provides a marketplace for trading shares and other derivatives and financial products. Today, investors are usually represented by stockbrokers who buy and sell shares of a wide range of companies on the exchanges. Buying or selling stocks is referred to as **trading**. The price of a share of a company is determined by the law of supply and demand. Like all commodities in the market, the price of a stock is sensitive to demand. When you sell the shares for more money than you initially paid, you make profit on this sale, called a **capital gain**. (There can also be a capital loss by selling for less than what you paid.)

The value of a share of a company at any given moment is determined by all investors voting with their money. If more investors want a stock and are willing to pay more, the price will go up. If more investors are selling a stock and there are not enough buyers, the price will go down. Investing in stocks may be a risky endeavor, which means that you may either lose all or part of your assets or gain some profit. To minimize a risk of investments, people buy stocks of different companies. In finance, a **portfolio** is an appropriate mix or collection of investments held by an institution or an individual.

There is another kind of investment you may consider—bonds. A **bond** is a debt investment in which an investor loans money to an entity (corporate or governmental) that borrows the funds for a defined period of time at a fixed interest rate. Thus a bond is like a loan: the issuer is the borrower (debtor), the holder is the lender (creditor). The **coupon** or coupon rate of a bond is the amount of interest paid per year (usually twice) expressed as a percentage of the face value of the bond. It is the interest rate that a bond issuer will pay to a bondholder. Bonds provide the borrower with external funds to finance long-term investments, or, in the case of government bonds, to finance current expenditure. Certificates of deposit (CDs) or commercial paper are considered to be money market instruments and not bonds. Bonds must be repaid at fixed intervals over a period of time.

Bonds and stocks are both securities, but the major difference between the two is that stockholders have an equity stake in the company (i.e., they are owners), whereas bondholders have a creditor stake in the company (i.e., they are lenders). Another difference is that bonds usually have a defined term, or maturity, after which the bond is redeemed, whereas stocks may be outstanding indefinitely.

Exercises for Section 3.6.

In problems 1 – 4, use the add-on method to determine the monthly payment.

1. $12,000 loan at 9% interest rate compounded monthly for one year.

2. $5,500 loan at 8% interest rate compounded monthly for one year.

3. $11,000 loan at 9.5% interest rate compounded monthly for two years.

4. $14,000 loan at 10.5% interest rate compounded monthly for three years.

In Exercises 5 –8, give the add-on interest rate.

5. A $3,400 loan for one year at 4.5% APR with a monthly payments of $290.27.

6. A $9,600 loan for one year at 5.25% APR with a monthly payments of $822.93.

7. A $19,000 loan for 2 years at 5.74966% APR with a monthly payments of $839.95.

8. A $15,500 loan for 3 years at 6.1494% APR with a monthly payments of $472.59.

9. Alyssa is 61 years old, is in the 33% marginal tax bracket, and has $450,000 in her traditional IRA. How much money will she have after taxes if she withdraws all the money from the account?

10. Connor is 65 years old, is in the 28% marginal tax bracket, and has $270,000 in his traditional IRA. How much money will he have after taxes if he withdraws all the money from the account?

11. If you are 22 years old, deposit $4000 each year into a traditional IRA for 48 years at 4.2% interest compounded annually, and retire at age 70, how much money will be in the account upon retirement?

12. If you are 25 years old, deposit $5000 each year into a Roth IRA for 45 years at 3.75% interest compounded annually, and retire at age 70, how much money will be in the account upon retirement?

13. Suppose that you are in the 35% marginal tax bracket for the duration of the account. You deposit $5000 into a traditional IRA that earns 5% interest compounded annually. Assuming no additional deposits are made, how much money will be in the account after 35 years? Suppose you withdraw all the money in the IRA after 35 years (assuming that you are eligible to do this). How much will you have after you pay the taxes on the money?

14. Repeat the previous exercise for a Roth IRA when you deposit an amount after paying taxes on $5000.

15. Suppose a lender gives you a choice between the following two 30-year mortgages of $220,000:
    Mortgage **A:**   5.75% interest compounded monthly, three points.
    Mortgage **B:**   5.92% interest compounded monthly, one point.

    Determine the break-even points for each mortgage if the original rent is $1326.09.

16. Suppose a lender gives you a choice between the following two 20-year mortgages of $180,000:
    Mortgage **A:**   6.2% interest compounded monthly, two points.
    Mortgage **B:**   6.4% interest compounded monthly, one point.

    Determine the break-even points for each mortgage if the original rent is $1352.65.

17. Suppose you want to save $5000 (before taxes) every year. You consider two options: either deposit this money into a systematic savings account paying 4.6% interest compounded annually, or contribute to a traditional IRA plan that earns 2.8% interest per year. After 10 years of investment, which plan is more profitable? Assume that you are in 28% marginal tax bracket.

18. Consider a 20-year mortgage of $230,000 at 6.15% interest compounded monthly with two discount points. Find the APR for this mortgage.

19. Consider a 30-year mortgage of $450,000 at 5.85% interest compounded monthly with two discount points. Find the APR for this mortgage.

20. A person deposits $3,500 each year for 30 years into an IRA plan that earns interest compounded annually. When she retires immediately after making the 30th deposit, the IRA account is worth $172,015.21.

    (a) Find the interest rate earned by the IRA plan over the 30-year period leading up to retirement.

    (b) Assume that the IRA plan continues to earn the interest rate found in part (a). How long can the retiree withdraw $30,000 per year?

21. A man deposits $4,000 in an IRA plan on his 25th birthday and on each subsequent birthday up to, and including, his 40th (26 deposits in all). The account earns 4.5% compounded annually.

    (a) If the man will not make any contributions to the account, how much will he have on his 60th birthday, assuming the account continues to earn the same rate of interest?

    (b) How much would be in the account if he continued to deposit the same amounts on each birthday until he reaches 60 years old?

22. If you are in the 28% marginal tax bracket, how much would you save in taxes if you deposited $7,500 into a 401(k) plan every year during 20 years?

23. If you are in the 33% marginal tax bracket, how much would you save in taxes if you deposited $9,500 into a 401(k) plan every year during 35 years?

Chapter 3 Review

Important Terms, Symbols, and Concepts

3.1 Review of Calculations

The word percent means per hundred. The percent of change is always in relationship to a previous, or **base amount**. The percent of change is computed in regard to a **new amount** as

$$\text{percent of change} = \frac{\text{new amount} - \text{base amount}}{\text{base amount}}.$$

Sales tax amount = tax rate × item's cost.

3.2 Simple Interest

When money is borrowed, a fee, called the **interest**, is charged for the use of that money for a certain period of time. An interest rate is the cost stated as a percent of the amount borrowed per period of time, usually one year. When you lend money, you expect to be paid back the initial investment, called the **principal** (amount of money that was borrowed) and the interest or, alternatively, the profit you earn. The amount of interest depends on the interest rate, the amount of money borrowed (principal), and the length of time that the money is borrowed. The formula for finding simple interest is:

$$\text{Interest} = \text{Principal} \times \text{Rate} \times \text{Time}.$$

The amount deposited or borrowed is the principal or present value. The total amount that is paid back (that is, the principal and the interest) is called the **future value**.

Discounted loans deduct the interest, called the discount, from the loan amount at the time the loan is made.

3.3 Compound Interest

Interest is the "rent" paid for the use of money. Compound interest has been called the eighth wonder of the world. It is the interest paid on the principal and reinvested interest:

$$F = P \left(1 + \frac{r}{n}\right)^{nt},$$

where F is the future value (the initial investment and the interest earned), P is the principal, or present value, the initial investment, r is annual interest rate, t is time (in years), n is the number of periods per year.

If a principal, P, is invested at an annual interest rate, r, earning **continuous compound interest**, then the amount after t years, called the future value, is given by

$$F = P e^{rt}.$$

The **growth time** of an investment is the time it takes for a given principal to grow to a particular amount. The **annual percentage yield** APY (also called the effective rate or the true interest rate) is the simple interest rate that would earn the same amount as a given annual rate for which interest is compounded. It is calculated as

$$\text{APY} = \left(1 + \frac{r}{n}\right)^n - 1.$$

If a principal is invested at the annual rate r compounded continuously, then the annual percent yield is given by

$$\text{APY} = e^r - 1.$$

The **rule of 72** says that to find the number of years required to double your money at a given interest rate, you just divide the interest rate into 72. The rule is only an approximation that is suitable for interest rates from 6% to 10%.

A **zero coupon bond** is a bond that is sold now at a discount and will pay its **par value** (= face value) at some time in the future when it matures.

3.4 Systematic Savings/Annuities

A sequence of payments or deposits made at equal periods of time is called an annuity or systematic savings plan. An increasing annuity is an investment that is earning interest, and into which regular payments of a fixed amount are made. If payments are made at the end of each time period, then the annuity is called an **ordinary annuity**. The time between payments is the **payment period**, and the time from the beginning of the first payment period to the end of the last period is called the **term** of the annuity. The **future value of an annuity** or the **value of an annuity** is the total lump sum of money including all deposits and interest earned:

$$F = D\frac{n}{r}\left[\left(1 + \frac{r}{n}\right)^{nt} - 1\right],$$

where F is the future value after t years, D is the amount of each deposit, r is the annual interest rate, n is the number of periods per year, and t is time (in years). A savings plan in which payments are made at the beginning of each period is called an **annuity due**.

An account that is established to accumulate funds to meet future obligations is called a **sinking fund**. The sinking fund payment is

$$D = \frac{Fr/n}{\left(1 + \frac{r}{n}\right)^{nt} - 1}.$$

3.5 Amortized Loans/Mortgages

A fixed installment loan is paid off with a sequence of equal periodic payments. A mortgage is a long-term loan to buy a real estate property, and for which the property is pledged as security for payment. The term of the mortgage is the number of years or months until the final payoff. The down payment is the upfront payment of the sale price that the buyer initially makes to secure the deal.

A decreasing annuity is an investment that is earning interest, and from which regular withdrawals of a fixed amount are made. If equal payments are made from an account (decreasing annuity) until the amount in the account is zero, the payment, called **rent**, R, and the present value, P, are related by the following formula:

$$P = R \frac{n}{r} \left[1 - \left(1 + \frac{r}{n} \right)^{-nt} \right].$$

Amortizing a debt means that the debt is paid off in a given length of time by equal periodic payments that include compound interest. The rent, R, is related with the principal (the borrowed amount), P, by the formula:

$$R = \frac{Pr/n}{1 - \left(1 + \frac{r}{n} \right)^{-nt}}.$$

An **amortization schedule** is a table that shows the interest due and the balance reduction for each payment of a loan. The amount that you owe on the loan at a given point in time is called the **loan balance**.

3.6 Personal Financial Decisions

Investment is the act of purchasing a monetary asset or resource in order to receive future income or a capital gain from its sale. A **bond** is a formal contract to repay borrowed money with interest at fixed intervals. The **coupon** of a bond is the amount of interest paid per year expressed as a percentage of the face value of the bond. **Stocks** are one of the most popular financial instruments in the world. Investors who have stocks of a company share the ownership of the company. A shareholder (or stockholder) is either an individual or a company (including a corporation) that legally owns one or more shares of stock in a joint stock company. When the company distributes all or part of its profits to shareholders, we say that stockholder earns **dividends**. In finance, a **portfolio** is an appropriate mix or collection of investments held by an institution or an individual.

Individual retirement accounts or IRAs were introduced in 1974 to provide some tax advantages for retirement savings in the United States. Traditional and Roth IRAs are established by individual taxpayers, who are allowed to contribute 100% of compensation up to a set maximum dollar amount. Contributions to the traditional IRA may be tax deductible depending on the taxpayer's income, tax filing status, and coverage by an employer-sponsored retirement plan. Roth IRA contributions are not tax-deductible, but all withdrawals are.

The **annual percentage rate** or **APR** is expressed as a single percentage number that represents the actual yearly cost of funds over the term of a loan. This includes any fees or additional costs associated with the transaction. Therefore, the APR gives a more accurate picture of the cost of borrowing than the interest rate alone in order to compare various loans you are considering.

Equity of the house is the difference between its current net market value and unpaid loan balance.

Dream Home Project. Task Description:

1. Conduct research on the Internet and identify your dream home that is currently being listed for sale. You may select any house that you wish, regardless of the asking price.

2. Provide a brief description of this home (1–2 paragraphs), including the asking price. You may include a picture or a web link to this dream home listing.

3. Conduct research on the Internet and identify a mortgage lender with the lowest current interest rate on a 30-year fixed rate mortgage that you would use to finance this home. Provide the lender's name and the interest rate of your mortgage.

4. Assume that you will be able to put down the 20% down payment on the home purchase. List the total mortgage cost (i.e., 80% of the asking price). You may include taxes and closing costs in the mortgage total if you wish.

5. Calculate the monthly payment and provide the amortization table for the first 6 months of the mortgage. The amortization table must clearly identify the following things for each of the first 6 monthly payments: month, payment amount, principal paid, interest paid, total interest, remaining mortgage balance.

Car Depreciation Project. Task Description:

1. Conduct research on the Internet regarding five car models that have been around for the last five years (for example, Honda Accord, Toyota Camry, BMW M3, etc.). You may select any car model that you wish, provided that the same model has been in existence for at least five years and is still currently on the market.

2. Using any valid Internet resource (such as Kelley Blue Book, Edmunds, NADA, etc) identify the current market value for each of the five selected car models. Assume that these five car models are five years old (i.e., originally manufactured five years ago from now) with a mileage of 100,000 and are still in decent condition.

3. Research online the current market value of each of the five car models, but this time identify their market value as new cars, not used (i.e., what do these five models cost as brand new this year models).

4. Calculate the total depreciation (the dollar amount and the percentage of the original price depreciated) for each of the five car models. Assume the cars were originally worth in 2005 what they are currently worth now as 2010 models and calculate the depreciation for each of the five car models.

5. Calculate the average annual depreciation for each car model for the past five years (both the dollar amount depreciated and the percentage of the original price depreciated).

6. Which of the five car models depreciated the most (in terms of dollars)? Which depreciated the most in terms of percentage of value lost? Which car model depreciated the least in terms of dollars? Which depreciated the least in terms of percentage of value lost?

7. Which of the five car models would you buy (assuming you had the money to purchase any of the five car models)?

Review Exercises.

1. Solve for the unknown variable x and round your answers to eight decimal places.

   (a) $x^{22} = 22$;   (b) $x^{37} = 37$;   (c) $x^{1.6} = 25$;   (d) $x^{0.23} = 23$;

   (e) $x^{-18} = 18$;   (f) $x^{-31} = 31$;   (g) $x^{0.42} = 3.275$;   (h) $x^{0.73} = 7.601$;

   (i) $\dfrac{1}{x^{0.28}} = 1.75$;   (j) $\dfrac{1}{x^{0.34}} = 5.8$;   (k) $(1+x)^{77} = 66$;   (l) $(2+x)^{31} = 2.8$;

   (m) $5^x = 41$;   (n) $7^x = 83$;   (o) $4^{2x} = \dfrac{1}{7}$;   (p) $7^{5x} = \dfrac{1}{13}$;

   (q) $(2.7)^{-8x} = 48$;   (r) $(1.7)^{-7x} = 77$;   (s) $9^{4x-3} = 5$;   (t) $2^{3x-7} = 7$.

2. Using a calculator, evaluate the following expressions within eight decimal places.

   (a) $\left(\dfrac{8}{7}\right)^{21}$;   (b) $\left(\dfrac{7}{9}\right)^{17}$;   (c) $\left(\dfrac{9}{4}\right)^{2.34}$;   (d) $\left(\dfrac{7}{5}\right)^{0.17}$;

   (e) $(1.125)^{60}$;   (f) $(0.571)^{13}$;   (g) $(1.263)^{0.15}$;   (h) $(0.539)^{0.11}$;

   (i) $\dfrac{5 \times 6}{7 \times 13}$;   (j) $\dfrac{19 \times 21}{16 \times 31}$;   (k) $\dfrac{1.8723 \times 0.4867}{2.7474 \times 0.1678}$;   (l) $\dfrac{8.8105 \times 0.8248}{8.3416 \times 0.2514}$;

   (m) $\dfrac{\ln 4.3197}{12 \ln 0.053}$;   (n) $\dfrac{\ln 3.705}{52 \ln 0.93}$;   (o) $\dfrac{\ln 0.6015}{12 \ln 7.15}$;   (p) $\dfrac{\ln 0.9792}{365 \ln 0.61}$.

3. In the 2009–2010 regular season, the Vancouver Canucks (NHL) won 49 games, lost 28 games, and lost 5 in overtime. What was their winning percentage for the 2009–2010 regular season?

4. The employment in the U.S. footware manufacturing industry drops from 44,800 employees in 1997 to 16,200 employees in 2007. What is the percentage of drop in employment during these ten years?

5. In 1985, the United States imported 36,407 thousand pairs of footwear from China. Twenty-two years later, in 2007, the country imported from China 2,004,322 thousand pairs of footwear. What is the percent of increase of footwear imports from China during 22 years?

6. The population of Ukraine was 45,905,341 in 2010. Ten years before it was 48,760,474 people. What was the percent of decrease?

7. The price of a color printer is reduced by 20% of its original price. When it still does not sell, its price is reduced by 20% of its already reduced price. What is the actual percent reduction from the original price?

8. If a dealer buys a Honda Odyssey LX 2010 van from the manufacturer for $25,007 and then sells it for $27,515, what is his markup (i.e., percentage increase from the original value)?

9. If a new Lincoln Town Car costs $46,925 and depreciates at a rate of 15% per year, what will be the value of the car in 4 years?

10. The price of a Chrysler Sebring sedan drops from $20,120 to $7,475 in 5 years. What is the annual percentage drop in its price?

11. A computer retailer buys a multimedia computer for $910 and then sells it for $999.99, what is her markup on the computer (i.e., percentage increase from the original value)?

12. How much income tax will a single male pay in 2010 if his gross income is $106,000? He contributed $7,000 to a tax-deferred IRA (Individual Retirement Account), paid $34,000 in mortgage interest, $7,800 in property taxes, and donated $3,500 to charitable institutions.

13. How much income tax will a single female pay in 2010 if her gross income is $98,000? She contributed $5,200 to a tax-deferred 401(k) plan, paid $21,000 in mortgage interest, $4,700 in property taxes, and made $3,400 in charitable contributions.

14. If you are not self-employed and earn $58,000 in 2010, what are your FICA taxes?

15. If you are not self-employed and earn $96,000 in 2010 and have a $24,000 bonus coming to you, what are your FICA taxes?

16. If you are self-employed and earn $170,000 in 2010, what are your FICA taxes?

17. In the following exercises, find the simple interest, I, earned on the given principal, P, for the time period, t, and interest rate, r, specified.
    (a) $P = \$12,200,\ r = 3.4\%,\ t = 6$ years;   (b) $P = \$3,450,\ r = 6.27\%,\ t = 4$ years;
    (c) $P = \$820,\ r = 2.8\%,\ t = 8$ months;   (d) $P = \$670,\ r = 4.7\%,\ t = 22$ months;
    (e) $P = \$4,500,\ r = 3.12\%,\ t = 40$ days;   (f) $P = \$7,510,\ r = 5.46\%,\ t = 80$ days.

18. In the following exercises, find the future value, F, of the given principal, P, earning simple interest for the time period, t, and interest rate, r, specified.
    (a) $P = \$35,000,\ r = 4.375\%,\ t = 8$ years;   (b) $P = \$7,300,\ r = 8.57\%,\ t = 3$ years;
    (c) $P = \$6,600,\ r = 5.8\%,\ t = 8$ months;   (d) $P = \$690,\ r = 5.7\%,\ t = 9$ months;
    (e) $P = \$340,\ r = 8.1\%,\ t = 9$ weeks;   (f) $P = \$750,\ r = 9.6\%,\ t = 35$ weeks;
    (g) $P = \$4,800,\ r = 10.5\%,\ t = 35$ days;   (h) $P = \$865,\ r = 12.5\%,\ t = 25$ days.

19. In the following exercises, find the present value, P, under simple interest given the future value, F, time period, t, and interest rate, r, specified.
    (a) $F = \$42,000,\ r = 5.3\%,\ t = 6$ years;   (b) $F = \$6,300,\ r = 4.32\%,\ t = 7$ years;
    (c) $F = \$700,\ r = 7.5\%,\ t = 25$ months;   (d) $F = \$800,\ r = 8.7\%,\ t = 10$ months;
    (e) $F = \$7,000,\ r = 9.5\%,\ t = 12$ weeks;   (f) $F = \$650,\ r = 10.2\%,\ t = 28$ weeks;
    (g) $F = \$5,700,\ r = 11\%,\ t = 40$ days;   (h) $F = \$6,350,\ r = 12\%,\ t = 35$ days.

20. In the following exercises, find the time, t, in years, for the given principal, P, to reach the future value, F, under simple interest for the interest rate, r, specified.
    (a) $F = \$4,500,\ P = \$4,200,\ r = 5.7\%$;   (b) $F = \$8,800,\ P = \$8,000,\ r = 8.5\%$;
    (c) $F = \$700,\ P = \$650,\ r = 7.825\%$;   (d) $F = \$850,\ P = \$800,\ r = 9.175\%$;
    (e) $F = \$7,200,\ P = \$7,000,\ r = 6.5\%$;   (f) $F = \$8,700,\ P = \$8,500,\ r = 7.6\%$.

21. In the following exercises, find the annual interest rate, r, for the given principal, P, to reach the future value, F, under simple interest for the time period, t, specified.
    (a) $F = \$47,000,\ P = \$45,000,\ t = 2$ years;   (b) $F = \$800,\ P = \$700,\ t = 3$ years;
    (c) $F = \$1000,\ P = \$900,\ t = 9$ months;   (d) $F = \$520,\ P = \$500,\ t = 8$ months;
    (e) $F = \$770,\ P = \$750,\ t = 40$ days;   (f) $F = \$660,\ P = \$650,\ t = 20$ days.

22. A loan of $3,167 has been made at a simple interest rate of 7.5% for four month. Find the loan's future value.

23. You borrow $3,400 from a friend and promise to pay back $3,500 in six months. What simple interest rate will you pay?

24. You plan to save $3,800 for a trip to France in two years. Should you purchase a certificate of deposit (CD) from a trade union bank that pays a simple interest rate of 8.7%? How much must you put in this CD now in order to have the $3,800 in two years?

25. In 1626, Peter Minuit, the first director-general of New Netherlands province, convinced the Wappinger Indians to sell him Manhattan Island for trinkets and clothes valued at about $24. If the Native Americans had put the $24 into a bank account at a 6% simple interest rate, how much would the account be worth after 384 years, in 2010?

26. Suppose a woman borrows $800 from a friend agreeing to pay it back in 18 months with 7.9% simple interest. How much will the woman owe her friend at the end of 18 months?

27. A $12,000 investment in the American Honda Company in 2000 rose in value to $14,000 in 2009. What was the rate of return, figured as an annual interest rate, for this investment?

28. A man borrowed $1,850 from his parents, agreeing to pay them back $2,000 in 3 months. What annual simple interest rate did he pay?

29. To complete the sale of a car, the seller accepts a 180-day note for $6,000 at 6% simple interest. Wishing to be able to use the money sooner, the seller sells the note to a third party for $6,080 after 90 days. What annual interest rate will the third party receive for the investment?

30. A woman received a salary bonus of $7,700. She needs $8,000 for the down payment on a boat. If she can invest her bonus into an account paying a simple interest of 6.125%, how long must she wait until she has enough money to make the down payment?

31. A client receives a $1,200 RAL (refund anticipation loan), which is paid back in 20 days. What is the annual rate of interest for this loan if the client paid a $29 fee?

32. You borrow $40,000 on a 12% discounted loan for a period of 1 year. Determine the loan's discount and the actual interest rate.

33. In order to pay for soccer uniforms, a school takes out a 10.2% discounted loan for 4 months of the amount of $350. Determine the loan's discount and the actual interest rate.

34. In order to pay for wedding expenses, you borrow $8,500 on a 10% discounted loan for six months. What is the loan's discount? What is the loan's actual interest rate?

35. You buy a T-bill for $1982.1 that pays $2,000 in 13 weeks. What simple interest rate, to the nearest tenth of a percent, does this T-bill earn?

36. To borrow money, you pawn your mountain bike. Based on the value of the mountain bike, the pawnbroker loans you $400. One month later, you get the mountain bike back by paying the pawnbroker $500. What annual interest rate did you pay?

In Problems 37 – 38, use the unpaid balance method to find the finance charge on the credit card account. Last month's balance, the payment, the annual interest rate, and any other transactions are given.

37. Last month's balance, $597.32; payment, $368; interest rate, 18.2%; bought a hockey ticket $60; returned shoes, $79.99.

38. Last month's balance, $1803.74; payment, $560; interest rate, 19.52%; bought clothes, $248.16; returned books, $90.

In Problems 39 – 42, use the average daily balance method to find the financial charge on the credit card account. The starting balance and transactions on the account for the month are given. Assume an annual interest rate of 19% in each case.

39. Month: June (30 days); previous month's balance: $814.34.

Date	Transaction
June 9	Made payment of $390
June 12	Charged $17.66 for gasoline
June 18	Charged $19.99 for T-shirt
June 26	Charged $26.95 for sandals

40. Month: July (31 days); previous month's balance: $739.04.

Date	Transaction
July 2	Charged $12.99 for a cap
July 11	Made payment of $440
July 17	Charged $145 for sunglasses
July 25	Charged $78.93 for food

41. Month: August (30 days); previous month's balance: $650.59.

Date	Transaction
August 3	Charged $300 for hotel
August 7	Charged $70 for a theater ticket
August 15	Made payment of $340
August 23	Charged $42.75 for restaurant

42. Month: September (31 days); previous month's balance: $592.87.

Date	Transaction
September 5	Made payment of $250
September 10	Charged $100 for spa
September 14	Charged $76.89 for perfume
September 28	Charged $120 for massage

Use the following commission schedule

Principal	Commission
Under $2,500	$19 + 1.9% of principal
$2,500 – $10,000	$27 + 1.3% of principal
Over $10,000	$50 + 0.8% of principal

to find the annual rate of interest earned by each investment in Problems 43 and 44.

43. An investor purchases 340 shares at $17.48 a share, holds the stock for 34 weeks, and then sells the stock for $19.37 a share.

44. An investor purchases 470 shares at $21.08 a share, holds the stock for 26 weeks, and then sells the stock for $23.71 a share.

45. In the following exercises, find the future value, F, of the given principal, P, earning compound interest for the time period, t, and interest rate, r, specified.

   (a) $P = \$62,000$, $t = 7$ years, $r = 8.31\%$ compounded annually.

   (b) $P = \$28,000$, $t = 3$ years, $r = 7.56\%$ compounded annually.

   (c) $P = \$37,000$, $t = 5$ years, $r = 7.75\%$ compounded quarterly.

   (d) $P = \$7,600$, $t = 9$ years, $r = 6.725\%$ compounded quarterly.

   (e) $P = \$11,710$, $t = 6$ years, $r = 8.21\%$ compounded monthly.

   (f) $P = \$14,170$, $t = 5$ years, $r = 8.18\%$ compounded monthly.

   (g) $P = \$16,200$, $t = 8$ months, $r = 10.25\%$ compounded weekly.

   (h) $P = \$33,000$, $t = 4$ months, $r = 9.8\%$ compounded weekly.

   (i) $P = \$15,800$, $t = 3$ years, $r = 5.76\%$ compounded daily.

   (j) $P = \$21,000$, $t = 4$ years, $r = 6.28\%$ compounded daily.

46. In the following exercises, find the present value, P, under compound interest given the future value, F, time period, t, and interest rate, r, specified.

   (a) $F = \$5,300$, $t = 9$ years, $r = 3.52\%$ compounded annually.

   (b) $F = \$5,700$, $t = 4$ years, $r = 3.89\%$ compounded annually.

   (c) $F = \$6,500$, $t = 5$ years, $r = 8.175\%$ compounded quarterly.

   (d) $F = \$2,500$, $t = 11$ years, $r = 3.725\%$ compounded quarterly.

   (e) $F = \$880$, $t = 6$ years, $r = 6.42\%$ compounded monthly.

   (f) $F = \$350$, $t = 5$ years, $r = 6.78\%$ compounded monthly.

   (g) $F = \$9,000$, $t = 7$ months, $r = 7.48\%$ compounded weekly.

   (h) $F = \$8,500$, $t = 5$ months, $r = 10.22\%$ compounded weekly.

   (i) $F = \$3,700$, $t = 5$ years, $r = 4.25\%$ compounded daily.

   (j) $F = \$7,200$, $t = 4$ years, $r = 5.76\%$ compounded daily.

47. In the following exercises, find the time, t, in years, for the given principal, P, to reach the future value, F, under simple interest for the interest rate, r, specified.

   (a) $F = \$8,200$, $P = \$8,000$, $r = 3.52\%$ compounded annually.

   (b) $F = \$7,300$, $P = \$6,500$, $r = 4.27\%$ compounded annually.

   (c) $F = \$37,000$, $P = \$35,000$, $r = 7.28\%$ compounded quarterly.

   (d) $F = \$25,900$, $P = \$25,000$, $r = 3.54\%$ compounded quarterly.

   (e) $F = \$820$, $P = \$790$, $r = 10.3\%$ compounded monthly.

   (f) $F = \$570$, $P = \$500$, $r = 9.28\%$ compounded monthly.

   (g) $F = \$17,300$, $P = \$16,500$, $r = 3.26\%$ compounded weekly.

   (h) $F = \$14,000$, $P = \$13,000$, $r = 4.16\%$ compounded weekly.

   (i) $F = \$25,000$, $P = \$22,000$, $r = 3.85\%$ compounded daily.

   (j) $F = \$35,000$, $P = \$32,500$, $r = 4.53\%$ compounded daily.

48. In the following exercises, find the annual interest rate, r, that will yield the given future value, F, starting from the given principal, P, over the specified time period, t.

   (a) $F = \$28,000$, $P = \$27,000$, $t = 5$ years compounded annually.
   (b) $F = \$16,200$, $P = \$15,000$, $t = 6$ years compounded annually.
   (c) $F = \$62,000$, $P = \$60,000$, $t = 3$ years compounded quarterly.
   (d) $F = \$38,000$, $P = \$35,000$, $t = 7$ years compounded quarterly.
   (e) $F = \$700$, $P = \$660$, $t = 9$ months compounded monthly.
   (f) $F = \$500$, $P = \$470$, $t = 11$ months compounded monthly.
   (g) $F = \$70,000$, $P = \$60,000$, $t = 10$ years compounded weekly.
   (h) $F = \$65,000$, $P = \$55,000$, $t = 15$ years compounded weekly.
   (i) $F = \$48,000$, $P = \$40,000$, $t = 8$ years compounded daily.
   (j) $F = \$58,000$, $P = \$48,000$, $t = 9$ years compounded daily.

49. In the following exercises, find the annual percentage yield (APY) for the given interest, r, compounded as specified.

   (a) $r = 7.33\%$ compounded quarterly; (b) $r = 5.64\%$ compounded quarterly;
   (c) $r = 5.23\%$ compounded monthly;   (d) $r = 4.71\%$ compounded monthly;
   (e) $r = 3.85\%$ compounded daily;   (f) $r = 2.715\%$ compounded daily.

50. For an account with an annual interest rate of 4.825%, find the annual percentage yield (APY) when it is compounded
   (a) quarterly,   (b) monthly,   (c) daily,   (d) continuously.

51. For an account with an annual interest rate of 3.25%, find the annual percentage yield (APY) when it is compounded
   (a) quarterly,   (b) monthly,   (c) daily,   (d) continuously.

52. Find the future value of an account after 11 years if $1200 is deposited initially and the account earns 4.12% interest compounded
   (a) annually,   (b) quarterly,   (c) monthly,   (d) daily.

53. Find the future value of an account after 13 years if $36,000 is deposited initially and the account earns 5.36% interest compounded
   (a) annually,   (b) quarterly,   (c) monthly,   (d) daily.

54. Find the future value of an account after 30 months if $820 is deposited initially and the account earns 4.48% interest compounded
   (a) quarterly,   (b) monthly,   (c) weekly,   (d) daily.

55. Find the future value of an account after 66 months if $30,000 is deposited initially and the account earns 5.72% interest compounded
   (a) quarterly,   (b) monthly,   (c) weekly,   (d) daily.

56. Bank A offers an interest rate of 5.73% compounded quarterly, whereas bank B offers 5.70% compounded daily. If you want to invest in a 1-year certificate of deposit, which bank should you choose?

57. Tuition at a state university grew by about 2.5% per year between 1999 and 2009. In 2009, the tuition was $4,800 for residents. Assuming that it will continue to grow at a rate of about 3.4% per year, what will the tuition be at the university in 2019?

58. Car prices grew on average by about 0.8% per year between 2007 and 2009. In 2010, the price for a new Chrysler 300 car was $27,260. Assuming that it will continue to grow at a rate of about 0.8% per year, what will be the cost of this car in 2012?

59. One-hundred-year railroad bonds were issued in 1909 when interest rates on railroad bonds averaged 7.8%. How much would a bond purchased for $1,000 in 1909 be worth in 2009 assuming

    (a) simple interest of 7.8% is paid?

    (b) the interest is compounded monthly with an annual rate of 7.8%?

60. Twenty-year recreation bonds were issued in 1989 when interest rates on recreation bonds averaged 5.64%. How much would a bond purchased for $1,000 in 1989 be worth in 2009 assuming

    (a) simple interest of 5.64% is paid?

    (b) the interest is compounded monthly with an annual rate of 5.64%?

61. Suppose you lend $24,500 to a friend, and your friend agrees to pay the loan back in 2 years at 6% interest compounded quarterly. How much will the friend owe you at the end of 2 years?

62. Suppose you borrow $9000 from your friend, and you agree to pay back the loan in 10 months at 7% interest compounded monthly. How much will you owe to your friend at the end of 10 months?

63. A house in 1969 cost $34,200; as of 2008, the median home price was $214,000. What is the rate of inflation based on the real estate growth?

64. A house was bought in 1989 for $80,500; in 2009, the house was sold for $214,000. What is the annual rate of growth (which is the inflation rate) in the cost of the house from 1989 to 2009?

65. Some Swiss banks pay negative interest: they charge the investor for keeping his or her money in a private account without asking where it comes from. Suppose a person deposited $100,000 in such account that charges 2% interest compounded annually. How much will be in the account after 5 years?

66. During Zachary's first year at college, his parents had been sending him $200 per month for incidental expenses. For his sophomore year, his parents decided instead to make a deposit into a savings account on August 1 and have their son withdraw $200 on the first of each month from September 1 to May 1. If the bank pays 2.75% interest compounded monthly, how much should Zachary's parents deposit?

67. Peru went through its worst inflation from 1988 to 1990. In September 1988, monthly inflation went to 132%. If this rate had continued, compounding monthly, for a year, how much would an item that cost 10 intis (the Peruvian currency) in 1988 be expected to cost next year? (In 1991, the government was forced to abandon the inti and introduce the nuevo sol as the country's new currency.)

68. A bank offers a variable rate for a three-year certificate of deposit, compounded monthly. For the first year, the bank will pay 3.0% interest, for the second your the rate will be 3.25%, and the last year it will be 4.0%. What is the blended APY for this three-year CD?

69. If a 2010 Jaguar XF car costs $51,150 and depreciates at a rate of 16% per year, what will be the value of the car in 4 years?

70. A promissory note will pay $32,000 at maturity 8 years from now. How much money should you be willing to pay now if money is worth 4.75% compounded continuously?

71. Is it more profitable to receive $8,500 now or $11,700 in 10 years? Assume that money can earn 3.6% interest compounded monthly.

72. Four banks offer different rates for a certificate of deposit:

Bank	Rate	Compounded
Alpha	3.856%	quarterly
Beta	3.855%	monthly
Gamma	3.854%	daily
Delta	3.853%	continuously

Find the APY (expressed as a percentage, correct to four decimal places) and the doubling times for each of the four banks and compare these CDs.

Problems 73 – 76 refer to zero coupon bonds. A zero coupon bond is a bond that is sold now at a discount and will pay its face value at some time in the future when it matures—no interest payments are made.

73. A zero coupon bond with a face value of $25,000 matures in 22 years. What should the bond be sold for now if its rate of return is to be 5.25% compounded annually?

74. A zero coupon bond with a face value of $15,000 matures in 12 years. What should the bond be sold for now if its rate of return is to be 4.75% compounded annually?

75. If you pay $5,276 for a 20-year zero coupon bond with par value of $10,000, what is your annual compound rate of return?

76. If you pay $15,762 for a 10-year zero coupon bond with par value of $25,000, what is your annual compound rate of return?

77. In the following exercises, find the future value, F, of a systematic saving plan with the given deposits, D, made at the end of the compounding periods for the time period, t, and interest rate, r, specified.

    (a) $D = \$600$, $t = 12$ years, $r = 6.75\%$ compounded annually.

    (b) $D = \$7200$, $t = 8$ years, $r = 3.125\%$ compounded annually.

    (c) $D = \$650$, $t = 25$ years, $r = 4.125\%$ compounded quarterly.

    (d) $D = \$775$, $t = 21$ years, $r = 3.39\%$ compounded quarterly.

    (e) $D = \$175$, $t = 5$ years, $r = 2.9\%$ compounded monthly.

    (f) $D = \$33$, $t = 4$ years, $r = 3.75\%$ compounded monthly.

    (g) $D = \$52$, $t = 17$ years, $r = 6.67\%$ compounded biweekly.

    (h) $D = \$25$, $t = 20$ years, $r = 5.2\%$ compounded biweekly.

78. In the following exercises, find the amount of each deposit, D, for a systematic saving plan with the deposits made at the end of the compounding periods given the future value, F, time period, t, and interest rate, r, specified.

    (a) $F = \$250,000$, $t = 35$ years, $r = 3.9\%$ compounded annually.

    (b) $F = \$120,000$, $t = 22$ years, $r = 5.25\%$ compounded annually.

    (c) $F = \$400,000$, $t = 38$ years, $r = 4.48\%$ compounded quarterly.

    (d) $F = \$75,000$, $t = 18$ years, $r = 5.41\%$ compounded quarterly.

    (e) $F = \$35,270$, $t = 25$ months, $r = 4.715\%$ compounded monthly.

    (f) $F = \$6,700$, $t = 8$ months, $r = 5.31\%$ compounded monthly.

    (g) $F = \$2,000,000$, $t = 36$ years, $r = 4.784\%$ compounded biweekly.

    (h) $F = \$1,000,000$, $t = 27$ years, $r = 6.448\%$ compounded biweekly.

79. In the following exercises, find the time, t, in years, for a systematic saving plan with the given deposits, D, made at the end of the compounding periods to reach the given future value, F, under the specified interest rate, r.

    (a) $F = \$270,000$, $D = \$2,000$, $r = 4.48\%$ compounded quarterly.

    (b) $F = \$320,000$, $D = \$3,500$, $r = 5.26\%$ compounded quarterly.

    (c) $F = \$12,350$, $D = \$125$, $r = 7.25\%$ compounded monthly.

    (d) $F = \$6,170$, $D = \$95$, $r = 5.12\%$ compounded monthly.

    (e) $F = \$41,000$, $D = \$250$, $r = 5.3\%$ compounded biweekly.

    (f) $F = \$53,000$, $D = \$200$, $r = 6.125\%$ compounded biweekly.

80. In the following exercises, use a calculator to find a minimum interest rate that guarantees achieving the given future value, F, for a systematic saving plan with the given deposits, D, made at the end of the compounding periods for the specified time period, t.

    (a) $F = \$7436.85$, $D = \$500$, $t = 14$ years compounded quarterly.

    (b) $F = \$288,881.09$, $D = \$2,000$, $t = 22$ years compounded quarterly.

    (c) $F = \$2501.42$, $D = \$120$, $t = 20$ months compounded monthly.

    (d) $F = \$10,078.73$, $D = \$200$, $t = 46$ months compounded monthly.

    (e) $F = \$20,920.46$, $D = \$40$, $t = 15$ years compounded biweekly.

    (f) $F = \$88,401.84$, $D = \$85$, $t = 20$ years compounded biweekly.

81. Suppose a 27-year-old woman bought an annuity (a systematic saving plan) that calls for deposits of $200 at the end of each month and pays an interest rate of 4.25% compounded monthly.

    (a) How much will the annuity be worth when the woman is 62 years old?

    (b) How much of the future value will be from deposits?

    (c) How much of the future value will be from interest?

82. A man has $72 deducted from his paycheck every month and the company matches this amount in its 401(k) plan. Suppose he continues to work for this company and contributes such payments for 22 years. The pension plan earns 3.28% interest compounded monthly.

    (a) How much will the plan be worth after 22 years?

    (b) How much of the future value will be from deposits?

(c) How much of the future value will be from interest?

83. A couple predicts that they will need to save $110,000 for their child's university education. Suppose their investment will earn 3.74% interest, compounded monthly.

   (a) If they make deposits at the end of each month for 18 years, how large must each deposit be?

   (b) If they deposit at the end of each month for 10 consecutive years, how large must each deposit be?

84. A man needs $2,994 for a down payment to purchase a car. Suppose that he can deposit $30 every month into a savings account that pays 4.85% interest compounded monthly.

   (a) How long should he wait until the investment will be worth $2,994?

   (b) How much of the future value will be from deposits?

   (c) How much of the future value will be from interest?

85. When Aaron takes a new job, he is offered a $1200 bonus now or the option of an extra $100 each month for the next year. If interest rates are 3.5% compounded monthly, which choice is better and by how much?

86. A city has a debt of $2,500,000 due in 12 years. How much money must it deposit at the end of each half-year into a sinking fund at 3.25% interest compounded semi-annually in order to pay off the debt?

87. How much money must you deposit into a savings account at the end of each quarter at 2.75% interest compounded quarterly in order to earn $350 interest during a 3-year period?

88. A philanthropist set up a 10-year trust fund to buy Christmas presents for needy children. The fund will provide $35,000 each year. How much money should the philanthropist put into the fund if the money earns 3.4% interest compounded annually?

89. In the following exercises, find the principal, P, of a loan with the given payment, R, time period, t, and interest rate, r.

   (a) $R = \$1,350$, $t = 8$ years, $r = 6.4\%$ compounded quarterly.

   (b) $R = \$1,245.49$, $t = 17$ years, $r = 5.785\%$ compounded quarterly.

   (c) $R = \$650$, $t = 14$ years, $r = 7.5\%$ compounded monthly.

   (d) $R = \$484.59$, $t = 18$ years, $r = 8.1\%$ compounded monthly.

   (e) $R = \$75$, $t = 35$ years, $r = 6.4\%$ compounded biweekly.

   (f) $R = \$67.49$, $t = 19$ years, $r = 5.25\%$ compounded biweekly.

90. In the following exercises, find the amount of each payment, R, for a loan with the given principal, P, time period, t, and interest rate, r.

   (a) $P = \$2,500,000$, $t = 43$ years, $r = 7.22\%$ compounded quarterly.

   (b) $P = \$2,720,580$, $t = 32$ years, $r = 8.25\%$ compounded quarterly.

   (c) $P = \$650,000$, $t = 19$ years, $r = 9.15\%$ compounded monthly.

   (d) $P = \$479,900$, $t = 21$ years, $r = 7.125\%$ compounded monthly.

(e) $P = \$8,250$, $t = 18$ months, $r = 6.45\%$ compounded monthly.

(f) $P = \$9,999$, $t = 22$ months, $r = 7.315\%$ compounded monthly.

(g) $P = \$75,200$, $t = 12$ years, $r = 8.24\%$ compounded biweekly.

(h) $P = \$69,790$, $t = 14$ years, $r = 9.35\%$ compounded biweekly.

91. In the following exercises, find the time, t, in years, required to pay off a loan with the given principal, P, payment, R, and interest rate, r.

(a) $P = \$1,750,000$, $R = \$30,000$, $r = 5.34\%$ compounded quarterly.

(b) $P = \$1,390,000$, $R = \$26,800$, $r = 6.22\%$ compounded quarterly.

(c) $P = \$765,000$, $R = \$5,500$, $r = 8.25\%$ compounded monthly.

(d) $P = \$849,500$, $R = \$6,200$, $r = 7.215\%$ compounded monthly.

(e) $P = \$275,800$, $R = \$2,200$, $r = 8.2\%$ compounded biweekly.

(f) $P = \$920,000$, $R = \$3,250$, $r = 7.24\%$ compounded biweekly.

92. An online lending service recently offered 4-year loans at 7.65% compounded quarterly. If you can afford to make payments of $450 each quarter, how much can you borrow?

93. How much money can you borrow at 5.75% interest compounded quarterly if you agree to pay $500 at the end of each quarter-year for 8 years and in addition a balloon payment of $1000 at the end of the eighth year?

94. How much money can you borrow at 9% interest compounded semiannually if the loan is to be repaid at half-year intervals for 4 years and you can afford to pay $1500 per half-year?

95. Three years ago you borrowed $25,000 at 11% interest compounded monthly, which was to be amortized over 6 years. Now you have acquired some additional funds and decide that you want to pay off this loan. What is the unpaid balance after making equal monthly payments for 3 years?

96. Find the monthly payment on a $360,000, 30-year mortgage at 6.25% interest compounded monthly.

97. You unexpectedly inherit $35,000 just after you have made the 48th monthly payment on a 30-year mortgage of $550,000 at 6.2% compounded monthly. If you use this money to reduce the loan principal, how it will affect the length of your mortgage?

98. Fill out the amortization schedule for the first 4 months of the 30-year mortgage at 5.875% compounded monthly.

Payment Number	Payment	Interest Paid	Principal Paid	Balance
				$685,900
1				
2				
3				
4				

99. Fill out the amortization schedule for the first 4 months of the 30-year mortgage at 6.72% compounded monthly.

Payment Number	Payment	Interest Paid	Principal Paid	Balance
				$281,300
1				
2				
3				
4				

100. Fill out the amortization schedule for the first 4 months of the 30-year mortgage at 7.16% compounded monthly.

Payment Number	Payment	Interest Paid	Principal Paid	Balance
				$319,440
1				
2				
3				
4				

101. Fill out the amortization schedule for the first 4 months of the 30-year mortgage at 6.27% compounded monthly.

Payment Number	Payment	Interest Paid	Principal Paid	Balance
				$388,150
1				
2				
3				
4				

102. Assume that you have taken out a 30-year mortgage for $341,100 at an annual rate of 6.12%. Assume that you have decided to pay an extra $125 per month to amortize the loan quicker. How much interest will you save compared with the original mortgage?

103. A family has a $275,000, 30-year mortgage at 6.75% compounded monthly. Find the unpaid balance after
(a) 10 years,   (b) 15 years,   (c) 20 years.

104. In terms of paying less in interest, which is more economical for a $240,000 mortgage: a 30-year fixed rate at 7.75% or a 20-year fixed rate at 8.75%? How much is saved in interest?

105. In terms of paying less in interest, which is more economical for a $187,500 mortgage: a 30-year fixed rate at 5.5% or a 22-year fixed rate at 6.5%? How much is saved in interest?

106. A couple buys a house for $478,000. They put 20% for a down payment and then finance the rest at 6.45% interest compounded monthly for 30 years.

   (a) Find the monthly payment.
   (b) Find the total amount to be paid for the house.
   (c) Find the total amount to be paid in interest.
   (d) Find the balance after 17 years.

107. A couple buys a house for \$382,600. They put 20% for a down payment and then finance the rest at 7.375% interest compounded monthly for 30 years.

    (a) Find the monthly payment.

    (b) Find the total amount to be paid for the house.

    (c) Find the total amount to be paid in interest.

    (d) Find the balance after 16 years.

108. A family takes out a \$256,900 mortgage at 5.9% interest compounded monthly with monthly payments for 30 years.

    (a) Find the monthly payment.

    (b) Find the total amount to be paid for the house.

    (c) If they decide to pay an extra \$120 each month on the house payments, how long will it take to amortize the loan?

    (d) How much will they save in payments over the life of the mortgage by paying \$120 extra each month?

109. A family takes out a \$524,600 mortgage at 6.49% interest compounded monthly with monthly payments for 30 years.

    (a) Find the monthly payment.

    (b) Find the total amount to be paid for the house.

    (c) If they decide to pay an extra \$227.62 each month on the house payments, how long will it take to amortize the loan?

    (d) How much will they save in payments over the life of the mortgage by paying \$227.62 extra each month?

110. A mortgage at 7.48% interest compounded monthly with a monthly payment of \$1369.17 has an unpaid balance of \$115,446.86 after 240 months. Find the unpaid balance after 241 months.

111. A mortgage at 6.375% interest compounded monthly with a monthly payment of \$1152.29 has an unpaid balance of \$86475.73 after 264 months. Find the unpaid balance after 265 months.

112. Consider a loan for 11 years at 7.77% interest compounded quarterly and a payment of \$1234.71 per quarter-year.

    (a) What is the total amount of interest paid on the loan?

    (b) Compute the unpaid balance after 6 years.

    (c) How much interest is paid by the end of the sixth year?

113. Consider a loan for 8 years at 6.66% interest compounded quarterly and a payment of \$1014.11 per quarter-year.

    (a) What is the total amount of interest paid on the loan?

    (b) Compute the unpaid balance after 4 years.

    (c) How much interest is paid by the end of the fourth year?

114. The price of a home is \$375,800. The bank requires a 20% down payment. The buyer is offered two mortgage options: a 20-year fixed rate at 7.34% or 30-year fixed rate at 6.95%. Calculate the amount of interest paid for each option. How much does the buyer save in interest with a 20-year option?

115. The price of a home is \$487,000. The bank requires a 20% down payment. The buyer is offered two mortgage options: a 15-year fixed rate at 7.8% or 30-year fixed rate at 6.6%. Calculate the amount of interest paid for each option. How much does the buyer save in interest with a 15-year option?

116. Michael is taking a car loan for \$18,716 for a term of 4 years at 4.6% interest compounded monthly.

   (a) Find his monthly payment.
   (b) Find the total amount he pays for the loan.
   (c) Find the total amount of interest he pays.
   (d) What is his unpaid balance after 3 years of payments?

117. Caitlin is taking a car loan for \$12,183 for a term of 4 years at 3.99% interest compounded monthly.

   (a) Find her monthly rent.
   (b) Find the total amount she pays for the loan.
   (c) Find the total amount of interest she pays.
   (d) What is her unpaid balance after 3 years of payments?

118. A couple is considered two mortgages. One of them is a 30-year loan with a 6.25% interest rate compounded monthly. Another 20-year mortgage has an interest rate of 6.36%. Calculate the equilibrium points for each mortgage and determine the percent of interest paid compared with the principal by the time it equals the equilibrium point.

119. Your parents take out a private loan of \$30,000 to cover your expenses during your four years in college. They agreed to pay it off within 10 years at the interest of 8.5% compounded quarterly. What will be their quarterly payment on this loan?

120. Suppose that your Stafford student loans plus accumulated interest sum to the total of \$22,000 at the time that you start repayment. If you select the standard repayment plan of a fixed rent payable every month during 10 years at 5.6% interest, what is your monthly payment?

121. A winner of the Powerball lottery can choose either an immediate lump sum of money which is worth 55% of the total prize or else an annuity. In the latter case, the advertised jackpot amount is paid in 20 annual payments, including one immediate payment. To keep up with inflation, each payment is 3.5% more than the previous year.

   (a) If the prize amount is 10 million, what will be the first payment? What is the amount of the last payment?
   (b) If the winner of the lottery deposits 55% of the jackpot amount into a savings account that pays 4% interest compounded monthly, how much will he or she have after 20 years?

122. Suppose that you have an investment that earns 0% in the first year, but 12% in the second year. What is the APR of your investment?

123. Suppose that you have an investment that earns 9% in the first year, loses 15% the second year, and earns 21% the third year. What is the APR of your investment?

124. Suppose that you make annual contributions of $4000 for 20 years into a systematic savings plan paying 5.5% interest compounded annually.

    (a) What is the value of the increasing annuity at the end of the 20th year?

    (b) Suppose that after 20 years, you decide to use the lump sum of money accumulated in the savings plan as a decreasing annuity over 10 years. What will be your annual rent?

    (c) Suppose that after you receive payments for 5 years, you withdraw the remainder of your decreasing annuity in a lump sum. How much will you receive?

125. Economists predict that the average interest rate for a 30-year mortgage will increase from 5.5% to 5.8% next year. However, home prices will decline by 11% during the same period of time. A couple is considering purchasing a $350,000 house. What would be their monthly payment today and in a year when they would buy the house?

126. Mario is going to buy a new car, but is not sure what maker to choose. He visited one dealership where he was offered a new $21,345 car that he could finance with a 4-year loan at 5.7% interest compounded monthly. Another dealership offers him a $20,875 car that he can finance with a 4-year loan at 5.8% interest compounded monthly. Which of these offers has the least monthly payment?

127. A couple has a 30-year mortgage of $375,000 borrowed at 7.25% interest compounded monthly. After 12 years of paying the rent on this loan, the couple decided to refinance it. They want to take a new 18-year mortgage to save $100 on their monthly payment. What interest rate are they looking for?

128. Erica deposits $2,500 annually into a Roth IRA that earns 2.75% compounded annually. Due to a change in employment, these deposits stop after 12 years, but the account continues to earn the same interest until Erica retires 38 years after the last deposit was made. How much is in the account when Erica retires?

129. Suppose a lender gives you a choice between the following two 25-year mortgages of $320,000:
    Mortgage **A:** 6.45% interest compounded monthly, two points.
    Mortgage **B:** 6.65% interest compounded monthly, one point.

    Determine the break-even points for each mortgage if the original rent is $2270.09.

130. Suppose a lender gives you a choice between the following two 30-year mortgages of $350,000:
    Mortgage **A:** 6.25% interest compounded monthly, three points.
    Mortgage **B:** 6.42% interest compounded monthly, one point.

    Determine the break-even points for each mortgage if the original rent is $2216.84.

131. A couple purchased a house 12 years ago for $150,000. The house was financed by paying 20% down and signing a 30-year mortgage at 7.6% on the unpaid balance. Equal monthly payments were made to amortize the loan over a 30-year period. After 144 payments, the couple wishes to refinance the house because of a need for additional cash. If the loan company agrees to a new 30-year mortgage of 80% of the new appraised value of the real estate, which is $164,000, how much cash (to the nearest dollar) will the couple receive after repaying the balance of the original mortgage?

References and Further Reading

[1] Barnett, Raymond A., Ziegler, Michael R., and Byleen, Karl E., *Finite Mathematics for Business, Economics, Life Sciences and Social Sciences*, Prentice Hall; 12th edition, Upper Saddle River, NJ, 2010.

[2] Blitzer, Robert, *Thinking Mathematically*, Prentice Hall; 5th edition, Upper Saddle River, NJ, 2010.

[3] Buchanan, J. Robert, *An Undergraduate Introduction to Financial Mathematics*, World Scientific Publishing Company; 2nd edition, Singapore, 2008.

[4] Gilbert, George T. and Hatcher, Rhonda L., *Mathematics Beyond the Numbers*, John Wiley & Sons, Inc., New York, 2000.

[5] Goldstein, Larry J., Schneider, David I., and Siegel, Martha J., *Finite Mathematics and Its Applications*, Prentice Hall; 10th edition, Upper Saddle River, NJ, 2009.

[6] Pirnot, Tom, *Mathematics All Around*, 4th Edition, Addison Wesley, Boston, 2010.

[7] Stefanica, Dan, *A Primer for the Mathematics of Financial Engineering*, FE Press, 2008.

[8] Zima, Petr and Brown, Robert, *Schaum's Outline of Mathematics of Finance*, McGraw-Hill; 2nd edition, New York, 1996.

The Internet Sites

http://www.answers.com

http://banking.about.com/

http://financial-dictionary.thefreedictionary.com/

http://www.investopedia.com

http://www.investorwords.com/

http://banking.about.com/library/calculators/bl_APR_calculator_code.htm
* APR calculator

http://www.debtconsolidationcare.com/calculator/apr.html
* APR calculator

http://www.bankrate.com/calculators/mortgages/mortgage-apr-calculator.aspx
* Mortgage APR calculator

http://www.interest.com/content/calculators/aprcalc.asp
* Mortgage APR calculator

Answers to Odd Problems

Chapter 1

§1.1, page 7

Number of votes:	2	3	1	1	1
A	1	3	4	4	2
B	2	4	2	2	4
C	3	1	3	1	1
D	4	2	1	3	3

1.

3. (a)

Number of votes:	2	1	2	3	1	4	2	1
First choice	A	A	E	C	C	B	E	C
Second choice	B	D	C	A	E	D	C	E
Third choice	C	E	D	B	D	A	B	D
Fourth choice	D	B	B	E	B	C	A	A
Fifth choice	E	C	A	D	A	E	D	B

(b) 16. (c) 9. (d) C. (e) E.

5. (a) 25; (b) 13; (c) Gennert; (d) Gennert.

7. (a) A, B, and C are eliminated. (b)

Votes	2	3	2	2	1	4	2	3
D	4	5	2	4	3	1	3	1
E	5	1	1	5	2	4	2	2

(c) D is the winner with 11 votes.

9. 23% $SD \succ F \succ G$; 31% $G \succ F \succ SD$; 46% $F \succ SD \succ G$.

11. 12.

§1.2, page 31

1. Italian.

3. Arriaza — 12; Barrett — 9; Coenen — 13.

5. Damato – 16; Erickson – 19 Foray – 17, Gamba – 12.

7. Lopes — 14; Mateo — 17; Renzi — 7; Songin — 17.

9. Andraka — 14; Beagen — 13; Frias — 16; Klotz — 17; Romano — 19.

11. 77.

13. 42.

15. **(a)** Winner: Faiola; loser: tie between Bertoncini and Lombardi.   **(b)** Winner: Bertoncini; loser: Lombardi.   **(c)** Winner: Bertoncini; loser: Lombardi.   **(d)** Winner: Faiola; loser: tie between Bertoncini and Lombardi.

17. **(a)** Winner: tie between Reis and Waldman; loser: tie between Reis and Sherman.   **(b)** Winner: Reis; loser: Reis.   **(c)** Winner: Reis; loser:Reis.   **(d)** Winner: Waldman; loser: Sherman.

19. Winner: Plurality – Chicken;   runoff – Steak;   Hare – Steak;   Coombs – Pasta.
    Loser: Plurality – Chicken;   runoff – Chicken;   Hare – Chicken;   Coombs – Steak.

21. Winner: Plurality – Steak;   runoff – Steak;   Hare – Steak;   Coombs – Fish.
    Loser: Plurality – Chicken;   runoff – Chicken;   Hare – Chicken;   Coombs – Chicken.

23. Winner: Plurality – Soccer and Basketball, a tie;   runoff – Soccer;   Hare – Basketball; Coombs – Volleyball.
    Loser: Plurality – Tennis;   runoff – Soccer;   Hare – Soccer;   Coombs – Tennis.

25. Winner: Plurality – M and T, a tie;   runoff – T;   Hare – T;   Coombs – W.
    Loser: Plurality – M;   runoff – F;   Hare – F;   Coombs – F.

27. Winner: Plurality – A;   runoff – B;   Hare – B;   Coombs – C.
    Loser: Plurality – A;   runoff – A;   Hare – A;   Coombs – D.

29. Winner: Plurality – F;   runoff – E;   Hare – E;   Coombs – B.
    Loser: Plurality – F;   runoff – F;   Hare – L;   Coombs – E.

31. Winner: Plurality – B;   runoff – A;   Hare – A;   Coombs – B.
    Loser: Plurality – A;   runoff – D;   Hare – D;   Coombs – C.

33. Winner: Plurality – B;   runoff – A;   Hare – A;   Coombs – A.
    Loser: Plurality – tie between B,C,D;   runoff – inconclusive;   Hare – C;   Coombs – C.

35. Plurality – A;   runoff – B;   Hare – C;   Coombs – D.

37. Plurality – C;   runoff – B;   Hare – D;   Coombs – A.

39. Plurality – D;   runoff – B;   Hare – A;   Coombs – E.

41. Plurality – E,   runoff – C+E,   Hare – A,   Coombs – B.

42. Plurality – A,   runoff – B,   Hare – C,   Coombs – D.

43. B is the winner with 7 first-place votes and the loser with 8 last-place votes.

45. B is the winner with 10 first-place votes and the loser with 9 last-place votes.

47. C is the winner and the loser.

49. C is the winner and the loser.

51. B is the winner and the loser.

53. A is the winner and the loser.

§1.3, page 54

1. 230.

3. $11 \times 17 = 187$.

5. The number of votes in each columns: 15, 25, 40, 50, 65, 75, 100, 130. The Borda winner is inconclusive: there is a tie between A and C—162 points each.

7. **(a)** 10.   **(b)** 1750.   **(c)** 377.

9.

	First	Second	Third	Total
Sam Bradford, Oklahoma	300	315	196	1726
Colt McCoy, Texas	266	288	230	1640
Tim Tebow, Florida	309	207	234	1575
Graham Harrell, Texas Tech	13	44	86	213
Michael Crabtree, Texas Tech	3	27	116	179
Shonn Greene, Iowa	5	9	32	65

11. Atlanta.

13. **(a)** Decrease ;   **(b)** Expand.

15. Gyoza – 24 points.

17. Hamburgers.

19. Winner: Donna, loser: Clara.

21. Winner: Potanin, loser: Berezovsky.

23. **(a)** D;   **(b)** C;   **(c)** D;   **(d)** C with $43\,11/12$ points;   **(e)** D;   **(f)** A.

25. **(a)** A;   **(b)** C;   **(c)** A;   **(d)** D with 93 points;   **(e)** A;   **(f)** A.

27. **(a)** A;   **(b)** C;   **(c)** A;   **(d)** C with 20.25 points;   **(e)** C;   **(f)** A.

29. **(a)** Rebello;   **(b)** tie between Rebello and Tellstone;   **(c)** tie between Rebello and Tellstone;   **(d)** Tellstone with $7\,31/60$ points;   **(e)** Rebello;   **(f)** Rebello.

31. **(a)** B;   **(b)** tie between B, D, and E;   **(c)** E;   **(d)** E with $19\,11/12$ points;   **(e)** D;   **(f)** D.

33. **(a)** E;   **(b)** B;   **(c)** E;   **(d)** B with $15\,31/60$ points;   **(e)** B;   **(f)** D.

35. **(a)** Movies;   **(b)** Golf;   **(c)** Golf;   **(d)** Theater with $14\,1/3$ points;   **(e)** Theater;   **(f)** Movies;   **(g)** Theater;   **(h)** Theater.

37. **(a)** B;   **(b)** D;   **(c)** A;   **(d)** D;   **(e)** D;   **(f)** D with $13\,17/30$ points;   **(g)** D;   **(h)** D.

39. **(a)** C;   **(b)** B;   **(c)** E;   **(d)** B;   **(e)** A;   **(f)** C with 10.6 points;   **(g)** A;   **(h)** A.

41. **(a)** A;   **(b)** B;   **(c)** B;   **(d)** D;   **(e)** D;   **(f)** D with 9.7 points;   **(g)** D;   **(h)** D.

43. **(a)** A;   **(b)** A;   **(c)** B;   **(d)** C;   **(e)** tie between B and C;   **(f)** A with $12\,23/30$ points;   **(g)** B;   **(h)** B.

45. Kevin Love with 54 points.

47.

	First	Second	Third	Total
Tim Lincecum, San Francisco	11	12	9	66
Chris Carpenter, St. Louis	9	14	7	62
Adam Wainwright, St. Louis	12	5	15	61

49.

Player	First	Second	Third	Forth	Fifth	Total
Toby Gerhart	7	2	2	4	0	27
Mark Ingram	4	8	2	1	0	30
Ndamukong Suh	3	5	6	0	1	25

51. B is the winner, but there are three losers: A, C, and D, with Borda scores of 13 points.

53. Winner: Borda – Engineering.   Losers: Borda – Liberal Arts.

55. Winner: Borda – C, Coombs – C.   Loser: Borda - A, Runoff – A, Hare - A.

57. Borda winner is Los Angeles; Borda loser is New York. None of plurality methods supports the Borda winner.

§1.4, page 78

1. 276.

3. Condorcet winner: A;   Condorcet loser: C.   Plurality winner: C, and plurality loser is C.

5. No Condorcet winner and no Condorcet loser.   Plurality winner: Schi; plurality loser: Borsch.

7. Condorcet winner: Beckham;   Condorcet loser: Cahill.

9. Transportation service is the Condorcet winner, which is supported by runoff, hare, Cooms. Athletic facilities is the Condorcet loser, which is supported by plurality, runoff, and IRV.

11. No Condorcet winner, and no Condorcet loser.   A:F=9:12; A:I=16:5, A:N=10:11; F:I=15:6, F:N=7:14; I:N=12:9.

13. C is the Condorcet winner, and D is the Condorcet loser because A:C = 11:26, B:C = 14:23, C:D = 25:12, A:D = 21:16, B:D = 19:18, C:D = 25:12.

15. No Copeland winner because three candidates, A, D, and E, have the same number of wins (3). Their Condorcet scores (A has +14, D has +3, and E has −6) identify A as the winner. Candidate C is the Condorcet loser.

17. A tie between B and C (A:B = 11:5, B:C = 10:6, B:D = 9:7, A:C = 7:9, C:D = 9:7), both have 2 Copeland points and 0 Condorcet scores.

19. A tie between A, B, and D, all have 2 Copeland points (A:B = 16:8, A:C = 13:11, A:D = 9:15, B:C = 14:10, B:D = 14:10, C:D = 10:14). The Condorcet scores (A has 4, B has 0, and D has +6) identify D as the winner.

21. C and D are two weak Condorcet winners, with 3 Copeland points. The Condorcet scores (C has +8 and D has +'12) identify D as the winner. There is no Condorcet loser.

23. E is the Copeland winner with 3 points, and A and C both have least number of Copeland points (1). The Condorcet scores (A has −24 and C has −22) identify A as the loser.

25. Pacheco is the winner under agenda Ramos, Loureiro, Pacheco. Loureiro is the winner under agenda Pacheco, Ramos, Loureiro.

27. D wins the election under agenda A, B, C, D. Alternative C wins the election under agenda D, B, A, C.

29. D wins the election under agenda A, B, C, D. Alternative B wins the election under agenda A, C, D, B.

31. A wins the election under agenda D, C, B, A. Alternative D wins the election under agenda A, B, C, D.

33. Plurality - A Hare - C, Borda - B, Sequential pairwise - D.

35. Plurality - C, Hare - B, Borda - A, Sequential pairwise - D.

37. Plurality - C, Hare - A, Borda - B, Sequential pairwise - D.

39. Plurality - A, Hare - B, Borda - E, Sequential pairwise - D.

41. Condorcet winner: KFC is supported by the Coombs method. Condorcet loser: Subway is supported by the Coombs. method.

43. Copeland winner: D is not supported by any of plurality methods. Copeland loser: B is supported by all four plurality methods.

45. Copeland winner: Singing, with 2 points, which is supported by none of plurality methods. Copeland loser: Shopping with 1 point, which is supported by none of plurality methods.

47. Copeland winner: Guitar with $2\frac{1}{2}$points, which is confirmed by the Coombs method. Weak Condorcet loser: Tambourine, which is confirmed by plurality, runoff, and IRV.

49. Condorcet winner: Horse, which is confirmed by IRV and Coombs. Copeland loser: Falcon (with Condorcet score -18), confirmed by plurality, runoff, Hare.

51. Ashley is the Condorcet winner, which supported by every point distribution method. Beatrice is the Condorcet loser, not confirmed by other methods.

53. The Condorcet winner is Football, which supported by all point distribution methods. The Condorcet loser is Auto Racing, not confirmed by other systems.

55. Rage Against the Machine is the Copeland winner, and Sublime is the Copeland loser. None of them is supported by the point distribution methods. Perl:Rage = 9:10, Perl:Red = 5:14, Perl:S = 12:7, Rage:Red = 12:7, Rage:S = 9:10, Red:S = 10:9.

57. H and D both have 2 Copeland points, but we declare Handzus the winner because the his Condorcet score is $+3$, whereas Doughty score is $+2$.

59. The Copeland winner is Bently, not supported by point distribution methods. The Copeland loser is Jaguar, supported by Borda.

61. A wins with the Condorcet score $+4$.

63. B wins with the Condorcet score $+3$.

65. Woonsocket wins with the Condorcet score $+9$.

67. Trix wins with the Condorcet score $+9$.

§1.5, page 97

1. McDonald's, with 8 approvals.

3. Ellis, with 7 approvals.

5. Twenty one, with 5 approvals.

7. Siberian, with 5 approvals.

9. Tony, with 45 approvals.

§1.6, page 110

1. A is the plurality winner, but B is the Condorcet winner.

3. The plurality winner is defeated by all others: A:B = 4:5, A:C = 4:5.

5. The plurality winner, A, is defeated by C when B is eliminated: A:C = 10:13.

7. B is the Here winner, which is defeated by C after elimination of A.

9. A is the IRV winner. If 2 voters switch for A in the second column with C, then B becomes the winner.

10. The Hare winner is A because it defeats B with score A:B = 9:6. If 3 voters in the second column switch A and B, then we will have

Ranking	Number of Voters (15)					
	4	3	2	3	2	1
First choice	A	A	B	C	D	E
Second choice	B	C	C	D	E	C
Third choice	C	B	A	A	C	D
Fourth choice	D	E	E	B	A	B
Fifth choice	E	D	D	E	B	A

From this table, it follows that C is the Hare winner.

11. A wins Hare, but when B drops, C becomes the winner.

13. E is the Hare winner. If A is dropped, C becomes the winner. So the IRV violates the IIA criterion.

15. A is the Coombs winner; however, the new election brings another winner—it becomes D. Therefore, the monotonicity criterion is violated.

16. C is the winner under IRV. Shifting up in the last column: D, A, C, B leads to A's win. So the monotonicity criterion is vilated.

17. "Reduce sports programs" is the Condorcet winner, but "Reduce art and music programs" is the Hare winner.

19. The Hare winner is C. However, making a clone of C, we get

Candidates	Number of Votes (20)							
	32	33	30	30	20	20	15	15
A	1	1	5	5	5	5	5	5
B	2	2	1	1	3	3	4	4
CC	3	4	2	3	1	2	2	3
C	4	3	3	2	2	1	3	2
D	5	5	4	4	4	4	1	1

which leads to another Coombs winner—B.

21. D is the Coombs winner; however, after adding one ballot $D \succ B \succ C \succ A$, the winner becomes B.

23. The Borda winner is A.

25. Borda winner is A with 51 points, but D defeats every others: D:A=17:9, D:B=14:12, D:C=14:12

27. Borda winner is Colorado, eliminating Arkansas and Kansas, Indiana becomes the winner.

29. The Borda winner is Ivana, the Nauru winner is Allison, and the Baldwin winner is Nadia. When Allison drops, the winner became Kristie.

31. Alternative A is the Copeland winner, but if B and C are dropped, D becomes the winner.

33. Monotonicity.

§1.7, page 128

1. Extended ranking: **(a)** $A \succ C \succ D \succ B$. **(b)** $A \succ D \succ C \succ B$. **(c)** $C \succ D \succ B \succ A$. **(d)** $C \succ B \succ D \succ A$. **(e)** $C \succ A \succ B \succ D$. **(f)** $C \succ B \succ A \succ D$. **(g)** $C \succ B \succ A \succ D$. Recursive ranking: **(a)** $A \succ C \succ B \succ D$. **(b)** $A \succ C \succ B \succ D$. **(c)** $C \succ B \succ A \succ D$. **(d)** $C \succ B \succ A \succ D$. **(e)** $C \succ B \succ A \succ D$. **(f)** $C \succ B \succ A \succ D$. **(g)** $C \succ B \succ A \succ D$.

3. Extended ranking: **(a)** $A \succ B \succ C \succ D$. **(b)** $B \succ A \succ C \succ D$. **(c)** $A \succ D \succ C \succ B$. **(d)** $A \succ D \succ B \sim C$. **(e)** $A \succ D \succ C \sim B$. **(f)** $A \succ C \succ D \sim B$. **(g)** $A \sim B \succ C \sim D$. Recursive ranking: **(a)** $A \succ C \succ D \succ B$. **(b)** $B \succ A \succ C \sim D$. **(c)** $A \succ C \succ D \sim B$. **(d)** $A \succ C \succ D \sim B$. **(e)** $A \succ D \succ C \sim B$. **(f)** $A \succ C \succ D \sim B$. **(g)** $A \succ C \succ D \sim B$.

5. Extended ranking: **(a)** $B \succ A \succ D \succ C$. **(b)** $D \succ B \succ A \succ C$. **(c)** $A \succ D \succ B \sim C$. **(d)** $D \succ A \succ B \succ C$. **(e)** $A \succ B \succ D \succ C$. **(f)** $A \succ D \succ C \succ B$. **(g)** $A \succ D \succ B \succ C$. Recursive ranking: **(a)** $B \succ A \succ D \succ C$. **(b)** $D \succ C \succ A \succ B$. **(c)** $A \succ D \succ C \succ B$. **(d)** $D \succ A \succ C \succ B$. **(e)** $A \succ D \succ C \succ B$. **(f)** $A \succ D \succ C \succ B$. **(g)** $A \succ D \succ C \succ B$.

7. Extended ranking: **(a)** $A \sim B \succ C \succ D$. **(b)** $B \succ A \succ C \succ D$. **(c)** $D \succ A \succ B \succ C$. **(d)** $D \succ A \sim B \succ C$. **(e)** $A \succ B \succ D \succ C$. **(f)** $B \succ D \succ A \succ C$. **(g)** $D \succ A \succ B \succ C$. Recursive ranking: **(a)** $A \sim B \succ D \succ C$. **(b)** $B \succ D \succ A \succ C$. **(c)** $D \succ B \succ A \succ C$. **(d)** $D \succ A \succ B \succ C$. **(e)** $A \succ D \succ B \succ C$. **(f)** $B \succ D \succ A \succ C$. **(g)** $D \succ A \succ B \succ C$.

9. Extended ranking: **(a)** $C \succ B \succ D \succ A$. **(b)** $B \sim C \succ D \sim A$. **(c)** $B \succ D \succ C \sim A$. **(d)** $B \succ C \succ D \sim A$. **(e)** $B \succ C \succ D \sim A$. **(f)** $B \succ C \succ D \sim A$. **(g)** $B \succ C \succ D \sim A$. Recursive ranking: **(a)** $C \succ B \succ D \succ A$. **(b)** $B \sim C \succ D \sim A$. **(c)** $B \succ A \succ C \sim D$. **(d)** $B \succ C \succ D \sim A$. **(e)** $B \succ C \succ D \sim A$. **(f)** $B \succ D \succ C \sim A$. **(g)** $B \succ C \succ D \sim A$.

Review Exercises for Chapter 1, page 132

1. Euler – 29, Gauss – 24, Kolmogorov – 26, Hilbert – 18, Leibniz – 27.

3. 77.

5. No majority winner, Arrighi wins the plurality.

7. Computer updates is the Hare winner.

9. Winner: plurality – a tie between A and B, runoff and IRV – B, Coombs – D. Loser: A using plurality, runoff, and Hare; Coombs loser is C.

11. Winner: A – plurality, runoff and IRV – C, E – Coombs. Loser: A – plurality and runoff, E – Hare, A – Coombs.

13. Winner: Soccer – plurality and runoff, IRV and Coombs – Basketball. Loser: Baseball – plurality and runoff, Basketball – Hare, Tennis – Coombs.

15. Winner: Tropicana – plurality and Coombs, Sweet Berry – runoff, Hare gives a tie between Lemon and Tropicana. Loser: Sweet Berry – all methods.

17. Cycling is the winner using all methods.

19. Tessa is the Borda, Nanson's, and Baldwin's winner, while Jess is the Nauru winner.

21. Banana is Borda's, Nanson's, and Baldwin's winner, Apple is Nauru winner.

23. A – 31, B – 34 (winner), C – 32, D – 33, E – 32.

25. A – 35, B – 26, C – 36 (winner), D – 27, E – 26.

27. D is the Copeland winner.

29. A is the Copeland winner.

31. Rondo wins with 3 points.

33. D wins.

35. New York is the winner.

37. Chernova is the winner.

39. Providence is the IRV winner; when New York drops, Denver becomes the winner. So the IIA criterion is violated.

41. E is the majority winner, but N is the Borda winner.

43. Hauser is the Coombs winner; however, if Johnson drops, the winner becomes Smith. Hence the IIA criterion is violated.

45. B is the Hare winner. If C drops, D becomes the winner.

47. B is the Hare winner. If 5 voters switch A with B (in the first column), D becomes the winner.

49. B is the Nauru winner with 30 $\frac{1}{2}$ points. If A drops, D becomes the winner. The IIA criterion is violated.

51. B is the Nanson winner. However, dropping the last option, A became the winner.

53. Ferrari is the Borda and Nauru winner. Second election: Lamborghini wins under Borda and Nauru systems.

55. A is the Condorcet/majority winner, but B is the Borda winner.

57. A is the Condorcet winner, while B is the Borda winner. The Condorcet and IIA criteria are violated.

59. Winner is Chicago Blackhawks with 2 $\frac{1}{2}$ points. However, dropping the last team (Columbus) leads to another winner: Nashville Predators.

61. C is Nanson's winner; but when D drops, B becomes the winner.

63. Extended ranking: **(a)** $A \sim B \succ D \succ C$. **(b)** $A \succ B \succ C \succ D$. **(c)** $A \succ D \succ C \succ B$. **(d)** $D \succ A \succ B \succ C$. **(e)** $A \succ B \succ D \succ C$. **(f)** $A \succ D \succ B \succ C$. **(g)** $A \sim D \succ B \succ C$. Recursive ranking: **(a)** $A \sim B \succ C \succ D$. **(b)** $A \succ D \succ B \succ C$. **(c)** $A \succ D \succ C \succ A$. **(d)** $D \succ A \sim B \sim C$. **(e)** $A \succ D \succ B \succ C$. **(f)** $A \succ D \succ B \succ C$. **(g)** $A \sim D \succ B \succ C$.

Chapter 2

§2.1, page 160

1. **(a)** States are bus routes and seats are buses. **(b)** 342. **(c)** A: $\frac{2770}{342} = 8.0994$; B: $\frac{3510}{342} = 10.2631$; C: $\frac{7630}{342} = 22.3099$; D: $\frac{1870}{342} = 5.4678$; E: $\frac{4920}{342} = 14.3859$; F: $\frac{5634}{342} = 16.4736$.

3. **(a)** 50; **(b)** 0.724; **(c)** $p_A \approx 3.4$; $p_B \approx 2$; $p_C \approx 2.7$; $p_C \approx 1.9$; $p_C \approx 3.7$.

5. **(a)** 10; **(b)** lower quotas: 0, 1, 3, 2, 1; upper quotas: 1, 2, 4, 3, 2; **(c)** 0.45 and 3.76.

7. 47257.

9. Indiana with absolute unfairness of 87262.4(4). The relative unfairness is 0.14.

11. Wisconsin with absolute unfairness of 108,768.5417. The relative unfairness is 0.183.

12. The absolute unfairness is 19.2; and relative unfairness is ≈ 0.57831.

14. The absolute unfairness is $\frac{1151762}{3} \approx 3.8392 \times 10^5$; and relative unfairness is ≈ 0.705.

15. Apportionment: 2+2+3+5+5=17 minimizes average constituencies, while 1+2+3+5+6=17 minimizes representative shares.

§2.2, page 170

1. Mississippi gets 4 seats, Colorado gets 6 seats, and Arizona gets 0 seats.

3. 0+0+2+4+4=10

5. 1+6+6+5+2=20

7. 0+2+6+9+13=30

9. $9+9+7+6=31$

11. Hamilton: $2+1+3+4=10$, Lowndes': $3+1+3+3=10$.

13. Hamilton: $15+10+4+1$; Lowndes' method is inconclusive.

15. Hamilton and Lowndes' are the same: $9+10+16+18+3=56$.

17. Hamilton: $34+24+24+8+0=90$; Lowndes': $33+24+24+8+1=90$.

19. **(a)** 123. **(b)** A – 15 seats, B – 28, E – 4, M – 71, S – 5. **(c)** The same as in part (b).

21. Hamilton: $69+15+36+24+48+22=214$; Lowndes': $69+16+35+24+47+23=214$.

23. Hamilton: $57+51+34+21+20=183$; Lowndes': $57+50+34+21+21=183$.

25. Hamilton: $107+24+93+31+37+31=323$; Lowndes': $106+25+92+32+37+31=323$.

27. Hamilton: $8+2+9+5+11+15+7+21+2+2+16+6+11+2+3=120$; \
Lowndes': $8+3+9+5+11+14+7+20+2+3+15+6+11+3+3=120$

29. Hamilton: $15+3+4=22$, $16+2+5=23$.

31. Hamiltons' allocation: $4+2+6+2=14$. With house size $h=15$: $5+1+7+2=15$.

33. $4+8+10+16+19=57$ and $4+8+9+17+20=58$.

35. Hamilton's method exhibits the Alabama paradox because its allocation is $24+5+59=88$ and $48+9+119=176$, while Lowndes' method allocates: $24+5+59=88$, $48+10+118=176$.

37. $3+0+8=11$ and $3+1+7=11$.

39. Hamilton: $14+0+3=17$ and $14+1+2=17$. Lowndes': $21+1+5=27$ and $22+1+4=27$.

41. State C lost 1 seat, but state B gained 1 seat.

§2.3, page 190

1. $12+11+9+14=46$, $2.538461538 < \mathbf{MD} \leqslant 2.\overline{54}$

3. $15+34+11+10=70$, $1192.085714 < \mathbf{MD} \leqslant 1210.4$.

5. Area: $3+3+2+7=15$, $8266.\overline{6} < \mathbf{MD} \leqslant 8466.\overline{6}$. Population: $5+4+6+0=15$, $368471.4286 < \mathbf{MD} \leqslant 401725$.

7. Impossible to apply.

9. $10+5+1+21+0=37$, $909402 < \mathbf{MD} \leqslant \frac{19490297}{21} \approx 928109.38095$.

11. $11+2+3+3+1=20$, $15.2500 < \mathbf{MD} \leqslant 17.\overline{27}$.

13. Hamilton: $7+6+7+7+6=33$; Adams: $7+6+7+7+6=33$, $3924.5 \leqslant \mathbf{MD} < 4150.1\overline{6}$; Jefferson: $7+6+7+7+6=33$, $3417.5 < \mathbf{MD} \leqslant \frac{24901}{7} \approx 3557.285714$.

15. $20+14+37+20+3+3+0=97$, $64877 < \mathbf{MD} \leqslant 66630.\overline{432}$.

17. Hamilton: $59+19+18+8+5+5+2+2+2+2+2=124$. Adams: $56+19+17+8+6+5+3+3+3+2+2$ $=124$ (violation of the quota property); $29567.58929 \approx \frac{1655785}{56} \leqslant \mathbf{MD} < 29769.2$. Jefferson: $61+20+18+7+5+5+2+2+2+1+1=124$ (violation of the quota rule); $26706.20968 \approx \frac{1655785}{62} \leqslant \mathbf{MD} < 26938.45000$.

19. Hamilton: 1+1+9+4=1=16. Adams: 1+2+8+4+1=16, $10235 \leqslant \mathbf{MD} < 10507$. Jefferson: 1+1+10+4+0=16, $7632.80 < \mathbf{MD} \leqslant 8188$.

21. 33+140+3+42+13+19=250, $94.4 < \mathbf{MD} \leqslant 49.54\overline{285714}$.

23. 8+7+1+6+8+0=30, $22 < \mathbf{MD} \leqslant 24.375$.

25. 17+28+29+10+18+18=120, $4.3793103448275862069 < \mathbf{MD} \leqslant 4.3\overline{8}$.

27. 8+2+9+4+11+15+7+22+1+2+16+6+12+2+3=120 with modified divisor: $28364.4 < MD \leqslant 28511$.

29. 1300+4234+2833+400+1233=10000 with modified divisor: $0.02999294284 < MD \leqslant 0.02999527633$.

31. 13+8+5+3+1=300 with modified divisor: $41.8\overline{3} < \mathbf{MD} \leqslant 42.\overline{6}$.

§2.4, page 211

1. 5+3+2=10 (initial allocation).

3. 4+9+10+13=36, $551.7142857 < \mathbf{MD} \leqslant 556.24$.

5. 2+3+1+1=7, $17.\overline{3} < \mathbf{MD} \leqslant 19.2$.

7. 1+4+4+4=13, $8.\overline{8} < \mathbf{MD} \leqslant \frac{32}{3.5} \approx 9.142857143$.

9. Hamilton: 5+4+8+4=21. Lowndes': 5+4+7+5=21. Jefferson: 5+4+8+4=21, $1062679.8 < \mathbf{MD} \leqslant 1098724.2$. Webster: 5+4+8+4, $1180755.333 < \mathbf{MD} \leqslant 1220804.\overline{6}$.

11. Hamilton: 3+3+6+2+3=17. Lowndes': 3+3+6+2+3=17. Jefferson: 3+3+7+1+3=17, $12 < \mathbf{MD} \leqslant \frac{87}{7} \approx 12.42857143$. Webster: 3+3+6+2+3=17 (initial allocation).

13. Hamilton: Lowndes': Jefferson: 6+1+4+0+1=12, $18.0 < \mathbf{MD} \leqslant 18.\overline{6}$. Webster: 5+1+4+0+2 =12, $22.\overline{6} < \mathbf{MD} \leqslant \frac{81}{3.5} \approx 23.14285714$.

15. Jefferson: 19+8+10+9+15+10=71, $3709.\overline{18} < \mathbf{MD} \leqslant 3716.875000$. Lowndes': 18+8+11+9+ 15+10=71. Webster: 18+8+11+9+15+10=71, $3864.\overline{216} < \mathbf{MD} \leqslant \frac{56244}{14.5} \approx 3878.896552$.

17. 25+26+5+3+29=88, $16.98113208 < \mathbf{MD} \leqslant \frac{418}{24.5} \approx 17.06122449$.

19. **(a)** Jefferson: 30+20+20+19+17+9+8+1=124, $31317.6 < \mathbf{MD} \leqslant 31329.1$. Webster: 29+20+ 19+19+17+10+9+1=124, $32197.53846 < \mathbf{MD} \leqslant \frac{628564}{19.5} \approx 32234.05128$. **(b)** Jefferson: 26+19+15+13+2=75, $24363.07407 < \mathbf{MD} \leqslant \frac{469792}{19} \approx 24725.89474$. Webster: 26+19+15+13 +2=75 (initial allocation). **(c)** Jefferson: 23+20+16+11+1+1=72, $5109.\overline{6} < \mathbf{MD} \leqslant 5166.4375$. Webster: 23+20+16+11+1+1=72, $5263.217391 < \mathbf{MD} \leqslant \frac{82663}{15.5} \approx 5333.096774$. **(d)** Jefferson: 4+4+3+3+2+1+0=17, $2391.4 < \mathbf{MD} \leqslant 2521.25$. Webster: 4+3+3+3+2+1 +1=17 (initial allocation). **(e)** Jefferson: 7+5+4+4+3+2=25, $1312.4 < \mathbf{MD} \leqslant 1342$. Webster: 7+5+4+4+3+2=25 (initial allocation).

21. Jefferson: 15+21+22+23+25=132, $11.208\overline{3} < \mathbf{MD} \leqslant \frac{237}{21} \approx 11.28571429$. Adams: 16+21+22 +23+24+26=132, $11.8\overline{3} \leqslant \mathbf{MD} < 11.85$. Webster: 15+21+22+23+25+26=132 (initial allocation).

23. Jefferson: 38+6+13+1+300+642=1000, $292.0256410 \approx \frac{11389}{39} < \mathbf{MD} \leqslant \frac{187723}{642} \approx 292.4034268$. Adams: 39+7+14+2+299+639=1000, $293.9297659 \approx \frac{87885}{299} < \mathbf{MD} < \frac{187723}{638} \approx 294.2366771$. Webster: 39+6+14+2+299+640=1000, $293.4390651 < \mathbf{MD} \leqslant \frac{187723}{639.5} \approx 293.5465207$.

25. Jefferson: (MN)311+(MS)433+(BN)433+(BS)420+(QN)298+(QS)295+(SI)71+(Bronx)502= 2763, $0.687872763 \approx \frac{346}{503} < \mathbf{MD} \leqslant \frac{205}{298} \approx 0.68691946$. Adams: 311+433+433+419+298+295+ 72+503=2763, $0.689737 \approx \frac{289}{419} < \mathbf{MD} < \frac{298}{432} \approx 0.689814$. Webster: 311+433+433+419+298+ 295+72+502=2763, $0.688915375 \approx \frac{289}{419.5} < \mathbf{MD} \leqslant \frac{298}{432.5} \approx 0.68901734$.

27. Jefferson: 26+15+13+19+22+29+10=134, $18.05 = \frac{361}{20} < \mathbf{MD} \leqslant \frac{527}{29} \approx 18.17241379$. Adams: 26+15+13+19+22+28+11=134, $19 < \mathbf{MD} \leqslant 19.04$. Webster: 26+15+13+19+22+28+11 =134 (initial allocation).

29. Jefferson: 61+48+47+35+28+18=237, $10.22448980 \approx \frac{501}{49} < \mathbf{MD} \leqslant \frac{186}{18} \approx 10.\overline{3}$. Adams': 61+48+47+35+28+18=237, $10.535714 \approx \frac{295}{28} \leqslant \mathbf{MD} < 10.55$. Webster: 61+48+47+35+28 +18=237 (initial allocation).

31. Jefferson: 3+2+6+3+4+2+2=22, $15337.5 < \mathbf{MD} \leqslant 15721$. Adams': 3+2+5+3+4+3+2=22, $20450 \leqslant \mathbf{MD} < 22000$. Webster: 3+2+6+3+4+2+2=22, $17600 < \mathbf{MD} \leqslant \frac{98690}{5.5} = 17943.\overline{63}$.

33. Jefferson: 106+103+97+77+50+43+37=513, $130746.3458 \approx \frac{13989859}{107} < \mathbf{MD} \leqslant \frac{12691043}{97} \approx$ 130835.4948. Adams: 106+102+96+77+51+44+37=512, $133142.3235 \approx \frac{13580517}{102} \leqslant \mathbf{MD} <$ $\frac{13989859}{105} \approx 133236.7524$. Webster: 106+103+96+77+51+43+37=513, $131780.7816 \approx \frac{5732464}{43.5}$ $< \mathbf{MD} \leqslant \frac{6662309}{50.5} \approx 131926.9109$.

35. Initial allocation: 120=8(CT)+2(GA)+9(MD)+5(NH)+11(NY)+14(PA)+7(SC)+21(VA)+ 2(DE)+2(KY)+16(MA)+6(NJ)+12(NC)+2(RI)+3(VT).

37. 153+127+37+39+20+22=398, $171292.8730 < \mathbf{MD} \leqslant 171413.5584$.

§2.5, page 238

1. **(a)** Arkansas;   **(b)** Colorado;   **(c)** Delaware;   **(d)** Florida.

3. 3+18+2=23, $376657.4286 < \mathbf{MD} < 376776.4000$.

5. 5+2+17=24, $1416709.910 \times 10^7 < \mathbf{MD} < 1421051.992 \times 10^7$.

7. (A)3+(B)2+(C)2+(D)1=8, $35.51760127 < \mathbf{MD} < 39.19183589$.

9. 2+7+49+69=127, $30.00153056 < \mathbf{MD} < 30.01539815$.

11. 36+32+7+5=80, $1.120132568 < \mathbf{MD} < 1.121237978$.

13. 56+51+64+38+74+19=302, $332.2430986 < \mathbf{MD} < 332.6697716$.

15. (ME)67+(NH)35+(VT)23+(MA)149+(RI)26+(CT)103=403, $0.9778046039 < \mathbf{MD}$ < 0.9780192939.

17. 2+7+2+2+2+2=17, $74.29862527 < \mathbf{MD} < 86.97413406$.

19. Hamilton: 11+16+1+5+13+9=55. Lowndes': 11+15+2+5+13+9=55. Jefferson: 11+16+1+5+13+9=55, $3.08\overline{3} < \mathbf{MD} \leqslant 3.\overline{1}$. Webster: 11+16+2+5+13+8=55, $3.294117647 < \mathbf{WMD} \leqslant 3.\overline{3}$. Hill-Huntington: 11+16+2+5+13+8=55, $3.299831644 <$ $\mathbf{MD} < 3.356585567$.

21. 192+10+7+274+4+513=1000, $127.2367928 < \mathbf{MD} < 127.2527434$.

23. Hamilton: 12+14+6+1=33. Lowndes': 12+14+6+1=33. Jefferson: 12+14+6+1=33, $1300 <$ $\mathbf{JMD} \leqslant 1308.\overline{3}$. Adams: 12+14+6+1=33, $1392.8571 \approx \frac{19500}{14} \leqslant \mathbf{AMD} < 1427.\overline{27}$. Webster: 12+14+6+1=33 (initial allocation). Hill-Huntington: (A)12+(B)14+(C)6+(D)1=33, $1345.627841 < \mathbf{MD} < 1366.509599$.

25. Hamilton: 1026+476+352+274+180+192=2500.
    Lowndes': 1026+475+352+274+180+193=2500.
    Jefferson: 1027+476+352+274+179+192=2500, $8053.5\overline{2} < \textbf{JMD} \leqslant \frac{3834340}{476} \approx 8055.3361$.
    Adams: 1026+475+352+274+180+193=2500, $8072.29473 \approx \frac{3834340}{475} \leqslant \textbf{AMD} < 8072.7092$.
    Webster: 1026+476+352+274+180+192=2500, $8063.683117 < \textbf{WMD} \leqslant \frac{3834340}{351.5} \approx$. Hill-Huntington: 1026+476+352+274+180+192=2500, $8063.710317 < \textbf{MD} < 8063.810978$.

27. Hamilton: 30+28+28+29+29=144. Lowndes': 30+28+28+29+29=144.
    Jefferson: 30+28+28+29+29=144, $60.27586207 < \textbf{JMD} \leqslant \frac{1750}{29} \approx 60.34482759$. Adams: 30+28+28+29+29=144, $62.42857143 \approx \frac{1748}{28} \leqslant \textbf{AMD} < 62.5$. Webster: 30+28+28+29+29 =144, $61.\overline{3} < \textbf{WMD} \leqslant \frac{1750}{28.5} \approx 61.40350877$. Hill-Huntington: 30+28+28+29+29=144, $61.34277432 < \textbf{MD} < 61.41296057$.

Review Exercises for Chapter 2, page 246

1. Absolute unfairness: 70244.2 = 728472.2 - 658228. Relative unfairness ≈ 0.106717.

3. Minimum of average constituencies: 2+3+3+4+4=16. Minimum of representative shares: 1+2+3+5+5=16.

5. 0+3+2+2+3=10.

7. 0+1+2+4+5+3=15.

9. Hamilton: 1+5+2+0=8; Lowndes': 1+4+2+1=8.

11. Hamilton: 26+5+11=42, 27+4+12=43.

13. Hamilton's allocation: 3+1+4+6=14. With house size $h = 15$, Hamilton's method gives: 2+1+5+7=15.

15. Hamilton: 17+2+9+4=32, 18+2+10+3=33. Lowndes: 12+2+6+3=23, 13+2+7+2=24.

17. 1+2+3+4+4=14, 1+2+2+5+5=15.

19. 14+8+5+2+2=31, 15+9+5+2+1=32.

21. Natural quotas: $q_1 = 13.35877863, q_2 = 11.26717557, q_3 = 10.54961832, q_4 = q_5 = 4.412213740$.

23. 14+12+11+6+4=47, 15+13+11+6+3=48.

25. 8+17+7+15+9=56, 7+18+7+16+9=57.

27. 13+13+4+23+5=58, 13+14+4+24+4=59.

29. Hamilton: 1+4+13=18 and 1+5+12=18. Lowndes': 2+8+22=32 and 3+8+21=32.

31. 7+16+4+8=35, $5041.400000 < \textbf{MD} \leqslant \frac{35491}{7} \approx 5070.142857$.

33. 3+3+8+13+18+14=59, $203095.75 < \textbf{MD} \leqslant \frac{3687050}{18} = 204836.\overline{1}$.

35. 7+3+4+40+27+4=85, $2659.8 < \textbf{MD} \leqslant 2675.\overline{3}$.

37. 34+25+20+11+7+3=100, while the initial allocation is 32+25+20+11+7+2=97.

39. 30+29+9+7+5+5+5=90 (initial allocation).

41. Initial allocation: 190+5+3+2+1+1=202. Final allocation: 188+5+3+2+1+1=200, $200.2387268 \approx \frac{37745}{188.5} < \textbf{MD} \leqslant \frac{37745}{187.5} \approx 201.30\overline{6}$.

43. Jefferson: 149+172+118+44+42+25=550, $42377.97674 \approx \frac{1822253}{43} < \mathbf{MD} \leqslant \frac{5004003}{118} \approx$ 42406.80508. Adams: 149+171+117+45+43+25=550, $42769.25641 \approx \frac{5004003}{117} \leqslant \mathbf{MD}$ $< \frac{6340534}{148} \approx 42841.44595$. Webster: 149+171+117+45+43+25=550 (initial allocation).

45. 0+7+5+3+2+1=18, $202.\overline{18} < \mathbf{MD} < 204$.

47. 10+5+2+3+1=21, $16.78094157 < \mathbf{MD} < 18.55202894$.

49. 25+9+24+15+27=100, $35.68090060 < \mathbf{MD} < 35.74221171$

51. 26+19+18+18+17+2=100, $4408.907857 < \mathbf{MD} < 4426.607147$

53. Hamilton: inconclusive. Lowndes: 18+25+28+13+13+3=100. Jefferson: 18+25+29+13+13 +2=100, $1.357142857 < \mathbf{MD} \leqslant \frac{40}{29} \approx 1.379310345$. Adams: 18+24+28+13+14+3=100, $1.458\overline{3} \leqslant \mathbf{MD} < 1.461538462$. Webster: 18+25+28+13+14+2=100, $1.403508772 < \mathbf{MD} \leqslant$ 1.407407407. Hill-Huntington: 18+25+28+13+14+2=100, $1.403724813 < \mathbf{MD} < 1.408373701$.

55. 9+6+6+2+1+1=25, $128.8432972 < \mathbf{MD} < 130.4612011$.

57. Hamilton: 22+30+68+36+28+16 = 200. Lowndes: 22+30+67+36+28+17 = 200. Jefferson: 22+30+68+36+28+16=200, $203.1764706 < \mathbf{MD} \leqslant \frac{5743}{28} \approx 205.1071429$. Adams: 22+30+67+36+28+17=200, $209.358209 \approx \frac{14027}{67} \leqslant \mathbf{MD} < \frac{6163}{29} \approx 212.5172414$. Webster: 22+30+67+36+28+17=200, $207.8074074 < \mathbf{MD} \leqslant 208.8\overline{36}$. Hill-Huntington: 22+30+67+36 +28+17=200, $207.8131088 < \mathbf{MD} < 208.8708906$.

59. Hamilton: 23+13+10+7+44+3=100. Lowndes: 23+12+10+8+44+3=100. Jefferson: 24+12 +9+7+45+3=100, $72.15384615 < \mathbf{MD} \leqslant 72.\overline{3}$. Adams: 23+13+10+8+43+3=100, $76.6744186 \approx \frac{3297}{43} \leqslant \mathbf{MD} < 78.1\overline{6}$. Webster: 23+13+10+7+44+3=100 (initial allocation). Hill-Huntington: 23+13+10+7+44+3=100, $74.86801185 < \mathbf{MD} < 75.10010416$.

61. Hamilton: 105+90+53+83+88+81=500. Lowndes: 105+90+53+83+88+81=500. Jefferson: 105+90+53+83+88+81=500, $9.235955056 < \mathbf{MD} \leqslant \frac{767}{83} \approx 9.240963855$. Adams: 105+90+53+83+88+81=500, $9.34\overline{09} \leqslant \mathbf{MD} < 9.346153846$. Webster: 105+90+53+83+88+81 =500, $9.288135593 < \mathbf{MD} \leqslant 9.29\overline{69}$. Hill-Huntington: 105+90+53+83+88+81=500, $9.288283834 < \mathbf{MD} < 9.297140447$.

63. Hamilton: 288+425+198+592+243+328+136=2200. Lowndes: 278+425+198+591+244+328 +136=2200. Jefferson: 288+426+197+592+243+328+136=2200, $0.89\overline{39} < \mathbf{MD} \leqslant \frac{381}{426} \approx$ 0.8943661972. Adams: 278+425+198+591+244+328+136=2200, $0.8970588235 \leqslant \mathbf{MD} <$ 0.8971193416. Webster: 288+425+198+592+243+328+136=2200 (initial allocation). Hill-Huntington: 288+425+198+592+243+328+136=2200 (initial allocation).

Chapter 3

§3.1, page 265

1.
(a)	.201417238;	(b)	.151900653;	(c)	1.17378981;	(d)	1.01902398;
(e)	1888.05771;	(f)	.0223948894;	(g)	1.07835102;	(h)	.962723314;
(i)	.565656566;	(j)	1.67647059;	(k)	.482129880;	(l)	6.19443158;
(m)	-.0525116007;	(n)	-.129972786;	(o)	-.134008405;	(p)	.00625397702.

3. Tax: $2,073.5, total cost: $33,973.5.   5. 16.45986%.   7. 26.996%.   9. −3.629%.

11. 19%.   13. $1,383.   15. 52.53%.   17. $1 - x^2$.   19. Males – 648, females – 552.

21. $9,518.75.   23. Drop: 90.8858%; growth: 146.82%.   25. $8539.2.

§3.2, page 276

1. (a) $1440; (b) $250; (c) $39.06; (d) $126.56; (e) $14.64; (f) $14.13.

3. (a) $17886.18; (b) $5894.74; (c) $453.17; (d) $710.34; (e) $4945.32; (f) $717.90; (g) $7475.04; (h) $7607.79.

5. (a) $r = 10\%$; (b) $r = 10\%$; (c) $r = 15\%$; (d) $r = \frac{3}{26} \approx .1071428571$; (e) $r = 73.\overline{73}\%$; (f) $r = 101.3\overline{8}\%$.

7. $1774.37. **9.** 1.419950302 years or 17 months. **11.** 5.3%. **13.** $749.39 million.

15. $6333.94. **17.** $I = \$2625$, $P = \$10,000$, $r = 0.0136$, $t = 2$. **19.** $r = 0.08\overline{3}$.

21. $r = 0.0214$ or 2.4%. **23.** $1724.33. **25.** $1171.28. **27.** $7.06. **29.** $19.38.

31. $I = \$606.89$, monthly payment = $376.5. **33.** $I = \$824.25$, monthly payment = $240.95.

35. $42.54. **37.** $665.21. **39.** $I = \$134.75$, $r = 4\%$. **41.** 54.3%.

43. $I = \$385.42$, $r = 8.35\%$. **45.** 3.69%. **47.** She pays $5207, rate = 16.56%.

49. Double is more profitable. **51.** 16 years. **53.** 1%.

§3.3, page 291

1. (a) $36272.86; (b) $40871.39; (c) $61574.79; (d) $16863.54; (e) $15791.89; (f) $19116.45; (g) $17853.59; (h) $46679.28; (i) $26603.16; (j) $46178.85.

3. (a) 1.087 years or 13 months; (b) 1.668 years or 20 months; (c) 3 years or 37 months; (d) 2.19 years or 26 months; (e) 2.2979 years or 27.57 months; (f) 1.24 years or 15 months; (g) 4.71 years or 56.58 months; (h) 1 year or 12 months; (i) 14 years or 170 months; (j) 5 years or 61.5 months.

5. (a) 7.5339561%; (b) 4.8145617%; (c) 4.7295267%; (d) 5.5777076%; (e) 2.2754386%; (f) 3.6135756%.

7. (a) 5.9117127%; (b) 5.9408782%; (c) 5.9551135%.

9. (a) $134,676.70; (b) $135,927.25; (c) $136,213.97; (d) $136,353.61.

11. (a) $52572.05; (b) $52597.51; (c) $52607.36; (d) $52,609.92.

13. $10477.71. **15.** (a) $62,200.00; (b) $447,844.85. **17.** $2761.21.

19. 4.8150838%. **21.** $30,695.66. **23.** $326,888.58. **25.** 169.

27. $999,676.86. **29.** 5.42075%. **31.** 4.1623870%. **33.** 3.445760369 years or 41 months.

35. 23.17442708 years or 278 months. **37.** $3049.27. **39.** $24669.44. **41.** 12.34567.

§3.4, page 300

1.
(a) $26341.77;   (b) $81429.03;   (c) $19004.10;   (d) $28357.81;
(e) $2281.89;   (f) $1291.86;   (g) $1805.66;   (h) $3650.21.

3.
(a) 18.3729 years or 220 months;   (b) 7.72767 years or 93 months;
(c) 7.6224 years or 91 months;   (d) 3.5739 years or 43 months;
(e) 5.3235 years or 64 months;   (f) 8.118 years or 97 months.

5. (a) $161433.41;   (b) $50,400;   (c) $111,033.41.

7. (a) $274.78;   (b) $599.37.   **9.** 2.7 years.   **11.** 26 months.   **13.** $175.93.

15. 6.11 years.   **17.** $12828.22.   **19.** $14,554.63.   **21.** $83372.36.   **23.** $81,817.32.

25. $370.09.   **27.** No, the annuity will worth $425019.18 while it is needed $441086.86.

29. 9.633206 years.   **31.** Annuity yields $32,242.21.   **33.** Extra payment yields $3653.27.

§3.5, page 315

1.
(a) $33954.12;   (b) $57369.35;   (c) $54772.74;   (d) $90689.87;
(e) $27943.07;   (f) $24941.74.

3.
(a) 5.832566 years or 70 months;   (b) 6.3483 years or 76 months;
(c) 28.523 years or 342 months;   (d) 30 years or 364 months;
(e) 8.9372 years or 107 months;   (f) 34 years or 410 months.

5. $7310.30.   **7.** $9557.33.   **9.** $73595.47.

11.

Payment Number	Payment	Interest Paid	Principal Paid	Balance
				$220,500
1	$1357.66	$1148.44	$209.22	$220290.78
2	$1357.66	$1147.35	$210.31	$220080.47
3	$1357.66	$1146.25	$211.41	$219869.06
4	$1357.66	$1145.15	$212.51	$219058.55

13.

Payment Number	Payment	Interest Paid	Principal Paid	Balance
				$375,500
1	$2336.50	$1987.02	$349.48	$375150.52
2	$2336.50	$1985.17	$351.33	$374799.19
3	$2336.50	$1983.31	$353.19	$37446
4	$2336.50	$1981.44	$355.06	$374090.94

15. Monthly rent is $1320.65, and interest paid is $17186.8.   **17.** Save $30463.8.

19. (a) $2026.22.   (b) $729439.2.   (c) $354439.2.   (d) $214715.27.

21. (a) $2616.80.   (b) $942048.   (c) 23.725 years.   (d) Save $125870.04.

23. $8159.37.   **25.** (a) $7362.4.   (b) $8512.59.   (c) $6628.59.

27. Interest paid: $210083.6 for 15-year mortgage and $441718.4 for 30-year mortgage.

29. (a) $208.77, (b) $10020.96, (c) $569.96, (d) $2466.33.

31.

Payment Number	Payment	Interest Paid	Principal Paid	Balance
				$327400
1	$2121.33	$1838.9	$282.43	$327117.57
2	$2121.33	$1837.31	$284.02	$326833.55
3	$2121.33	$1835.72	$285.61	$326547.94
4	$2121.33	$1834.11	$287.22	$326260.72

(b) $$242510.4; (c) 314 months; (d) save $23522.4.

33. $1124.08. **35.** Save $39871.44. **37.** $R = \$185.34$, Interest paid $5240.8.

39. $R = \$175$, Interest paid $4500.

41. (a) $529.03, (b) $528.59, (c) you pay less on the loan with 6.175% interest rate.

§3.6, page 331

1. 4.942%. **3.** 5.097%. **5.** 2.45%. **7.** 3.05%. **9.** $301500.

11. $590978.49. **13.** $293541.

15. Break-even point for mortgage A: 13 years, and for mortgage B: 10 years.

17. The IRA account is more profitable: $40891.85 vs $25920 of the annuity.

19. 5.03145697%. **21.** (a) $458.907.16. (b) $1,534,874.13. **23.** $109725.

Review Exercises for Chapter 3, page 337

1.

(a) 1.150851300; (b) 1.102513283; (c) 7.476743906; (d) 832828.9414;
(e) 0.8516529178; (f) 0.8951412243; (g) 16.85363491; (h) 16.09430535;
(i) 0.1355211159; (j) 0.005683600555; (k) 0.055918602; (l) −0.966228744;
(m) 2.307372058; (n) 2.270834864; (o) -.7018387305; (p) .2636246446;
(q) −0.48718778; (r) −1.169450798; (s) 0.93312169; (t) 3.2691183.

3. 59.756%. **5.** 5405% **7.** 36%. **9.** $24,495. **10.** 18%. **11.** 9.9%.

13. $11,431.25. **14.** $4437. **16.** $15076.

17. (a) $2488.8; (b) $1514.2; (c) $15.3; (d) $57.73; (e) $15.38; (f) $89.87.

19. (a) $31866.46; (b) $4837.22; (c) $699.82; (d) $745.92; (e) $6849.83; (f) $616.16; (g) $5632.11; (h) $6277.76.

21. (a) 2.222%; (b) 4.76%; (c) 14.81%; (d) 6%; (e) 24.33%; (f) 28.08%.

23. 5.88%. **25.** $576.96. **27.** $1.\overline{6}$%. **29.** 4.934%. **31.** 44.1%.

33. Discount is $36.75 and the rate is 35.2%.

35. 3.61%. **37.** $218.39. **39.** $8.63. **41.** $12.75. **43.** 13.9%.

45.
(a) $108410.55; (b) $34842.58; (c) $54310.18; (d) $13851.12;
(e) $19132.06; (f) $213000.68; (g) $17344.53; (h) $34094.75;
(i) $18780.07; (j) $26996.32.

47.
(a) 8.5 months; (b) 2.77 years or 33.3 months;
(c) 0.77 years or 9.24 months; (d) 1 year;
(e) 4.3 months; (f) 1.41 years;
(g) 16 months; (h) 1.78 years;
(i) 40 months; (j) 19.6 months.

49.
(a) 6.8386%; (b) 5.76%; (c) 5.357%;
(d) 4.813%; (e) 3.925%; (f) 2.752%.

51. (a) 3.2898%, (b) 3.2988%, (c) 3.30324%, (d) 3.30339%.

53. (a) $70972.05, (b) $71929.87, (c) $72151.44, (d) $72259.78.

55. (a) $40999.88, (b) $41060.63, (c) $41084.23, (d) $41090.32.

57. $6144.4.   **59.** (a) $8800, (b) $2,379,770.46.   **61.** $27599.06.   **63.** 4.68%.

65. $90392.08.   **67.** 2.8×10^{50}.   **69.** $25466.

71. Investment with 3.6% interest yields $12176.73.   **73.** $8110.61.   **75.** 3.25%.

77.
(a) $10576.45; (b) $64309.52; (c) $92442.10; (d) $94349.33;
(e) $11284.77; (f) $1706.10; (g) $85356.23; (h) $45693.68.

79.
(a) 10.67 years; (b) 15.1 years; (c) 6.47 years;
(d) 4.79 years; (e) 2.92 years; (f) 4.44 years.

81. (a) $192,813.22. (b) $84,000. (c) $108,815.22.

83. (a) $357.7. (b) $757.32.

85. Bonus yields $1242.65, while the monthly payments earn $1219.43.   **87.** $753.8.

89. (a) $39617.86, (b) $53680.96, (c) $67487.36, (d) $71791.11, (e) $27216.15, (f) $21084.57.

91. (a) 28.436 years or 341 months, (b) 26.61 years or 319 months, (c) 20.676 years or 248 months, (d) 24.135 years or 290 months, (e) 23.76 years or 205 months, (f) 23.769 years or 285 months.

93. $11,752.89.   **95.** $14534.78.   **97.** 22.16 years or 266 months.

99.

Payment Number	Payment	Interest Paid	Principal Paid	Balance
				$281,300
1	$1818.90	$1575.28	$243.62	$281056.38
2	$1818.90	$1573.92	$244.98	$280811.40
3	$1818.90	$1572.54	$246.36	$280565.04
4	$1818.90	$1571.16	$247.74	$280317.30

Payment Number	Payment	Interest Paid	Principal Paid	Balance
				$388,150
1	$2394.96	$2028.08	$366.88	$387783.12
2	$2394.96	$2026.17	$368.79	$387414.33
3	$2394.96	$2024.24	$370.72	$387043.61
4	$2394.96	$2022.30	$372.66	$386670.95

101.

103. (a) $234577.12, (b) $201561.86, (c) $155336.71.

105. 20-year mortgage is more economical: you save $45683.76.

107. (a) $2114.02, (b) $761047.2, (c) $378447.2, (d) $201627.29.

109. (a) $3312.38, (b) $1192456.8, (c) 24.98 years or 300 months, (d) save $130456.8.

111. $85782.84.   **113.** (a) $7451.52; (b) $14141.81; (c) $5367.57.

115. 20-year mortgage: interest paid $272508.4, 30-year mortgage: interest paid $506155.6.

117. (a) $275.03; (b) $13201.44; (c) $1018.44; (d) $3230.12.

119. $783.14.

121. (a) First payment is $353610.77 and the last one is $679817.17.   (b) $12,051,177.29.

123. 4.0355%.   **125.** This year the monthly rent is $1987.26; next year it would be $1827.74.

127. 6.7%.   **129.** Mortgage A: 4.46 years, Mortgage B: 3.36 years.   **131.** $31627.89.

Index

CPSIA information can be obtained
at www.ICGtesting.com
Printed in the USA
LVOW02s2339260816
501758LV00004B/19/P